# CURRENT RESEARCH ON OCCUPATION AND PROFESSIONS

*Volume 10* • 1998

## JOBS IN CONTEXT: CIRCLES AND SETTINGS

# CURRENT RESEARCH ON OCCUPATIONS AND PROFESSIONS

## JOBS IN CONTEXT: CIRCLES AND SETTINGS

*Editor:* HELENA Z. LOPATA
*Department of Sociology and Anthropology*
*Loyola University Chicago*

VOLUME 10 • 1998

 JAI PRESS INC.

Greenwich, Connecticut                                      London, England

# CONTENTS

# INTRODUCTION:
## JOBS IN CONTEXT: CIRCLES AND SETTINGS

Helena Z. Lopata

The overall theme, differentially central to the various chapters in this volume, is the importance of the social circle and the setting within which jobs are located. Jobs organize work in the economic institution into social roles, that is, sets of negotiated, mutually interdependent, social relations between a social person and a social circle, involving duties and rights (see previous volumes in this series, especially volume 9, co-edited by me and Anne Figert; see also Lopata 1971, 1991, 1994; Znaniecki 1965). Such social roles have two major components. They contain tasks fitted into a system of tasks in cooperation with other workers in order to accomplish the purposes of production and distribution of goods or of provisions of services. They also fit into a system of social roles located in industrialized societies in organized social groups in a particular locale or setting or combinations of such (Lopata and Figert 1996).

The social person, or the job holder, is that "package" of abilities and characteristics with which an individual enters the role center. These are the traits deemed necessary by an established social circle, or one brought together by the person, for the performance of the duties and the receipt of the rights of the role. The social circle contains all the people or other social units toward whom the duties are directed and from whom rights are received by the person. Circle members can,

Current Research on Occupations and Professions, Volume 10, pages 1-10.
Copyright © 1998 by JAI Press Inc.
All rights of reproduction in any form reserved.
ISBN: 0-7623-0034-5

but do not necessarily, relate to each other because they are part of that role. For example, nurses, receptionists, and the physician's patients relate to each other, but all the patients may not even know each other, or meet only while in the waiting room. The concept of social circle has the advantage over Merton's (1957, 1968) role set in that members are seen as possibly interacting. They often depend upon each other as providers of the person's rights and resources, or as co-recipients of the duties and obligations. The role may need mutual cooperation among at least some of such people, as in the case of a an officer's platoon in the armed services. The important point is that jobs can not be performed without the cooperation of circle members with the social person. Such cooperation may be direct, as in case of a class of students and the professor, or indirect, as when the board of directors grants the professor rights to perform duties in the negotiated settings. A job is, thus, a particular social role involving the person and the circle working and relating to each other in a setting (Lopata 1994) or workplace (Abbott 1988).

The concept of occupation helps to organize the vast range of jobs in a complex society such as America. Occupations are aggregations of jobs, grouped on the basis of their similarity in content—similarity in the tasks they involve, the duties and responsibilities they entail, and the conditions under which they are performed. The DOT (*Dictionary of Occupational Titles* of the United States Department of Labor) combined the several million jobs held by the 100 million members of the U.S. labor force into 12,099 occupations (Cain and Treiman 1981; Lopata, Miller, and Barnewolt 1984, p. 4). An occupational title is a "summary label, covering many jobs and variations of content, duties and conditions" (Lopata, Miller, and Barnewolt 1984, p. 5). An additional label of "occupational category" that aggregates occupations considered to be similar is used by the DOT and occupational sociologists. Professions form one of these categories, service occupations form another.

This volume contains descriptions of jobs that fall into a variety of occupations involving many social persons within social circles located in one or a combination of settings. A related concept used in some of the chapters is that of career, seen as a series of jobs held by an individual in horizontal, vertical, or nonlineal sequences. In each of these jobs, the individual packages herself or himself as a different kind of social person, using abilities and personal characteristics required, or negotiated with circle members, for that job. Each job is located with its person and the circle of cooperating contributors in a setting—a school, hospital, motel, home, or office.

## THE MATCH BETWEEN PERSON AND CIRCLE

One of the fascinating questions about occupations is the manner by which social persons and social circles find each other. After all, there are millions of persons needing circles and millions of circles needing social persons in order for jobs to be carried forth. A major part of the problem of the search procedures is the matching of requirements. Intermediary matching agencies, such as employment bureaus, have developed in most locales and some matches are made through them, but even they rarely succeed in ideal matches. Matching is complicated by the wish on everyone's side for the perfect other. Social circles have images of the ideal social person for their social role, and each worker has an image of the ideal setting and social circle within which to work. These images are developed historically in the cultures of societies and communities and contain explanations, justifications, or rationales for the need for specific qualifications. Each culture or circle contains assumptions about characteristics of categories of people that would best fit or mismatch the needs of a job. For example, discriminatory practices in rejecting prospective workers may be explained as, "Those people (women, blacks) can not possibly do such precise work" or "We would hire (women, blacks) but our clients would not work with them" (Lopata, Barnewolt, and Miller 1986; in this volume, see Slevin and Wingrove; Calasanti and Smith; Genovese and Fava). As Myrdal (1944) found when examining racism in America, such rationales are especially needed if the country or social circle claims democratic ideals negated by discriminatory practices.

People looking for a future career may reject many occupations on the assumption that they demand unwelcome tasks, or they may reject a particular organization because of a belief such as "It is run by Germans and they discriminate against people like us."

Many chapters in this series on occupations and professions analyze some of the ways in which persons or circles try to prevent a mismatch (see, in volume 9, Erdmans 1996; Tang 1996; Lawson 1996; Ward and Mendoza 1996). One way in which an obvious mismatch can be prevented is through avoidance of the relationship, generally possible only when either circle or person has other choices. The circle simply does not hire a person with "inappropriate" characteristics of, for example, skin color, or gender, as all the literature on discrimination documents, including the chapters by Slevin and Wingrove, and by Calasanti and Smith. The potential candidate may seek another organization or setting, or may create his or her own circle, as in the case of Asian entrepreneurs in Chicago (Zhang) or the Asian Indian motel owners (Assar), all of whom face or fear discrimination in white "WASP" organizations, or women who apply or finally enter less prestigeful law firms, being unwilling to meet the demands of male-dominated elite firms (Schrimsher).

Although employing organizations have developed procedures for trying to find a fit between their needs and the qualifications of potential persons, much occupational matching is quite sporadic and haphazard, even in cases where the worker has great personal responsibility. The Kejing and Dempsey chapter details the problem of finding matches between worker and employer in the case of rural women, who are unfamiliar with life and people in cities, finding employment, and the converse case of women household managers finding personal service workers. The women come to the cities for many reasons. They do not necessarily want jobs as houseworkers but usually settle for such because they do not have the skills for other jobs. They then worry about finding a good social circle—about the kind of work they will be asked to do and the people in the circle within which they must work. The employer has images of an ideal worker always performing duties pleasantly and honestly and worries about the abilities and personalities of potential workers. Some matching contacts are so brief that women on both sides of the situation are really taking chances that things will work out. Using matching services has disadvantages for both the employer and the worker. As usual in even allegedly rational bureaucratic settings of cities, women wanting to work and women wanting workers often depend on personal recommendations.

The fact is, as will be evident from the studies included here, that most circles or persons do not have the power or resources to attract their ideal counterparts. How, then, do these circles and persons negotiate cooperation and handle imperfect matches? The next question answered by occupational sociologists is how social circles and potential social persons actually work out relationships of social roles. Let us examine some of the way circles of different complexity and social persons adjust to their own and each other's imperfections.

## SOCIAL CIRCLES AND LESS-THAN-PERFECT PERSONS

Large and complex employing organizations have developed complex procedures for preventing serious mismatch between their needs and the qualifications of employees. Powerful organizations and professions have spent a great deal of effort over time in ensuring their rights to selection of candidates, training them in their esoteric knowledge, and keeping out those deemed unacceptable at any stage of assimilation.

Law, medicine, science and academia, and real estate are occupations that have been able to draw tight boundaries over whom, and how, social persons can enter and practice their work (Abbott 1986) In each of these cases, the occupation has been dominated by social persons of different characteristics than those who are now trying to find a social circle and location, and much of sociological literature of the past decades deals with this situation. Not only has the American legal profession been dominated by white males but the whole structure of the legal system has been built around the male model of work in it during the height of the two-sphere world imagery which assumed that each professional had a back-up person at home. This profession, and most other high-prestige occupations with socialization and cooperation power over workers exerted by employing organizations and supported by professional associations, were established when American society was focused on its economic expansion so that its values predominated and evolved into what Coser (1974) described as "greedy" institutions. Those groups that obtained power in that period of American history demanded of the occupational co-members total commitment, at the cost of devotion to family, religion, recreation, and even education that did not meet professional goals. Members of minority groups with devalued racial or ethnic identification were purposely kept from the high power professions on the assumption that their status would lower that of the occupation As stated above, the leaders of such groups justified this exclusion in a society claiming democracy through a set of stereotypes as to the presumed inferiority of the minorities and women (Lopata, Miller, and Barnewolt 1984, 1986). One of the justifications used to prevent women from entering has been that they could never be as committed to their work as required.

One way in which established occupation holders and their associations handle impingement by persons lacking desirable characteristics is through what Abbott (1988) defined as segmentation of the field, especially through segmentation of client or setting. Goldschmidt and Wharton show one way that professions have handled jurisdictional conflict not only with nearby occupations, but also with social persons in their own occupation. Less desirable practitioners who could not be kept out of legal or real estate jobs have been assigned to less desirable clients or subspecialties. Men in real estate segmented the clientele when large numbers of women entered the field: men kept corporate sales, women developed almost a monopoly over sales of private property. Such a division fits the imagery of the two-sphere world: sales of private homes are less prestigous and less profitable than those of public buildings, so they are still considered all right for less prestigous persons to control. Public buildings require complex relations with high-status corporations. The legal profession also tends to be segmented by client and type of problem (Goldschmidt, Schrimsher): men handle corporate law with large firms, women usually end up dealing with private cases, such as divorce. Interestingly enough, real estate is one occupation in which the boundaries of law are encroached, although attorneys are first to declare real estate as really lacking professional status, as evidenced by the dearth of abstract knowledge. Abbott's description of the way in which professions control impingement by other occupations through assignment of subservient status is evident in the attempts of the law profession to insist that attorneys supervise paralegals, and in the real estate associations' insistence that brokers supervise agents.

The Calasanti and Smith study shows some of the consequences upon the job arising from the presence of social persons with less than desired characteristics. Minority women in academia are discriminated against in organizations that hire them, and found in less prestigious fields of specialization, at lower salary and rank levels. Those who are hired as tokens often face visibility problems which make their work difficult, or isolation from segments of the social circle whose interaction they need. Slevin and Wingrove emphasize the fact that the same occupation may have different meaning for different social persons due to social circle relations.

One way that circles can continue functioning is to accept imperfect workers on the assumption that better choices are not available and that the decision will not be fatal because others will pitch in, and can be changed through dismissal. Social circles may also assume that the new hire can be taught to perform the role adequately, and segments of these circles can be delegated the task of such training. The pre-existing circle of lobstermen can try to teach newcomers the skills and etiquette of their occupation and simply decrease cooperation with those beginners who refuse to learn (Travisano).

Another way of handling imperfect matches on the part of the circle is to decrease the rights that normally go with the job. The Slevin and Wingrove chapter illustrates the situation in which the social circles—dominant culture schools—hiring African-American teachers can refuse, or simply fail, to provide them with the resources necessary to carry forth teaching in the manner of the educational system. Best resources are saved for favorite teachers. Neither side meets the ideal, so negotiations result in the job being performed as best as it can.

Sometimes the lack of resources is inevitable because the social circle lacks power in the larger system. Not all communities can provide a new surgeon with all the resources with which she or he trained in medical school (Goodrick, Flood, and Fremont) or offer a teacher the equipment and students necessary for wished-for job development (Calasanti and Smith). Even the powerful Arizona bar association can not fully prevent encroachment on the work of its attorneys (Goldschmidt). Not all candidates match the picture of the perfect lawyer, physician, engineer, or household worker. Even physicians professionally trained for autonomous work must adjust to the organizational culture when they join a group practice (Goodrich, Flood, and Fremont). At other times, the absence of resources is due to the unwillingness of some or many circle members to grant these rights to social persons whom their construction of reality defines as inferior. The need to negotiate compromises happens so often that sociologists can not use inflexible definitions of social role if they wish to understand what happens in the actual world of work (or in other roles, for that matter).

## SOCIAL PERSONS IN LESS-THAN-PERFECT SOCIAL CIRCLES

There are many ways that social persons may deal with imperfect circles, ones lacking desired characteristics or unwilling or unable to provide necessary rights. What is done, of course, depends on the amount of power the person has to change the circle. Those persons with power can get rid of the circle or segments of it, substituting people with whom they can work. They can bring in their own resources and change the culture within which they work. Those without the influence to change the circle can perform their duties or meet their obligations at a minimal level, or differentially toward differently cooperating circle segments. Taking the job with the assumption that one can find a better one soon, is another means of dealing with the situation. Organizations can hire temporary workers as a means of avoiding commitments, and workers can treat organizations the same way. Commitment to an occupation does not necessarily transfer to commitment to an employer, as Schrimsher found in her study of lawyers.

Most social persons do not have much power vis-à-vis established social circles and must do as best as they can in the situation. Women of color who teach in integrated schools must learn how to deal with challenges from white students. Those teaching in inferior schools may feel obliged to do both "gender and race work," providing resources that should have been supplied by the school system and helping students from disadvantaged backgrounds to work in the system (Slevin and Wingrove; Calasanti and Smith). Minority and women academics in any field and setting in which they are relatively powerless outsiders have to learn to live within the restrictions provided by the system or find ways of getting their work done in independent ways, as evidenced in the case of woman physicists studied by Genovese and Fava. A change in the system may even remove whatever advantages the person developed, as illustrated in the experiences of African-American teachers after desegregation (Slevin and Wingrove).

The problems of finding and keeping a job by migrants within their own country, typical now of the whole world due to the tremendous amount of urbanization and unequal economic development in regions, are exacerbated in the situation of immigrants crossing national and cultural boundaries. Such job seekers are often totally unfamiliar with the language and culture of the host society, lack skills in new types of jobs or contacts with cooperative, more knowledgeable job holders. In such situations, they must take jobs that social persons with greater qualifications would not accept. Such jobs are usually undesirable for many reasons, including the types of tasks, negative aspects of relations with circle members, and lack of adequate resources and rewards. Such a situation was described by Erdmans studying Polish illegal immigrants in personal care jobs in America (volume 9). The alternative is to create a new job or a social circle that will accept qualifications the dominant circles define as inadequate. Assar found that a segment of the Asian Indian immigrant group in America went into the motel business, and Zhang found other Asians creating their own businesses in Chicago. Both of these groups selected such self-employment (or, actually, self- and family-employment) partly in response to discrimination in the American corporate and professional worlds.

Of course, an excellent way in which a person can function in an inadequate or uncooperative social circle is to develop cooperative relations with at least a part of that circle, obtaining necessary information for existing and getting the job done. Travisano stresses the importance of developing mutually respecting relations with other lobstermen, a process that includes the acceptance of advice and the giving of help in times of trouble or when apparently acceptable. In this way, persons lacking necessary qualifications learn them and become self-sufficient in new jobs. New lobstermen must depend upon a network of people doing the same work, who have greater familiarity with the locale. The lobstermen's earnings, even their very lives, are dependent upon the cooperation of the social circle. Some of such cooperative interdependence would be helpful to rural women in urban China, but it is hard for them to form networks because of their isolated work settings during their work time in the city.

Even the selection of the location of work may be helped by insider knowledge obtained from people already in that type of job or work organization. Some homes, communities, or organizations are better to work in than others.

The matching of person and job may require special training in centers for such preparation, or from current or future circle members. A vital need prior to the functioning of a social role may be for economic or emotional support for the transition period, as when someone is trying to start up a business.

Choosing an occupation and then a job in a particular circle and setting location does not guarantee continued involvement. The turnover of workers in any setting is a good indicator of mismatches, because of any combination of factors on the part of the worker: the assumption that "the grass is greener on the other side," definition of present social relations or tasks as lacking sufficient support for career committed lawyers (Schrimsher), or the wish to move into another stage of the occupational career, as opening a self-employed business in Chicago (Zhang), or a new "farm" in a housing community (Wharton). One reason for rapid turnover may be the circles' finding that new hires, assumed to be trainable, cannot do the work.

Of course, a wide choice of settings may not be available at any time or place or for any set of social persons. We have in this volume documentation of four situations in America in which characteristics of the social person limit choices of occupations or jobs: women physicists of past decades (Genovese and Fava), women and minority faculty in academia (Calasanti and Smith), African-American professional women (Slevin and Wingrove), and married women lawyers (Schrimsher).

A mismatch in which either the circle or the person does not have the necessary resources or qualifications to enable the necessary relations for the achievement of the role's goals, but in which both sides are willing to negotiate adjustments, results in changes in the role. The types of negotiation and balance in relationships on all sides are clearly seen in jobs within small social circles, as studied by some of our researchers.

## JOBS IN SMALL SOCIAL CIRCLES

Most jobs, even in complex societies, are not actually performed in large, bureaucratized organizations, but rather in relatively small circles within small social units. This is particularly true of service jobs:

> the service sector has been defined to include wholesale and retail trade, finance, insurance and real estate, general government (including the military in most instances) and the services traditionally so designated, including professional, personal, business and repair services (Fuchs 1968, p. 16; see also Lopata, Miller and Barnewolt 1986, p. 78).

The rapidly expanding service sector of American and global economies involves two layers of jobs. One layer, at the bottom of the occupational ladder, consists of relatively unskilled or semiskilled jobs in building maintenance, food, health, protection, and personal service. The higher layer involves professionals who belong to occupations generally identified as professional. For example, nurses aides are service workers, physicians are professionals, and nurses are still considered semiprofessionals striving for full professional status (Etzioni 1969). Here, we find analyses of household workers who provide personal services, maintain housing, prepare food, and provide child care (Kejing and Dempsey). Real estate sales workers also fall into the service sector. They are attempting to move their occupation toward increasing professionalization (Wharton). Lobstermen of the kind described by Travisano provide goods after much individual work, the "Potel" motel owners studied by Assar provide services, and Zhang's Asian business entrepreneurs contribute either services or products to customers.

One major difference between jobs in large bureaucracies and in small groups is the influence of informal controls and methods by which duties and rights are established and crossed over. Assar makes an important point that economic contributions, within the group she studied, do not necessarily lead to shared decision making and increased status, as often assumed by sociologists. Both the decision making and prestige in work do not follow for women, since such rights go against the traditional, culturally ingrained gender hierarchy that contains a devaluation of work defined as "women's."

## CONCLUSIONS

Volume 10 of *Current Research on Occupations and Professions* is devoted to the study of *Jobs in Context: Circles and Settings*. One of the main themes contained in the various chapters is the importance of an acceptable match between a social person wishing to enter a job and the social circle with whom he or she must obtain necessary cooperation. The job, as a social role, is performed in a specific setting and only through cooperation between the person and the circle. Various means have been developed by social circles in the attempt to prevent a mismatch—the hiring of a person with inadequate or inappropriate characteristics, as defined by the circle, for the job. However, neither the circle nor the person usually has the ability to obtain a perfect match. The various studies reported here analyze the consequences for the circle and for the person of at least partial mismatch. This can mean training and learning the needed qualities, and/or negotiating compromises in the relationships. In most job situations in larger social circles, the person does not have much power to change the job under given conditions. Cooperation from parts of the circle are even then necessary, although they may vary in importance and manner. What is surprising is that so many jobs are carried forth at least well enough that societies, circles, and persons are able to function.

## REFERENCES

Abbott, A. 1988. *The System of Professions*. Chicago: University of Chicago Press.

Cain, P.S., and D.J. Treiman. 1981. "The DOT as a Source of Occupational Data." *American Sociological Review* 46: 253-278.

Coser, L. 1974. *Greedy Institutions*. New York: Free Press.

Coser, L., and R.L. Coser. 1974. "The Housewife and her 'Greedy Family'." Pp. 89-100 in *Greedy Institutions*, edited by L. Coser. New York: Free Press.

Etzioni, A. 1969. *The Semi-Professions and their Organization*. New York: The Free Press.

Fuchs, V.R. 1968. *The Service Economy*. New York: National Bureau of Economic Research

Erdmans, M.P. 1996. "Illegal Home Care Workers: Polish Immigrants Caring for American Families. Pp. 267-292 in *Current Research on Occupations and Professions*, Volume 9: *Getting Down to Business*, edited by H.Z. Lopata and A.E. Figert. Greenwich, CT: JAI Press.

Lawson, H. 1996. "Car Saleswomen: Expanding the Scope of Salesmanship." Pp. 53-71 in *Current Research on Occupations and Professions*, Volume 9: *Getting Down to Business*, edited by H.Z. Lopata and A.E. Figert. Greenwich, CT: JAI Press.

Lopata, H.Z. 1971. *Occupation: Housewife*. New York: Oxford University Press.

Lopata, H.Z. 1991. "Role Theory." Pp. 1-11 in *Social Roles and Social Institutions: Essays in Honor of Rose Laub Coser,* edited by J.R. Blau and N. Goodman. Boulder, CO: Westview.

Lopata, H.Z. 1994. *Circles and Settings: Role Changes of American Women*. Albany, NY: SUNY Press.

Lopata, H.Z., D. Barnewolt, and C.A. Miller. 1986. *City Women in America: Work, Jobs, Occupations, Careers*. New York: Praeger.

Lopata, H.Z., and A.E. Figert. 1996. *Current Research on Occupations and Professions,* Volume 9: *Getting Down to Business*. Greenwich, CT: JAI Press.

Lopata, H.Z., C.A. Miller, and D. Barnewolt. 1984. *City Women: Work, Jobs, Occupations, Careers*, Volume 1: *America*. New York: Praeger.

Merton, R. 1957. "The Role Set." *British Journal of Sociology* 8: 106-120.

Merton, R. 1968. *Social Theory and Social Structure*. New York: Free Press.

Myrdal, G. 1944. *An American Dilemma*. New York: Harper and Bros.

Tang, J. 1996. "To Be or Not to Be Your Own Boss? A Comparison of White, Black and Asian Scientists and Engineers." Pp. 129-165 in *Current Research on Occupations and Professions*, Volume 9" *Getting Down to Business*, edited by H.Z. Lopata and A.E. Figert. Greenwich, CT: JAI Press.

United States Department of Labor. 1997. *Dictionary of Occupational Titles*, Volume 1: *Definition of Titles*. Washington, DC: U.S. Government Printing Office.

Ward, J.L., and D.S. Mendoza. 1996. "Work in the Family Business." Pp. 167-188 in *Current Research on Occupations and Professions*, Volume 9: *Getting Down to Business*, edited by H.Z. Lopata and A.E. Figert. Greenwich, CT: JAI Press.

Znaniecki, F. 1965. *Social Relations and Social Roles*. San Francisco, CA: Chandler.

# WORKING SISTERS FROM OUTSIDE:
## RURAL CHINESE HOUSEHOLD WORKERS
## IN BEIJING

Dai Kejing and Paula R. Dempsey

## ABSTRACT

In 1989, there were three million rural women employed as household workers in Chinese cities. They left country villages to see the outside world and to earn money as domestic servants. Many of the younger women hoped to change their fate in life by marrying and settling down in the city, but most became integrated into both settings in intricate ways. The paper is based on in-depth interviews and field observations of rural women looking for work as housemaids in official and informal labor markets, as well as an extended case study of one household worker. This evidence is set in the larger context of economic reform and rural/urban migration patterns in China. The situation of the "working sisters from outside" casts light on the gap between the poor and the rich under improving economic conditions for the country as a whole.

Current Research on Occupations and Professions, Volume 10, pages 11-29.
Copyright © 1998 by JAI Press Inc.
All rights of reproduction in any form reserved.
ISBN: 0-7623-0034-5

# INTRODUCTION

In 1989, there were three million rural women employed as household workers[1] in Chinese cities. These women are known to urban residents as *dagong mei*, "working sisters from outside" (Dai 1993). They leave their country villages for the purpose of earning money and to see the outside world. Most of the younger women hope to change their fate in life by marrying and settling down in the city. Once they establish a home in the city, the total course of migration is complete. If they do not find the opportunities they hope for within three to five years, they return home to the village and get married. Sometimes after marrying and having children, the young mother leaves her children with their grandmother and returns to the city to try her luck again.

The history of rural/urban migration reveals this pattern as one common throughout the world (Hill 1994). Researchers have traced the causes and consequences of the international labor market for household workers (Heyzer, Nijeholt, and Weerakoon 1994), particularly as it has affected Filipina women in the Pacific Rim (Armstrong 1996; French and Lam 1988; Tan and Devasahayam 1987), Sri Lankan women in the Middle East (Gamburd 1995), and women of color in the United States (Bakan and Stasiulis 1995; Romero 1988a; Rollins 1985). The situation of household workers in China brings to this body of literature new insights into the operation of these forces, in view of China's ongoing transition from a centrally controlled economy to a modified socialist market system. It also reveals a pattern within a single country that reinforces the concept of "transnational networks of workers who are neither temporary migrants nor permanent immigrants" (Ozeki 1995).

This paper is based on in-depth interviews and field observations of *dagong mei* looking for household work in official and informal labor markets. This evidence is set in the larger context of economic reform in China. After an overview of rural/urban migration, an extended case study is presented of one household worker, Fan Dajie. Her situation casts light on the gap between the poor and the rich under improving economic conditions for the country as a whole.

# SCOPE AND CONDITIONS OF CHINESE
# RURAL/URBAN MIGRATION

Presently, there are three large population groups flowing to the urban labor market in China. According to labor department statistics, in 1993 there were 4.8 million unemployed workers (2.9% of all workers), 15 million redundant employees (12% of all workers), and 130 million rural surplus laborers (30% of the 440 million rural laborers). In order to find themselves new means of subsistence, in 1993 more than 50 million rural laborers left the countryside for work in the cities. It is estimated that by the end of this century, the number of rural surplus laborers will exceed 200 million. The speed of the flow of rural laborers to the cities will increase each year (Lee 1995).

According to the Research Institute of Rural Development of the Chinese Academy of Social Sciences, about 130 million rural residents moved to urban areas for work between 1978 and 1988. Of these, 33 million were women, accounting for 25% of the total rural migrants. These migrants work in the city either throughout the year or for a regular period each year. Because three times as many rural men migrate to cities as women, Chinese social scientists are concerned that this may result in less technical improvement in farming and reduced agricultural production. Therefore, the increasing role of women in agriculture is perhaps overemphasized in the literature.

In addition, the rapid increase of this floating population has aggravated urban traffic, worsened urban housing shortages, and further polluted the environment. Facing these problems, official opinion holds that it may be inappropriate to have a massive number of temporary workers settling in the city, and they must be encouraged to return to their homelands. Therefore, workers must obtain temporary resident cards giving them the right to look for work in the city. Local public security guards check for these cards regularly and fine or detain workers who do not possess them.

However, it is evident that the migration of rural youth into urban areas will help alleviate the problem of the aging urban population, a result of the low birth rate. A survey of 100,000 women in 74 small cities and towns conducted in 1991 by the Institute of Population under the CASS indicated that these migrants have fewer children than those still living in the countryside. Five causes were noted for the decrease in births:

1. The family planning policy is more strictly implemented in urban areas;
2. People moving to urban areas usually are better educated and prefer to raise fewer children;
3. Most young migrants must save to establish their homes, delaying plans to bear children;
4. Married migrants, most likely living apart, have reduced chances for having children; and
5. It is more expensive to bring up a child in an urban area (Zhai 1991).

## Urban Living Conditions for Rural Workers

The living conditions of rural workers depend on the type of work they are engaged in: dormitories for employees of enterprises, houses rented by employers, houses owned by factories or plants, and houses owned by local residents that workers rent for themselves. Housing conditions range from low-quality flats with kitchens and showers to dormitories with four to eight persons sharing one room. There are also attic rooms with long lines of bed boards, and houses built of sheet iron. In order to reduce expenses, two to four workers often share a room. Sometimes two or more couples share a room with a piece of cloth draped between the beds to provide limited privacy. It is not surprising that some young workers become involved in love affairs or eventually get married.

Some employer-owned housing arrangements have dining halls for the employees, but most workers cook their meals separately on kerosene stoves or brick cooking ranges fueled by waste construction materials or straw. They eat simple, poorly prepared meals of rice and vegetables. Under such conditions, most rural workers find that "home" is only a place to stay and a "meal" merely something to fill the stomach.

Many young people find large cities like Guangzhou and Beijing stimulating and fascinating, but they do not feel comfortable because of their rural background and limited income. Therefore, in their leisure time they visit fellow villagers and relatives, with whom they feel more at home. Holiday activities include playing poker, watching free television programs in the shops, sightseeing, and watching performances by artists on the street. Some middle-aged workers lead monotonous lives, with nothing to entertain them except a small radio, and some do not have even that. They care only for earning money, and when they feel they have earned enough, they return home (Dai 1991).

### Income and Expenses of Rural Workers

The workers' income varies greatly depending on their occupation. Generally, labor contractors earn 40-75 RMB Yuan[2] per day; average workers in the plants, Y300-700 per month (with overtime, the maximum pay was approximately Y900 per month); full-time household workers, Y200-300 per month (or Y2.50-3.00 by the hour). The workers' expenses are also very different, but, in general, the main expenses are as follows: seasonal or annual temporary resident's certificate, approximately Y120-240 per year; housing rent, Y120-300 per room (10-20 square meters), shared by several persons; meals, Y150-200 per month; round-trip transportation from the home village to the city, Y100-250. Thus, the total yearly income for rural migrants to cities after expenses is roughly Y1,000-10,000 per year, depending on their occupation, how quickly they find their jobs, and how long they stay in the city (Research Institute 1995).

Rural workers have a conclusive consensus regarding the use of their earnings. Unmarried men save to build a house and support a wife; unmarried women save for their dowry. Married people save for a new house in the home village and their children's tuition. Tuition is approximately Y800-900 per year for children who go to Xiang for junior middle school; Y1,000 for those who study in the county junior middle school; and Y2,000 for senior middle school in the county town. For further study in the occupational middle school, tuition is about Y4,000 per year.

In addition, some young working brothers send part of their earnings back to their parents. Daughters are not obligated to send their earnings to their parents, due to the old idea that daughters' labor will eventually belong to their husband's family. Some daughters believe that they are already contributing to the family income by supporting themselves and earning their own dowries. However, some daughters do send back some money each year.

## Relations with the Rural Family

Concerning their relations with folks at home, there were differences between married and unmarried workers. Parents tended to worry about their unmarried daughters leaving home, but they would be reassured if the daughters were accompanied to the city by relatives or fellow villagers. Married men and women need not ask for their parents' opinion about going to the city for work, nor would their parents intervene. They could leave home if they wanted, but they were obligated to come back for seasonal farm work. They could also ask their parents to take care of their children as well as their farm work, or they could pay other villagers to take care of the farm work. Most of them paid their parents each month for their children's meals at home.

Many young women working in the city took the opportunity to find themselves a husband from the same province or the same county. Generally, when a woman reached marriage age, if she had not found a husband for herself in the city, her parents or her elder brother would find a fiancé for her in the village. Then she would be drawn back to the village like a kite on a string. Some women already had a boyfriend or fiancé before they came to the city. They kept in touch with each other, but the women did not exclude the possibility that they would find a better mate in the city.

The values of rural migrants in selecting husbands tended to change radically in the city. According to a survey of working sisters from outside by the Women's Federation of Guangdong Province in the delta area of Zhujiang, 20% of the respondents answered that their first priority in mate selection was money, power, and the man's ability to turn their rural household registration to an urban one. And 17.7% and 21.4%, respectively, answered that there was "no need to be surprised at" and "no need to talk about" a broken family caused by a third person. In some places, divorce cases increased. A 1993 study of 439 female bigamous marriages in Guangzhou revealed that 69% were peasants, of whom most were originally working sisters from outside (Liu 1995).

One 26-year-old household worker, already much older than the usual marriage age for rural women, was introduced to a boy from her village through a villager she knew in Beijing. She returned to her homeland to meet him and had a good impression of his appearance and his education at the senior middle school level. They became engaged, but she returned to Beijing and stayed for many years as a household worker. She worked for seven households a day, for two hours each. Her income was approximately Y1,200 per month. Her boyfriend, parents, and fellow villagers often urged her to go back to get married, but she did not feel like hurrying. First, she wanted to stay in Beijing to earn more money. Second, she hoped to find a more satisfactory mate on her own in Beijing. Third, she wanted to wait to see whether her boyfriend would make progress with his business and begin to earn more money, because she was not pleased with her boyfriend's family's two-room house. In June 1994, after a lot of effort, her boyfriend built a new house and was engaged in a prosperous wood trading business. She decided to marry him and leave behind her life as a household worker.

Another woman married a bricklayer from Hebei Province, whom she met through an acquaintance. Her husband was honest but not hard-working. She wanted him to do extra work in the evenings so they could earn more. His family was poor, and he had three unmarried siblings. Eventually, she ran away to Guangzhou to earn money. After her return, his father wanted the couple to provide Y2,000 to build the family home. She was reluctant, and her husband became unhappy, lying on the bed and refusing to do anything. They quarreled, and he threatened to beat her. She said she wanted to leave the family, and her husband hid her identity card and took her money. She was penniless and unable to buy a train ticket to leave the village.

## WOMEN'S URBAN EMPLOYMENT

Chinese women have made great progress since 1949. Their employment conditions have improved tremendously since the founding of the People's Republic. Employment is the basic form of women's participation in social progress and their only chance of winning economic independence. Since 1949, the population of employed women has risen constantly in China. Women now account for some 44% of the total number of employees. In 1992, employed women were more than 72% of all women over 15 years old. In the countryside, women laborers made up half the rural labor force. The number of urban working women increased from 600,000 in 1949 to 56 million in 1992 (China Information Office 1994). From 1982 to 1990, among the 63 million newly employed women, 23 million, or more than 36% of the total, were employed in cities and towns; 40 million, or 54%, were employed in rural areas.

The working sisters from outside are often brought to big cities like Beijing, Shanghai, and Guangzhou or introduced to potential employers by relatives, friends, and fellow villagers. They also find work by reading local recruitment notices, through advertisements, or by making inquiries in shops, enterprises, or plants. In addition, labor contractors recruit workers at railway stations. After briefly negotiating working conditions and pay, the job hunter follows the labor contractor to the place of employment. Usually workers with a specific goal find work after a couple of days or a week. The few who cannot find suitable work in this amount of time return home. Most work in one of four primary occupations:

- Low-skilled, manual labor in textile plants, light construction, or manufacturing;
- Serving in small restaurants, preparing vegetables and washing dishes;
- Selling clothing, shoes, and hats in small shops; or
- Domestic work in private households.

Bridget Hill found that the most common occupations for female rural/urban migrants throughout the world are household labor, street vending, and prostitution (Hill 1994). Among the cases of prostitution investigated by Beijing public security institutions, more than 90% are rural migrants. This is a source of social prejudice against *dagong mei*.

## REASONS FOR BECOMING A HOUSEHOLD WORKER

Rural women had only one reason in common for becoming household workers, which was to earn money. In 1986, a household worker could earn a monthly income of Y150-180, roughly twice the salary of an office worker. In 1993, the salary for full-time service of cooking, washing, and cleaning was Y200-300, in addition to meals and lodging. Care of a child under age two or a lying-in mother or sick elderly family member cost Y400-600, plus meals and lodging.

However, *dagong mei* had various reasons for needing money. One household worker from Heilonjiang Province said that her brother had spent Y2,000 on his wedding and fallen into debt. After consulting with her mother, she decided to come to Beijing to earn the money to pay off his debt and save for her own marriage. The father of another woman, from Henan, had died when she was only four years old. Her mother had remarried, and she had gone to live with her grandmother. They had nine Mu$^3$ of contracted plots, but they were unable to farm the land. Therefore, they turned it over to their neighbors for help. A friend had brought her to Beijing to see if she could earn some money to support her grandmother. A third woman, from Sichuan, said, "My family is not poor. We live in a large house with a new tile roof, and we eat well. Unluckily, my father wanted me to marry a carpenter. The matchmakers had come many times. I hated him, so I left home."

Money was not the only reason for coming to the city. One rural young woman who sought domestic employment did not carry clothing, a comb, or cosmetics in her netted bag, as did the other *dagong mei*. Instead, she carried a big, thick book, the *Chinese Language Encyclopedia*. This looked odd to many potential employers, and she became known as "Big Dictionary." She refused an employer who wanted her to care for his baby, another who needed someone to look after her elderly mother, and someone else who wanted her to cook particular dishes. The rejected employers were all surprised, and one was even angry. He looked at her thick book and asked, "Then what are you coming for?"

She answered, "Most people in Beijing are intellectual. I came to become a housemaid so I would have a place to live in the capital and be able to learn something from the people. I will certainly do some work for my host, but I hope I will also find the time to study. I plan to take a university entrance examination." She was very fortunate to be introduced to a professor of drama and to learn art and drawing (Anfeng 1990). This was a rare and special case, but it illustrates the broad range of reasons for becoming a domestic worker.

Other researchers have found a similar range of reasons among domestic workers in other parts of the world. Hill cites the three main attractions of domestic labor in the city for nineteenth-century rural British women, in addition to the "lure of the town," as being the chance to accumulate a dowry through higher relative earning, the potential for upward mobility through marriage to an urban man, and escaping an unwanted marriage (Hill 1994). Erdmans found that Polish immigrants to the United States preferred the lower risk of deportation in a private home and the chance it provided to read, study, and learn English (Erdmans 1996). Armstrong found that Malaysian Chinese women considered the benefits of household work to be relatively safe living conditions, independence from male dominance at home, and the chance to learn English. Although they had less privacy and control over their personal lives than factory workers, they had more control over scheduling and performing their daily tasks (Armstrong 1996, p. 157).

A relatively neglected area of inquiry is the way in which household labor in the city can provide opportunities for rural migrants to develop their skills in a service trade and earn more money. One woman from Wuwei County said, "More than 10,000 village girls from my county are now working in Beijing. Some of them have come several times, or for several years. They have not only earned a lot of money, but also have learned much. I also want to know what's happening in the outside world. It's fun!" What did they learn? They wanted to learn to cook new dishes, so they could run a restaurant when they returned home. They wanted to learn dress cutting and sewing, so they could open a tailor shop. They wanted to learn painting, so they could be teachers in village schools.

This effort is useful for the development of both individual rural women and the rural society, because *dagong mei* often return to develop service trades or township enterprises in their homeland (Gao 1993). In addition, household workers often function as amateur marketing consultants and active participants in economic activities. For example, based on her observations in several wealthy households, a woman might pass along to her brother in the village the information that a kind of specially preserved and dried duck meat would sell well in Beijing. Two years later, her brother might be able to build a two-story house to replace the family's old straw huts with the money earned in this business. The local Women's Federation of Lujiang County in Anhui Province sent 200 young women to Shanghai to learn toy-making skills through the information and contacts of a woman who worked in an official's home in Shanghai. The county now has a toy plant with yearly benefits and revenue of approximately Y4 million (Wang 1992).

## OFFICIAL AND INFORMAL MARKETS FOR DOMESTIC LABOR

China is a populous country with relatively slow socioeconomic development. In the past, the supply of labor surpassed its demand. With the gradual establishment since 1979 of a socialist market economy, the protection for unproductive enterprises that existed in the past is weakening. Competition is beginning to benefit the superior enterprises and eliminate the inferior ones. In this process, the structure of production in China is continuously under adjustment. Some enterprises have been forced to close, merged into other companies, or transferred to other types of production. The weakening of the work unit as a source of such benefits as cafeterias and child care has increased the demand for domestic help in urban areas.

In Beijing, 80,000 families—or one in 20—need domestic help with housekeeping or child care, but only 50,000 people offer the service. In 1983, the Beijing Women's Federation set up the March 8th Service Company, the largest of the five domestic employment agencies in Beijing. In the spring of 1984, it sent out a notice nationwide that healthy, well-behaved rural women over 17 years of age could apply for housekeeping jobs. Applicants were required to bring a health certificate from their local hospital and a letter of introduction from the local rural administration.

Since then, the company has recruited about 24,000 village women from 24 provinces through the help of local women's federations. As soon as the women arrive at the office from the railway station, they put down their luggage and talk with potential employers. If both sides agree, the woman follows the employer to his or her home. When they first arrive at the employer's home, the women often look pale and fatigued, with disheveled hair. Usually they have a very big appetite for staple foods at first. Later, they become ruddy-cheeked.

March 8th helps employers and potential employees negotiate and sign contracts, which usually stipulate the following mutual duties and obligations: Employees must stay for a trial period of at least one month, take care of employers' property, and compensate them for anything they destroy. Employers must be polite to employees, give them a day off every fortnight, provide meals and lodging, and pay them at least Y30 per month (in 1985). Once the contract is signed, each party pays March 8th Y5 for the service rendered. The staff of March 8th is supposed to visit the employer and the house to enforce the contract (Xiao 1986). However, this is not always done.

In the 1980s when it was new, March 8th was considered reliable, the wages were reasonable, and its organizing work suited the needs of many families in the city. Beijing residents queued at the office at Jianguomen Street to request employees, and the small staff and limited number of village women could hardly meet the demand. Later, the numbers grew as household workers brought country friends and relatives there to find jobs. As the employers and young women waited outside the company, they talked. If the women had not brought the necessary health certificates or letters of introduction, or if the employers wished to avoid the registration procedures or fees, the employers and employees would negotiate directly. Thus, a large informal labor market formed just outside the doors of March 8th.

In addition, there were three other large informal markets in the residential areas in the north, southeast, and west of Beijing, where five times as many household workers were available as through March 8th Service Company. Trading in the markets was brisk between 10:00 a.m. and 5:00 p.m. on Sundays. In the market, usually 200-300 women assembled, mostly young girls and some middle-aged women. There were also some male farmers, brothers and nephews of the women, who followed the example of their female relatives and came to try their luck. Most of the prospective household workers were from Anhui Province, with smaller numbers from Sichuan, Henan, Zhejiang, and Hebei.

Women in the informal markets demanded high prices, worked on an oral agreement, and often left their work at will. Generally, Beijing residents did not have much confidence in workers from the free markets. Likewise, honest workers did not have much faith in unscreened employers. Investigations revealed that numerous incidents of female abduction occurred in these markets. It is a matter of luck or chance when city residents manage to employ a reliable helping hand, either in the official or the free labor market, who can get along well with others in the household. Therefore, people often prefer to employ women introduced to them by a friend, neighbor, or acquaintance.

## Working Conditions for Household Workers

Because of the high demand, employees can be selective about their employers. Generally, *dagong mei* are unwilling to serve sick people, or care for a child under two years old. They want employment with a working couple or a small family. They prefer light work in a home with electrical appliances. In the 1980s, they demanded more than Y50 a month in addition to meals and a room of their own. One woman from Anhui refused two employers with babies who offered Y40 per month, a military man who needed someone to cook and do the washing for his elderly parents, and a working couple who needed someone to prepare lunch for their 7-year-old son. The employer she finally accepted was a self-employed entrepreneur in his thirties who needed someone to prepare supper for him and his wife every evening. There was to be no other work in the daytime except to stay at home and take messages if anyone visited them on business. She was pleased because the employer rented a counter and sold fashions in the biggest department store in Beijing. He owned a car, a fully carpeted three-room flat, a television set, and a refrigerator. He would pay her Y50 a month in addition to meals and a room of her own.

## Employers of Household Workers

According to the March 8th Service Company, 30% of the employers were factory or office workers, 15.3% were professionals, and 11.2% were self-employed entrepreneurs (Xiao 1986). The employers were mostly dual-wage earners. Some were retired couples where the wife was sick. Others were couples who needed child care while they did errands. There were also young couples who both wanted to devote time to their careers or were studying for degrees and had no time to prepare their meals. In addition, in the big cities, employers included families of foreign ambassadors and their staff or expatriate employees of foreign firms.

Very recently, it has been reported that peasant families in the Beijing suburbs have started to employ household workers from Sichuan and Hubei Provinces. The employers are mostly those who ran the township enterprises. According to March 8th, in Haidian District in western Beijing, plenty of suburban peasant families wanted to employ household workers, but few women wanted to work in such an inaccessible area, especially those who were already used to working in metropolitan Beijing. Only a few dozen women out of the 500-600 working in Beijing were employed by suburban peasant families (Ligang 1995). The most attractive employers were families of overseas Chinese who came back to do business, the bosses and staffs of joint ventures, and families of the staff of foreign organizations.

Researchers have found similar preferences among household workers in other areas. The job satisfaction of Filipina household workers in Hong Kong was most positively affected by a private room, good wages, the employer's nationality, length of residence in Hong Kong, and the number of their relatives living in Hong Kong. They preferred to work for Europeans, whose attitude toward servants was more informal and less hierarchical than that of Chinese families, who considered servitude normal (French and Lam 1988, p. 88). Armstrong found that Malaysian Chinese household workers preferred jobs that were convenient to transportation and near the places of employment of friends or relatives (Armstrong 1996, p. 157).

## Living In, Living Out

With an average income of under Y1,000 per month, many Chinese families cannot afford a full-time servant. Therefore, many *dagong mei* work for several families by the hour. In general, working hours are very flexible. Some work half-time for one family, with whom they live and take their meals, and work by the hour for other families. Others work for several families, cleaning house for one in the morning, cooking for another at noon, and washing for a third in the afternoon.

Some women work for each family for two hours, twice a week. This means that a woman working by the hour might work for 12-18 families each week, not including Sundays.

Over time, as inflation has risen in China, wages for household workers have also risen—from Y2 per hour in 1992, to Y2.50 in 1993, to Y3 in 1994, and to more than Y4 by 1996. Wages also varied depending on location and duties. For example, about 14% of families in the city of Nanjing in southeast China employed household workers by the hour to cook lunch or supper or to take care of a child. The pay for cooking was generally Y10 per hour. Some employers paid Y150 per month for one lunch each day. The employees were mostly from Chu county, Shou county in Anhui Province, or Huaiyin county in northern Jiangsu Province, as well as local Nanjing people who were unemployed, of whom women made up 86%.

Many researchers have analyzed the competing benefits of working by the hour over living in the employer's home full time. Because duties and hours for household workers are almost universally at the employers' discretion (French and Lam 1988; Romero 1988b; Tan and Devasahayam 1987), working by the hour gives workers slightly more control over their duties and money earned:

> Domestic service is not product-oriented; it calls for no specific skills on the part of the servant and no precise, evaluative standards on the part of the employer. Its personalized, diffuse, and capricious expectations fortify and legitimize the dependency and its ascription (Tellis-Nayak et al. 1983, pp. 72-73).

Working by the hour leads from this pseudo-kin relationship to a contractual one, which gives the household worker more leverage in negotiating working conditions (Armstrong 1996, p. 148). It also leads to the "casualization of work" (Sanjek and Colen 1990, pp. 183-184; Armstrong 1996, p. 157), in which household tasks are divided and workers specialize in either cleaning or child care. This leads to greater professionalization and higher wages, rather than the traditional expectation that household employees will work harder and longer as they become more attached to the family.

## CASE STUDY: FAN DAJIE

Fan Dajie[4] was 45 years old and came from Wuwei County in Anhui Province. She had no first name but was known by everybody as Dajie, which means "elder sister." She had married at age 28 and had a 15-year-old daughter who was in her third year of study at the junior middle school in Xiang. Her husband worked on the farm, tilling 4.5 Mu of the family's contracted plots. She had a younger sister who also worked as a household worker in Beijing. Another younger sister and brother-in-law sold raincoats in Beijing that were made in the township enterprise in their homeland. She had a younger brother who was involved in trading wood in Anhui Province. His wife took care of their four daughters and the household chores. Fan said that her brother and sister-in-law had wanted to have a boy, but one girl was born after another. Because there were so many children, her sister-in-law could not leave the village, otherwise she also would have come to Beijing, like Fan.

Fan said, "My father-in-law died very early, leaving two daughters and a small boy, my husband. My mother-in-law was already 73 years old, and she was not very competent even when she was young. My husband had grown up almost under the total care of his elder sister, who was engaged to her uncle's son. Her uncle was a cadre[5] in the village and supported [my husband's] sister until she graduated from senior middle school. Then she married her uncle's son, who was also of senior middle school background. So my family was for a long time supported by my husband's sister." They were considered one of the luckiest families in their village in Wuwei county.

Fan said, "My husband's village was originally much better off than the village where I was from." The village of her family of origin was rather poor, so many people left the village to make a living or married people from wealthier places. The people of her husband's village, including her husband, did not like to move, because life there was not too difficult. He had traditional ideas and was satisfied with working on the farm. "But my fellow villagers thought differently, maybe because they were too poor. Many of them left the village even in the early and mid-1970s. Not long after I got married, I discovered that I could not get along with him well. We quarreled a lot." In 1979, she decided to leave one day. "I got the address of one of my village sisters who had gone to Beijing. I just told a lie to my husband that I was going to visit one of his uncles in the city of Wuhu in Anhui Province, not far from Wuwei county." As soon as she left, her husband discovered her lie and wanted to chase after her and get her back. But the villagers laughed at him, saying, "She has already gotten to Wuhu, what's the use of chasing her?" Finally, he knew she was not at his uncle's when he called him on the phone.

When she arrived in Beijing, through the introduction of a fellow villager, she got a job as a full-time household worker in the family of a woman on the staff of the Capital Hospital and a man who worked in the Ministry of Foreign Trade. She stayed in Beijing for an entire year without writing to her husband. Later, she wrote a letter to her brother because she feared that her mother would worry too much. She never dreamed that the letter would be intercepted by her husband at the post office. Her husband rushed to Fan's village and threatened her family of origin that he would interfere with Fan's younger brother's marriage if she did not return home.

Fan's parents were very frustrated and had the administrative government of Xiang write to persuade her to return. She felt she had to do so to avoid trouble for her younger brother's marriage. Before she returned to the village, she asked her younger sister, who was only 15 at the time, to come to Beijing to take over her job, because her employer had four children who were then all in college, and the family was in great need of household help. Fan's brother got married without trouble from her husband. Later, Fan gave birth to a baby girl. She left the village for Beijing again when her daughter was five years old and her husband and mother-in-law were able to look after the child.

After Fan returned back to Beijing in 1982, she found a job in the family of an old military man who had a wife and a daughter. Fan worked half-time for this family. She spent the morning cleaning, washing (generally using the family's washing machine), and cooking both lunch and supper. She left the supper to be reheated in the evening. She earned Y250 per month plus food and lodging in a room of her own.

In the afternoon, Fan worked by the hour for families who lived in the neighborhood, or even in the same building. If there were no jobs nearby, she often went to work for families who lived farther away. But the other jobs had to be accessible by trolley or bus with a monthly bus ticket of Y7.50. She was usually able to arrange to work for one or two families in one afternoon or later in the evening for the whole week, so she could earn an extra Y200-250 per month.

After two or three years, the old military man retired and moved to a much larger house with five rooms, although the wife had died of breast cancer and it was now just the man and his daughter. The new house was very far away, in a development built specifically for retired military personnel. Fan hesitated, but the military man asked her to continue her service. The salary offered was satisfactory, the lodging and food were more than sufficient, the work was not heavy, and she had quite good relations with her employer's daughter. However, the difficulty was that Fan was not able to find enough work in the afternoon among the families in the neighborhood. This was definitely a disadvantage, and she finally decided to leave this family.

She returned to her village for the harvest and stayed for about a year. Her daughter had just started middle school, and she had to walk 30 minutes to Xiang. It took her an hour to get back and forth for lunch, and there was no one at home to prepare her lunch in time. Therefore, Fan stayed at home until she managed to arrange for her daughter.

When she came back to Beijing, she managed to find jobs in several households in the dormitory courtyard of the staffs of the organization of the *People's Daily*. Starting with these households as the center, she tried to find as many other households as possible nearby so she could save time and arrange her work for the whole week. She now works for two families twice a week, two hours each time. These employers are on the staff of the *People's Daily*. She works the same amount for two other families located only 10 minutes' walk away. These employers are researchers of the Chinese Academy of Social Sciences.

She also works every Tuesday, Friday, and Sunday morning for the family of a Singaporean businessman. He is about 40, married to a 27-year-old mainland woman, and they have twins, a boy and a girl. Fan does only the cleaning. There are other workers who take care of the twins during the daytime and at night. The Singaporean pays Fan Y250 a month.

She also prepares supper from Monday through Saturday for a Hong Kong man of nearly 50 who is engaged in advertising. He lives with a mainland woman of 25, from Guizhou Province in southwestern China, whom Fan calls "Miss." Fan said, "I don't know what kind of relationship they have, but the Mr. sometimes was wanted on the phone by his wife in Hong Kong. He also had two big children there. The Guizhou Miss is fairly pretty and looks young, but the Mr. is not at all handsome. He looks somewhat unpleasant. They go to work together; perhaps she is also his secretary. They pay me Y300 for the work, plus my supper. They dine at the restaurant on Sundays. I also have the responsibility of buying all the vegetables, groceries, and staple foods for the supper. My Miss used to give me, say, Y100 at a time. When I spent it all, I told her, and she gave me some more."

The Miss asked Fan to buy high-priced vegetables and rice from the supermarket, a German and Chinese joint venture. The rice was from Thailand and cost Y6.50 for a half kilo, or three times what middle-income people paid for rice. Fan said, "It seems too luxurious to buy ordinary vegetables and everyday groceries in this market. I used to save money, but my Miss asked me to buy more. She said the meals did not cost too much for them."

Fan often sought ideas from other employers about what kinds of good dishes she could learn to cook, because her Hong Kong employers never said what they wanted to eat, what they liked and did not like. "They do not know anything about cooking, they know only how to eat." She wanted to make new dishes and not always repeat the old ones, but she did not have the necessary cooking skills.

One evening when her employers returned for supper, it happened that Fan had not prepared enough food. She said, "I was very embarrassed. My Miss then called the restaurant of the apartment building to order a dish. It was just one sea crab. My Miss gave me Y100 to fetch the dish downstairs. Oh, my! It was not enough—two more Yuan were needed. I was really astonished!" This was perhaps not expensive to them, judging from the Y20,000 in rent they paid for their one-bedroom apartment with one sitting room, a bath, and a kitchen, and including furniture and cleaning services. Fan said, "The amount of money I spent for their supper was over Y500 per month." So it really meant nothing to them.

Every Saturday, Fan is employed by a 26-year-old woman who is the daughter of a former employer of Fan's, for whom Fan still works on Sunday afternoons. Fan does not know what company she works for, but she knows she earns a lot. She has her own car, and she had a mobile telephone until it was stolen. She also has a large house in suburban Changping County. The daughter drove her there when the house was new and asked her to help clean it. Now she employs Fan to do the cleaning every Saturday, all day, until it is time for Fan to go to the Hong Kong man's apartment to cook supper. At first, Fan did not feel like going there, because it was too far from the city and she might not be able to arrange time for other work in the city. However, the daughter was generous in paying for Fan to take a taxi to her house and back. In addition, the daughter pays Fan the full day's wage of Y24. Fan finally agreed, and now her work schedule was full for the whole week.

When Fan was asked, "Your work schedule is so full; do you feel tired?" she answered with a smile, "I am used to it. I fall asleep as soon as I touch my pillow. I get up at about 6:00 or 6:30 a.m. and cook and eat a big breakfast every morning." At noon she often buys a snack, such as biscuits, baked sweet potatoes, or pastries. She eats them on the way from one household to another for work. She said, "Later next year, I am not going to arrange my work schedule so full."

After cooking, eating her supper, and washing the dishes, Fan usually returns to her house between 9:30 and 10:00 p.m. Fan shares a room in a one-story house with another household worker, who is also from Anhui. They pay Y130 per month for the room, which measures about eight square meters. The room holds a bed that is two meters long and one meter wide, a table for their meals, and in winter a furnace for both cooking and heat.

One summer, Fan's brother-in-law, who often went between their village and Beijing for his raincoat business, brought Fan's daughter to see her mother and tour Beijing for the first time. Fan had never had time to see the sights in Beijing in the six years she had been there. She asked for a day off and took her daughter to the Forbidden City. Her brother-in-law took her daughter to some other scenic spots.

Her daughter stayed in Beijing for one week. When it was time for her to return home, she cried and insisted that her mother accompany her back to the village. Fan was very lucky, because it happened that the Hong Kong man and Miss had also taken a week's vacation outside Beijing. Therefore, she was able to go back to the village with her daughter. It took about three days by train to go round trip from Beijing to the village. Her brother-in-law easily bought a ticket on the black market, otherwise it would have been a problem. She stayed at home for just over two days. She waited until her daughter was not home to hurry back to Beijing. She purposely left without saying good-bye for fear that her daughter would feel sad and delay her departure. She had promised to be back to prepare supper for Miss the following Monday.

Fan is a strong-willed woman, diligent, quick, and daring. She said, "I was the first one in my husband's village who left for Beijing to work as a household worker. At that time, many fellow villagers gossiped a lot about me, criticized me as not caring for my family and my husband and wanting to divorce. In fact, I never thought of that, and I did not care, either. Now there were no such gossips any more. There were more and more people coming out to make a living, to earn money. In 1979 I earned Y30 a month, which was quite a sum. Generally, a cadre at that time earned only about Y70-80 per month. Besides, on Sundays I was allowed to go out to other households to work for a morning or so, and I was very pleased to earn Y0.25 per hour and one extra Yuan for the morning." She spent all she earned in the mid-1980s to build a one-story house of three rooms with a tiled roof. It cost approximately Y10,000.

## DISCUSSION

An old Chinese saying goes, "Man tends to go upward while water flows downward." This means that people would always like to find a higher position and better life. Rural youth, including women, are certainly no exception. From the point of view of the social sciences, the migration of rural laborers from agriculture to industry, from villages to the city, is a common phenomenon in the process of modernization (Hill 1994). However, for *dagong mei* engaged in household labor, migration does not represent a complete break with the past or an individualistic quest for personal betterment. Rather, in the context of the Chinese labor market, their migration serves as an ongoing survival strategy for the rural family unit (Ozeki 1995; Hill 1994).

Yanjie Bian analyzed the role of work units in social stratification among urban Chinese workers. He found that workers' labor was not a personal commodity; it belonged to the urban work unit, which controlled both material rewards and ideological and organizational incentives. "Just as peasant households are actors in the rural marketplace, work units are the direct producers in the cities" (Bian 1994, p. 16). Because rural migrants are restricted by law from joining urban work units or receiving food allotments (Bian 1994, p. 54), they remain integrated in the economic system of the rural family unit, even after many years of working in the city.

Fan remained tied to her family and local administrative units, returning to the village at their insistence to protect her brother's marriage plans. However, she demonstrated her commitment to her employers in Beijing by arranging for her 15-year-old sister to take over her duties with them. Her sense of responsibility to her employers is also illustrated by her overcoming her reluctance to accompany the military man and his daughter to an inconvenient suburb. Nonetheless, when she left this employer and returned to the village, she remained there a year until other arrangements could be made for her daughter's care.

Tan found a similar pattern among Filipina maids in Singapore, whose labor is considered to belong not to them but to their families. The family in the Philippines bears the costs of rearing and educating the household workers, and they are responsible for caring for them in old age (Tan and Devasahayam 1987). This is why even those *dagong mei* who come to the city hoping to transfer to an urban household registration keep up ties with their fiancés in the village until they are assured of marrying a city man. For their part, rural families maintain control over the behavior and earnings of *dagong mei* through such common tactics as hiding the women's identity cards or involving the local administrative authorities in family disputes.

The pattern of household workers' integration in both the rural family as an economic unit and the urban labor market reflects the perspective of "transnationalism" that researchers have adopted in analyzing international migration. Erino Ozeki defines the phenomenon as:

> a way of life whereby they lead their lives across cultural, economic, and political boundaries, but still maintain their ties to home, irrespective of the geographical distance between their countries of origin and destination (Ozeki 1995, p. 38).

Although in the case of China the migration is exclusively internal, household registration as a requirement for permission to work in the city sets up many of the same dynamics as international citizenship requirements. However, relatives engaged in trading between the village and the city provide opportunities, which rarely exist in the case of international migration, to maintain vital contact between the spheres. As household workers move from full-time employment in a pseudo-kin relationship with a single family to contractual arrangements by the hour with multiple families, the importance of sustaining relationships at home will increase.

## NOTES

1.  Household workers are defined as those hired from outside the family to engage in reproductive labor that maintains the residents on a daily basis and intergenerationally by cleaning the dwelling, preparing food, and caring for the young and elderly. Reproductive labor intersects with productive labor by freeing the household members for leisure, productive work or social contacts (Sanjek and Colen 1990).

2.  The exchange rate for RMB Yuan per U.S. $1.00 was Y5.76 in January 1993, Y5.51 in 1992, Y5.32 in 1991, Y4.78 in 1990, Y3.77 in 1989, and Y3.72 in 1988 (Central Intelligence Agency 1996).

3.  About 1/6 acre.

4.  This case study is based on conversations between Fan and Dai Kejing, who employed her twice a week for two hours over the course of almost six years.

5.  Communist Party official.

## REFERENCES

Anfeng. 1990. "Little Housemaids in Beijing." *Journal of Marriage and Family* (February) [in Chinese].

Armstrong, J. 1996. "Twenty Years of Domestic Service: A Malaysian Chinese Woman in Change." *Southeast Asian Journal of Social Science* 24(1): 64-82.

Bakan, A.B., and D.K. Stasiulis. 1995. "Making the Match: Domestic Placement Agencies and the Racialization of Women's Household Work." *Signs* 20(2): 303-335.

Bian, Y. 1994. *Work and Inequality in Urban China*. Albany, NY: State University of New York Press.

Central Intelligence Agency. 1996. *World Factbook*. [Online]. Available: http://portal.research.bell-labs.com/cgi-wald/dbaccess/411?key=52

China Information Office. 1994. "The Situation of Chinese Women." Beijing: State Council of the People's Republic of China [in Chinese].

Dai, K. 1991. "The Life Experience and Status of Chinese Rural Women from Observation of Three Age Groups." *International Sociology* 6(1): 5-23.

Dai, K. 1993. "The Impact of Changing Socio-economic Policy on Women's Choices and Challenges in Mainland China." Paper prepared for the International Sociological Association Committee on Family Research XXX Seminar, Annapolis, MD, November.

Erdmans, M.P. 1996. "Illegal Home Care Workers: Polish Immigrants Caring for American Elderly." Pp. 267-292 in *Current Research on Occupations and Professions*, Volume 9, edited by H.Z. Lopata and A.E. Figert. Greenwich, CT: JAI Press.

French, C., and Y.M. Lam. 1988. "Migration and Job Satisfaction: A Logistic Regression Analysis of Satisfaction of Filipina Domestic Workers in Hong Kong." *Social Indicators Research* 20(1): 79-90.

Gamburd, M.R. 1995. "Sri Lanka's 'Army of Housemaids': Control of Remittances and Gender Transformations." *Anthropologica* 37: 49-88.

Gao, X. 1993. "New Social Divisions." *Chinese Women News* (3, December) [in Chinese].

Heyzer, N., G.L. Nijeholt, and N. Weerakoon, eds. 1994. *The Trade in Domestic Workers: Causes, Mechanisms and Consequences of International Migration*. Atlantic Highlands, NJ: Zed Books/Kuala Lumpur: Asian and Pacific Development Centre.

Hill, B. 1994. "Rural-Urban Migration of Women and Their Employment in Towns." *Rural History* 5(2): 185-194.

Lee, Y. 1995. "Three Large Groups Pressing to the Labor Market." *Chinese Women News* (December 17) [in Chinese].

Ligang. 1995. "Housemaids Enter Suburban Rural Families." *Beijing Youth Daily* (December 17) [in Chinese].

Liu, X. 1995. "About Love and Marriage of Working Sisters." *Chinese Women News* (August 10) [in Chinese].

Ozeki, E. 1995. "At Arm's Length: The Filipina Domestic Helper: Chinese Employer Relationship in Hong Kong." *International Journal of Japanese Sociology* 4(September): 37-55.

Research Institute of Rural Development. 1995. "Study of Female Laborers." *Sociological Study*, Vol. 4. Beijing: Institute of Sociology, Chinese Academy of Social Sciences [in Chinese].

Rollins, J. 1985. *Between Women: Domestics and Their Employers*. Philadelphia, PA: Temple University Press.

Romero, M. 1988a. "Sisterhood and Domestic Service: Race, Class and Gender in the Mistress-Maid Relationship." *Humanity and Society* 12(4): 318-346.

Romero, M. 1988b. "Chicanas Modernize Domestic Service." *Qualitative Sociology* 11(4): 319-334.

Sanjek, R., and S. Colen, eds. 1990. *At Work in Homes: Household Workers in World Perspective*. Washington, DC: American Anthropological Association.

Tan, T., and T.W. Devasahayam. 1987. "Opposition and Interdependence: The Dialectics of Maid and Employer Relationships in Singapore." *Philippine Sociological Review* 35(3-4): 34-41.

Tellis-Nayak, V., K.T. Hansen, H.D. Lakshminarayana, J. Pettigrew, G.N. Ramu, S. Shahani, and L. Walter. 1983. "Power and Solidarity: Clientage in Domestic Service." *Current Anthropology* 24(1, February):67-74.

Wang, X. 1992. "New Interpretation of Housemaids." *Chinese Women News* (December 30) [in Chinese].

Xiao, M., et al. 1986. "The Househelpers in Beijing." *Journal of Chinese Women* (October) [in Chinese].

Zhai, F. 1991. "Rural Workers Needed in Cities." *China Daily* (June 26) [in Chinese].

# ASIAN ENTREPRENEURS IN CHICAGO

Jing Zhang

## ABSTRACT

This study examines business development in a recent Asian community in Chicago, which is known for multiple Asian cultures such as Chinese, Vietnamese, Thai, Cambodian, Laotian, and East Indian. The paper explains why certain Asian groups are more likely to concentrate on family business, their journey to start a business, and how they operate their businesses on daily basis. Through observations and interviews, it was found that the range of business types among Asians were much wider than previous sociological research suggested. Family businesses not only provide means of living for these groups but also play an crucial role in building emotional bonds among community residents, providing a cultural window to non-Asians outside the community, and providing a cultural base for Asians who live outside the community. In addition, business development further attracts external resources to the community for its overall development.

## OVERVIEW

Chicago School sociologists suggest the middlemen minorities are the exceptions to the assimilation process as most of them will eventually return to their home coun-

Current Research on Occupations and Professions, Volume 10, pages 31-66.
Copyright © 1998 by JAI Press Inc.
All rights of reproduction in any form reserved.
ISBN: 0-7623-0034-5

try. These minority groups often associate with special occupational niches by virtue of a combination of circumstances, and a cultural heritage that has been used as an adaptive mechanism (Lee 1960; Boggs and Fave 1984; Light 1972). Light's research (1972) found that most Asians operated laundries, restaurants, groceries, and import outlets, because they had the advantage of their customers' "unusual" demands. Asians either did not buy automobiles, lumber, clothing, and furnishings, or if they did, they purchased them from whites to avoid competition. Asian businesses are overrepresented in retail businesses. Lee's (1960) research suggests that the successful and highly competitive practice of ethnic entrepreneurs originates from the "sojourner" mentality—thrift, diligence, and willingness to stand longer hours are all consistent with the goals of saving enough money to eventually return home.

Other sociologists (Yoon 1990; Tsia 1986; Wong 1982; Nishi 1976; Loewen 1988; Boswell 1986; Doeringer and Poire 1971; Li 1988) argue that racial discrimination has forced many earlier Asian immigrants into self-employed businesses. For example, the Chinese Exclusion Act in 1882 and anti-Chinese sentiments forced Chinese out of manufacturing employment and pushed them into petty businesses catering to their own ethnic groups or providing services such as laundry and cooking for the wider community. Even Chinese-owned factories were forced to cease operations. By 1900, the only labor markets open to the Chinese were domestic service and self-employment. Sociologists from this tradition argue that Asian-American professionals are blocked from their upward mobility in the corporate world because of discrimination; therefore, they are more likely to enter into self-employed business as an alternative (Tang 1996; Lau 1988).

This chapter is focused on a most recent Asian community in Chicago and Asian entrepreneurs. This is a small geographic space centered around "the shopping strip" on Argyle Street between Broadway Street and Sheridan Road. Its northern border is Foster Avenue, its southern Lawrence Avenue. Sheridan Road forms the eastern boundary, and Magnolia Avenue the western one. The community is officially known as "New Chinatown." However, it has diverse populations and is also known by other names, such as "Vietnamese Refugee Camp," "International Shopping Center," and "Asian Village." Despite its multiple names and diverse populations, the community is best known for its Asian businesses and exotic Oriental food, arts, and crafts.

In the 1960s, the community was known as a most depressed areas on the north side of Chicago because of its poor housing facilities, high crime rates, high poverty rates among its residents, and lower property values. However, the community began to change its image in the 1970s, when a group of Chinese merchants from the south side of Chicago relocated there and refugees from Southeast Asia were resettled in the neighborhood. The businesses in the community have expanded far beyond this initial "shopping strip." Many stores have opened on Broadway, Sheridan, and Lawrence Avenues as well. Various businesses meet almost every need of the community. They include grocery stores, restaurants, bakeries, gifts, jewelers, clothing stores, laundromats, medical clinics, dental clinics, car dealerships, video shops, banks, law firms, accountants, travel agencies, real estate agencies, photo shops, electronic products, insurance companies, and a mortgage and loan company. In addition to these Asian businesses, there are also increasing numbers of non-Asian businesses.

At the beginning of the development of the Argyle area, Asian business owners used to go to banks, accountants, and attorneys in Chinatown on the south side of Chicago. In the 1990s, however, more and more Asian operated service businesses have been set up in the Argyle community. In 1991, the New Asian Bank of the south side opened its first branch office on Broadway. In 1992, the International Bank opened its business. Accountants, insurance companies, and law firms from the south side and other parts of the city have also opened branch offices to meet the needs of the Argyle community.

The growth of the "commercial strip" on Argyle street further attracted other businesses from outside the community to invest in the area. The community has increasingly become a self-sustainable entity and a regional shopping center. In the early 1980s, property values jumped about 60% as the Chinese first moved in to the area. By the end of the 1980s, the property values had risen by another 35%, according to a community business leader (Feyder 1977). Local officials indicated that the Argyle commercial area yields $77 million in city sales tax revenue and helps retain and/or create over 300 jobs annually. This development is seen by many as "a cure" for this distressed area in Uptown.

In this study, I examine the process of starting self-employed business among new Asian refugees and immigrants, how other non-Asian businesses responded to these trends and survived when the community became increasingly Asian, how businesses operate, and the role of self-employed businesses in the community. This study shows that Asian businesses are operated by many Asian groups and that the types of businesses are more varied than earlier literature suggests. For example, increasing numbers of Asian businesses in the Argyle community have expanded their businesses by starting their own wholesale businesses or combining several businesses at the same time. This study also describes the significant role businesses play in bringing people together and contributing to regional development.

## STARTING A SMALL SELF-EMPLOYED BUSINESS

The general public holds a number of "myths" about Asians and Asian businesses. First, Asians as a group are seen as more successful than other minority groups (Woo 1992). Second, Asians are also believed to be more likely to engage in small self-employed businesses than other ethnic groups (Light 1972). This section is not intended to dispel these myths. Nevertheless, the findings may help explain some misunderstandings or lack of understanding. This section focuses on the diverse Asian groups in the business community and the process by which they come to operate small self-employed businesses.

## Groups in the Business Community

Asian business owners in the Argyle community come from various back-grounds. For many Asians, starting a business means starting a family or an ethnic enterprise in their own community. Asian businesses in the Argyle community are dominated by four major groups: earlier Chinese who came before the arrival of Southeast Asian refugees; refugees from Southeast Asia; Chinese from Taiwan, Hong Kong, and mainland China; and other Asian groups. In addition, several older businesses predating the 1970s remain in operation, and other "non-Asian" businesses have opened up.

Chinese immigrants from Taishan (Toishan) were the first Asians to own busi-nesses in the Argyle community. Many of them were affiliated with the Hip Sing Association.[1] They often call themselves "old timers," or "old comers." As one "old timer" business owner said:

> Those were the pioneers here, they really had a hell of a time. We have to give them credit and we respect them for the hard time they went through. At the same time, those are basically like history. You have to catch up with the new trend.

In recent years, the influence of "old timers" in the community has gradually given away to other Asian groups, especially to the Southeast Asian refugee business people and entrepreneurs from Taiwan and Hong Kong. Some of the "old timers" sold their businesses to the newcomers and retired and some focus their attention on financing and property development in the community. Several shopping cen-ters and plazas and a row of townhouses on Ainslie and Winthrop were built by "old timers."

Mr. Kan is an "old timer" Chinese. He owned a restaurant on Argyle. In the 1930s, he followed his father and immigrated to the United States from rural Can-ton. Before he was employed in a Chinese restaurant in the downtown area, he worked as a "Chinese laundry man" for many years. He explained why most Chi-nese were involved in the laundry business during that time and how they started to run restaurants:

> At that time, only rich people among the Chinese ran restaurants. Most people ran laundries, because you could not find anything else. At that time, I earned $2.50 a week in a laundry.[2] Then the Jewish people came. They began to take over the laun-dry business. The Jews began to use machines. Before that, the Chinese did laundry by hand. Machine is more efficient than hand. Many Chinese lost their businesses. They thought what to do then. That's how the Chinese began to run restaurants. Everybody worked for restaurants or owned a restaurant.
>
> The Second World War provided opportunities for the Chinese restaurant business and it began to grow. Before the war, only 105 Chinese were allowed to enter the United States each year. After the war, 300 or 400 hundred. At that time the family only sent boys to America, because girls can't work and send money back home. So

Chinese went home for a year or two to have children and then brought them to the United States."'Do you have?" people asked each other. "Yes." "What do you have?" "Son." Thus, every one had a son. Some people made money out of it. People had boys and sold them to American fathers. After World War II, even the Germans brought their families into the United States. Only the Chinese still could not, even they served during the war.

In 1972, Mr. Kan quit his job at the downtown restaurant and bought his own restaurant on Argyle street. It was the first Chinese restaurant in the community. At that time, there was also a laundromat owned by a Chinese across the street. Mr. Kan said his business was quite good then. But recently there was not much business for him, because there were too many restaurants competing with one another. Occasionally, his old customers still came to eat. When they came, they asked about Kan's children who used to hang around in the restaurant. Mr. and Mrs. Kan sold their business and retired in 1995 after they were robbed on their way home from work.

This first group of business owners often compare their experience with the newer groups. As Mr. Kan said:

> When the Vietnamese came they had everything, not like us. They don't speak English, but we did. They have credit and money. If they don't have enough money, they buy secondhand cars. When we came to the U.S., we did not have credit. Even if we had money, we could not buy a car. It took us many years to have credit.

This comment points not only to the different experiences groups of immigrants have had, but also to a changed climate of race and ethnic relations. It has become easier for people of the same group who came to the United States more recently to do business than it was for older generations.

The second group of business owners are refugees from Southeast Asian countries like Vietnam, Cambodia, and Laos. Many of these people had previous experience of running a family business in their home countries, although those businesses might have been different in type and size. Among Southeast Asian refugee business owners, there is a significant number of ethnic Chinese (about 70% of businesses, according to a local business leader) who migrated to Southeast Asia mostly from Teo Chew and other regions in the south of mainland China.[3] Many of these ethnic Chinese intermarried with people in the Southeast Asian countries. The majority of current businesses in the Argyle commercial area are run by this group of business owners.

The other major group consists of Chinese immigrants from Taiwan, Hong Kong, and mainland China. Most of the Taiwanese businesses are professional and include a bookstore, an accounting firm, a bank, and several dental clinics. The rest of business community considers them as "successful." For example, a Taiwanese owns the only bookstore, an international franchise in the Argyle community. It has its American headquarters in New York and another branch store in

Chinatown on the south side. Mary Lee, a CPA, has a small office in Chinatown South but recently opened an accounting firm on Argyle street because many of her clients have businesses in this community. Her family also opened the International Bank on Broadway in 1992. As major shareholders, the family owns 70%-80% of the stocks. The president of the bank used to work at New Asian Bank in Chinatown on the south side. Most dentists are Taiwanese and Vietnamese. The Vietnamese who came in the first wave own comparatively more professional businesses compared to the rest of the same group.[4] They are medical doctors, dentists, CPAs, and pharmacists.

Within this group of Asian businesses owners are also immigrants from Hong Kong. Some escaped from mainland China and settled in Hong Kong. Others were born in Hong Kong but received educational degrees in the United States. For example, one of them is a partner in a dental office. A few medical professionals are from mainland China. The majority of these Chinese medical professionals work out of Southeast Asian grocery stores that carry herb medicines. Their earnings are split between the store owners and themselves. One of the doctors told me that her average monthly income is about $600. Some of these doctors are well-trained but do not have an American medical license, adequate English language skills, or permanent residency status. Thus, they have a hard time finding jobs outside their community. While their earnings are low by American standards, they are about three times higher than they were in China. In addition, they enjoy increased political freedom. In the Argyle community, these doctors are the closest to "sojourners," although their temporary status is not of their own choice and some have managed to stay.

The fourth group consists of a number of Asian businesses from a variety of other Asian ethnic backgrounds, including Korean, Pakistani, Thai, East Indian, and Filipino. There are four Korean businesses: a dry cleaner, a clothing store, a beauty supply store, and a fast food restaurant. Most of the Korean business owners received college degrees in their home country. In contrast to most other recent Asian business owners, these Korean business owners had also worked in areas other than family businesses in their home country. Thai business owners operate three Thai grocery stores and two Thai restaurants. Two Filipino doctors have offices in the community; three East Indians are operating businesses—one grocery, one furniture, and one laundry; and three or four Pakistani food businesses recently opened on Sheridan Road. According to a Pakistani business owner, more Pakistanis have been moving into the community recently. Most of them live in four apartment buildings on Sheridan Road and Foster Avenue.

## The Road to a Self-Employed Business

Earlier researchers attribute the success of Asian self-employed businesses to Asian cultural traits or heritage. They suggest that Asians and some other minorities start their businesses by rotating credit among a group of people who trust each other (Boggs and Fave 1984; Light 1972). Ordinary people believe that refugees receive government money or bank loans to help them start their businesses, or that they arrive with wealth to invest. However, the majority of refugee business owners in the Argyle community worked for several years in factories or at other jobs before starting their businesses.

To many Asians, owning a business becomes a family effort. Brothers and sisters, parents, children, or other relatives provide capital and labor.[5] The time taken to start the new business varies depending on family size, the scale of the business, and connections and credentials built up before coming to the United States. Some families had businesses and experiences in their home countries; relatives and friends are therefore likely to lend them money to start their businesses. For example, Minh's family is from Vietnam. There are six children in his family, four of whom are married. After working in factories for two years after they came to Chicago, the family was able to put enough money together to open a small grocery store. Since then, they have gradually expanded it. A young Cambodian man named Mike bought a fast food restaurant on Broadway in 1989 after he had worked in a factory here for eight years. However, even with eight years of savings, he still had to ask his uncle to help him financially.

When businesses begin to do well, some family members may leave and start their own business. Liang, a Vietnamese-Chinese beauty shop owner, first started a beauty shop with his sister-in-law in 1978. In 1979, he started his own business with his wife on Argyle street. In 1989, when the landlord raised the rent, Liang and his family moved to a building owned by Lakefront SRO (Single Room Occupancy) on Sheridan Road. According to Linda, a staff member at Lakefront SRO, the organization would like to rent space to business owners who stay and provide services to the community. Now, Liang and his wife and children own a second store on Argyle which they bought in 1980 when property values were still low. They seem happy with both locations. The one on Sheridan attracts customers from the east side of Sheridan road, such as the Jewish residents of Self Help Home on West Argyle and Marine Drive, or simply people who pass by on Sheridan Road. The shop on the Argyle shopping strip attracts customers shopping for groceries on Argyle street. Thus, Liang's family owns two beauty salons in Argyle and his sister-in-law owns one.

First-time business owners in the first generation of Asian immigrants and refugees are frugal and cautious. They started their businesses on a small scale, mostly with their own capital. According to a officer of a community bank, many borrowed money from banks only when they began to expand their businesses; the majority of the bank customers are individual commercial customers rather than consumer customers at this point. He explained:

> A lot of Asian-based businesses [in the community] are fairly new, there are a lot of small businesses within the community. Their credit needs are toward business growth and expansion. They are looking into different business opportunities. They

are getting into the import business and wholesale business. These are younger businesses, maybe around five or eight to ten years.

The other credit needs are on housing on the consumer side. It has been fairly active. They are all privately owned housing, the majority being single family residents. A lot within the community where you have two flats, three flats. Most of our customers at this point, have been Asian. Mostly Vietnamese, lots of first-time home buyers who may live within the family unit. They all work. They have obviously very good saving habits. They have accumulated enough cash for down payments and loans.

Most Asians are cautious when they borrow money from the bank, because for Asians, "to be in debt is not a proud thing indeed." As an officer from the International Bank said:

> The majority of the Asian population has good saving habits. They approach a bank, they will experience what they want in terms of business itself. They are concerned, not big risk takers. They are willing to commit their own capital. Not looking for the bank to lavish the whole business. They are looking for loans, say 200 dollar loans, their own expenditure could be 400 dollars. So they are at risk of their own capital.

Rosy, a social worker at the Indochinese Pastoral Center of St. Thomas Church, revealed that she was planning to build an apartment for Vietnamese elderly people close to the church. They were trying to get help from a nonprofit women's business development center in the city. She commented, "more or less, the American business people like to make things big. Asian people are cautious, concrete and much simpler." Asians have a very different definition of credit. For them, credit means they do not owe anybody anything. So most of time, they start their businesses on a small scale and then gradually expand. Charlie Soo, Director of the Asian Small Business Association, also made the same observation through helping Asian businesses invest in business improvement. He said:

> Years ago, I set up a reception at one of the restaurants, a free buffet, and nobody showed up. We tried to get into a city program, the facade rebate program, but none of the owners would apply. It's the old Chinese business philosophy. They don't spend money until it's absolutely necessary. They told me, "why should I redecorate my business facade? How does that make more business for me."

Nationwide data (see Table 1) show that Asians are more likely to use their own resources in their businesses. Like other groups, the vast majority of Asians start small, with 53% requiring less than $10,000 in startup capital. When capital is needed, Asians, like others, use a combination of personal savings, personal loans, and commercial loans to raise the startup capital (Ong and Hee 1994). However, Asians rely more on family, friends, personal savings and ethnic ties to help raise funds. The data also indicate that vertical linkages—trade to other firms—within

**Table 1.**   Characteristics of Business Owners

| | Asian Pacific American (%) | Latino (%) | African American (%) | White Males (%) |
|---|---|---|---|---|
| Have entrepreneur relative | 35.3 | 30.7 | 27.8 | 48.0 |
| Worked for entrepreneur relative | 17.0 | 12.1 | 10.0 | 23.7 |
| Required startup capital | 81.6 | 69.4 | 69.5 | 75.3 |
| Personal loans for startup* | 8.5 | 5.5 | 5.4 | 4.1 |
| Had commercial startup loan | 12.2 | 8.7 | 9.5 | 16.0 |
| Borrowed from relative(s) | 12.3 | 6.7 | 3.3 | 7.2% |
| Borrowed from friend(s) | 7.7 | 2.5 | 2.0 | 1.7 |
| Borrowed from prior owner | 3.9 | 0.9 | 0.6 | 1.9 |
| Purchased business | 19.1 | 9.9 | 9.2 | 15.4 |
| Sales to other firms | 18.0 | 17.0 | 12.1 | 25.0 |

**Note:** * Personal loans include loans from spouses, personal credit, and refinancing of homes.
**Sources:** U.S. Bureau of the Census (1992), Ong and Hee (1994).

the Asian community exist but are not extensive compared to the other three groups. This demonstrates that the vertical linkage is predominantly between wholesalers and retailers. There are fewer manufacturers and producers. Ong and Heé (1994) point out that the unique resources of Asian Americans explain their relative economic advantages but warn that we should not neglect the barriers that keep many operations marginal, even though some limitations may be overcome with time.

## Why Start a Self-Employed Business?

Since many Asian immigrants and refugees are well-educated and worked in mainstream professions and occupations in their home countries and even after immigration, why should they decide to invest in small self-employed businesses in the Argyle community? Asian business owners have offered a variety of explanations. It is true that many have a family tradition of being self-employed. Asians who quit mainstream jobs and invest in family businesses rely on pooling the resources of the whole family. Some business owners, especially those who have children, feel that self-employment allows them more autonomy and flexibility than salaried or waged employment. For many Asian business owners, the Asian residential community helped them succeed in maintaining and expanding their businesses. For some professionals, linguistic and cultural barriers restricted them from working outside the community or advancing within mainstream organizations.

### Family Tradition

Most business owners in Argyle, especially traditional and intermediate business owners, operate a family business because their parents had such a business in the home country. They grew up in a family business environment and observed their parents operating the business. Or they themselves had such a business before coming to the United States. It was their "dream" to have a business of their own in a new country. They were motivated to work hard and save enough to open their own enterprises. Nevertheless, they also recognize the differences between running a business here and in the home country. They may find themselves not as prosperous or respected as they would have been in their home country even if they have their own businesses and work very hard. Liang, the beauty shop owner from Vietnam, complained, "it was much easier to have the same business in a big city in Vietnam. There were people coming and going. Business was good. Here we do not have that many customers."

Another business person echoes this view. Nya, a Vietnamese refugee noted that her family had several businesses in Vietnam—a restaurant, an electric company, and a construction company. Talking about their American experience, she has mixed feelings:

> We are doing fine here. People all say we are most successful. But, we work much harder than we did in Vietnam, and we did not get much. Here there are too many businesses—American ones. No matter how hard you work, you can not get as much as you could in Vietnam. In Vietnam, our family business had good reputation, and we enjoyed prestige, but here although we work hard and we have money, we are just small business. United States is too big, Chicago is too big.

For most business owners, starting a business in a different culture also means they have to adjust to a different business environment. In Chicago, they have diversified customers; therefore, business owners have to identify and carter to diverse tastes and needs. They carry "a little bit of everything." They are not only ethnic businesses who sell to "their own kind" but businesses that need to serve a diverse market if they are to survive. For example, among their customers are people from Vietnam, Laos, Cambodia, Hong Kong, Thailand, mainland China, Taiwan, the Philippines, and the native born. In addition, their customers, while sharing a country, may be from different regions and therefore have differing tastes. At different seasons, they purchase different amounts of different goods. Because "many goods are transported from a long distance instead of nearby fields," shop owners have to pay more attention to cleanliness and the quality of food. The business owners also feel frustrated by the bureaucratic and technical procedures required in the United States, for instance, business regulations and tax requirements. In the home country, many jobs that were done by family members—for instance, book-keeping—require the services of a professional, such as a CPA. "Business operations became much more complicated," according to many Asian business owners.

*Financial Benefits from Working as a Family*

Family members explain that working in the factory or at other jobs only allows one or two family members to work, usually for low wages. When running a family business, the whole family can help and contribute to family income. In Asian businesses, older family members, parents, and grandparents often come to help during the day, cleaning the floor, or sorting and packing goods. Brothers and sisters and children help after work or school and on weekends. Mr. Lee, the Cambodian Chinese grocery store owner, said:

> I worked in a factory in Morton Grove, because my sponsor lives there. We decided to open a business here because we thought it could run around the family. My sponsor did not want us to come here. "It's not safe," he said. But if I work in the factory, I could not support family. I was the only one working and children were young, so we decided to come to Chicago and start my own business.

Similarly, the Korean clothing store owner has two children going to college. She works at a family business alone while her husband works in another store at another location in order to help put their children through college. Minh, a Vietnamese Chinese youth, whose family owns a grocery store on Argyle, said, "on week days, four family members are working in the store, including my mother and myself. The rest of family members are working at other jobs, but whenever they have time, they will come and work in the store, especially at weekends." Later, Minh left the family business to continue his college education.

Family members are favored at Asian businesses not only because they speak the same language but also because they are willing to work long and flexible hours. At times, their businesses do hire non-Asians from the community, but mostly as temporary workers for such tasks as delivering, carrying, shelving goods, or cleaning, depending on the seasons. As Nya, the Vietnamese clothing store owner, said:

> It is not that we can't pay for a worker. It is just because family members will work harder. If you hire someone, he will leave on time after seven or eight hours working, but family members can work longer time if necessary. We usually close at 7:00 in the evening. If there are still people coming in, we close a little late.

She and her sister operate two businesses in the community. Nya and her sister noted, "Our family work as a team between these two businesses. As soon as this store (clothing) closed in the evening, we all work in our restaurant."

Family business also becomes a way to take care of one's extended family. Phan, the Vietnamese pharmacist and also owner of the building, leased a space on the first floor to Dr. Marcos from the Philippines. Phan introduced a Vietnamese tenant in his building to help the doctor. Phan also hired his cousin, who was attending Truman College, to work part-time in his drug store. Phan explained, "I have to help him to go to school. I have to take care of my family. It is the priority."

Most of the businesses in the community close at 7:00 p.m., except restaurants and bars which are open from 11:00 a.m. until 11:00 p.m. However, most of the businesses would keep open until the last customer is gone. Sometimes, they even open up if customers show up after the store is closed for cleaning. The official "weekend" for Argyle is Tuesday. Stores stay open on Sundays and close on Tuesdays, but more and more businesses are staying open seven days a week, including one of the banks.

Among the few businesses that hire regular non-Asian employees are a Vietnamese dentist who has an American partner, and a Korean clothing store owner on Argyle who hires a Latino to help her sell clothes because she has Latino customers from the west side of Broadway. She says that "hiring depends on customers. We have non-English-speaking Latino customers, so we hired her. She speaks the language." Whether to hire local non-Asian residents depends on the customer bases of each store and family needs. According to a survey conducted by Korean American Community Services in 1992, 176 Korean business owners in African American neighborhoods hired 483 people, or 2.07 employees per store, of which 364 (80%) were African Americans. Although these business ventures are basically mom and pop operations, once they need outside help, they tend to rely on the labor force in the local community to improve customer relations. According to national data, Asian businesses hired more employees than other minority-owned businesses. Between 1972 to 1987, the number of businesses established and employees hired increased 973% and 411%, respectively, faster than other minority-owned businesses, although the increase in sales was less than for both African American and Latino-owned businesses. During this period of time, Latino-owned business sales increased 366%, African American businesses sales increased 176%, and Asian business sale increased 121% (Ong and Hee 1994).

## Language and Structural Barriers

Lack of proficiency in English is one of major reasons why most Asian professionals who recently moved to the United States work in the Argyle community. Phan, the Vietnamese pharmacist, came to the United States in 1975 and was resettled in Argyle. Six months later, he went to a university in Iowa to study for a degree in pharmacy. He was conditionally accepted by the university because he was not well prepared in English, but his background in pharmacy from Vietnam helped him to pass all the requirements. He came back to the community after graduation because he felt:

> It is hard to apply [for] jobs in other communities because of the language difference. I feel more comfortable to serve people who can understand me a little bit more.

Even though I have a college degree here, language is still a problem. I stayed here. It is OK.

Although language can be a barrier in finding work outside the community, it is an advantage in finding jobs in the community. A basic requirement of hiring for most of the businesses is fluency in at least one of the major local dialects or a foreign language such as Cantonese or Vietnamese.

Some business owners speak little English, but they can still manage to survive in such a multicultural community if they speak a major dialect or foreign language. Most of these business owners have their children or another family member to help them with English during busy hours, such as late afternoon, especially Fridays, and weekends. Customers who do not speak the same language or dialect find themselves feeling less comfortable, because they do not understand the community and cultural contexts of business transactions. They may know that they can bargain about the price at some Asian stores, but without speaking a major dialect or language, they will not know how.

Asians who have obtained professional degrees in the community often feel it is hard to get a job outside their own cultural community or to move up the ladder in the American organizational structure. Lau, in her doctoral dissertation (1988), suggests that Asian-American professionals are blocked in their attempts to gain recognition and rewards in the organizational world. As their career trajectories level off, members of this group become dual careerists, getting sideline ventures to supplement their careers, which are restricted by corporate immobility. Many people—including those facing mid-life challenges—reach limitations well below their expectations. They begin to think of ways to leave the structures of limited opportunities, developing "side bets" to carry them over into new lives. Richard Wong, an officer at one of the Asian Banks, obtained Masters degrees in both computer science and business management and initially worked in a bank outside the community. He remarked:

> We are minority in this society and disadvantaged. Although discrimination is illegal, it does exit. How many Asians are bank presidents? Can you say they are less capable or less intelligent? They are squeezed out and have no chance to move up the ladder.

Professionals like Richard and Phan felt more at home working in an ethnic community. Marcos, the Filipino doctor, indicated that it was difficult for him to work outside the community, because people would not trust him or come to see him. Phan and Richard contribute their business successes within the community to their understanding the language and life style of their own customers.

While some professionals, even middle-aged professionals who had experience working in mainstream occupations, came back to their own ethnic community because of language and structural barriers, other professionals looked for alternative career paths for themselves. Doctor Cheng is from mainland China. He has two medical degrees from China but here he could not find a job that he liked. Again, one of the main reasons was his lack of English proficiency. He is studying very hard to learn English and hopes some day to get an American medical license. He said, "without passing the language barrier, nothing else can be put on the agenda." However, he is still not optimistic about his future. He continued, "even if you get the MD, you will not necessarily have patients." Therefore, he is considering an alternative career. "I think since I am a Chinese and know Chinese medicine, I will develop my own advantage. So I use traditional medication and acupuncture." This is how he started to practice traditional Chinese medicine.

### Location: A Commercial Area with Residential Consumers

Most refugees settled in the community with the help of their sponsors, resettlement agencies, or churches when they first came to Chicago, and many still live in the same neighborhood. They feel at home and find it convenient to live in a community with businesses from their own culture. Others believe the community has an ideal cultural and business environment. This was the case for Dr. Marcos from the Philippines when he began to look for an opportunity to practice medicine by himself in 1989:

> I got this job from an advertisement. A doctor, a friend of mine, is also practicing, but I chose this place because it is also a business and commercial area. My friend does not work in a business area, so he did not grow very well. I got another place also, every other week I also work in Uptown National Bank Building. I see patients with another Filipino doctor there.

With the growth of the commercial area, Asian businesses from other parts of the city also began to invest in the area; these included the New Asian Bank, the International Bank, CPAs such as Mary Lee, the New Hong Kong Bakery, and Chiao Qiun Bakery, all of which are extensions of businesses from Chinatown on the south side of Chicago. A couple of restaurants moved to Argyle from New York and California. Similarly, some businesses on Argyle also opened branches at Chinatown on the south side of Chicago.

With more retail businesses in operation, modern service and entertainment businesses followed to provide services to the retail sector and the residential community. At the beginning, many Asian business owners went to south Chinatown for services in their own language and culture. Phan, the Vietnamese pharmacist, owns a drug store in Argyle. In addition, he manages a real estate business and recently opened a travel business, all in the same building as the pharmacy. Shortly after Phan opened his drugstore, Dr. Chen, a dentist from

Taiwan, came in to look for office space. Because Phan was looking for a CPA, Chen introduced him to Mary Lee, who had an office in Chinatown on the south side of Chicago. Later, the business people in Argyle encouraged Mary to open an office there. Several years later, she has opened not only an office of her accounting firm but also a bank, together with her family members.

## Autonomy and Flexibility

Some business owners enjoy the autonomy of a self-employed business and its flexible schedule. This pattern is exemplified by Hunan, one of the busiest grocery stores in Argyle because it sells groceries at a lower price. The owner took issue with the notion that Asian business families are more successful or makes more money than other immigrant families. He explained that he liked working in a self-employed business, not because it made him much more money but because of the autonomy if offered:

> People think we are making money. We make almost as much as working for someone else. Maybe only a little bit more, a little bit better than living from hand to mouth. The key is working hard and save. What is good about family business is that you do not have to take what other people give you. You are your own boss.

The professionals emphasized the same point. Dr. Marcos likes the single practice which enables him to spend more time with his children. He said, "my wife also works. I want to have more time with my kids. I want my time to be more flexible. Sometimes, I want to have a vacation. It's up to me to decide when."

However, for many self-employed business owners, self-employed business is simply a way of making living. Mike, the Cambodian fast food restaurant owner on Broadway, said he does not like anything about small business, "first I don't like selling food, second I don't like selling things. But it is a way of making money, that is the bottom line."

Almost all Asian business owners, except for the banks and book stores, started after saving enough capital from working somewhere else. They began their businesses on a small scale, relying on family resources, including savings and labor. Family included uncles and aunts, brothers and sisters, parents and children. The family received greater financial benefits from working together than just having one or two wage earners working in low-skilled and low-paid jobs. Many Asian business owners had experience working in similar businesses in their home country. For these businesses owners, being self-employed was a way of life that they already knew. The only difference is that they have to adjust to a more diverse market and a new environment. Business people with higher educational degrees feel that language and cultural differences are a greater barrier to them than they are to store owners. Some business owners feel that self-employment gives them more autonomy and flexible time. The "commercial strip" in Argyle was able to develop and attract investment from people who live outside the community because it has a residential customer base.

As this section has shown, there are several major groups within the Argyle area business community. These include "old timer" Chinese, refugees from Southeast Asia, and Chinese from Taiwan, Hong Kong, and the mainland, as well as other Asian groups. Many Southeast Asian refugees are ethnic Chinese. In addition, there are old non-Asian businesses which stayed in the community and new non-Asian business which are run most by newer immigrants. The groups represents different waves of immigrants who entered the United States at different times from different places. The "older timer" Chinese from mainland China in the 1930s or earlier encountered very different social, economic, and political conditions than did those who came to the United States in the 1970s. In addition, each of these immigrant waves shows intragroup variations, with some having greater educational, economic, and cultural resources than others.

## BUSINESS TYPES, OPERATIONS, AND CONNECTIONS

In contrast to conclusions from earlier sociological studies that showed self-employed Asians operating the retail stores to meet the needs of their customers with exotic tastes, the businesses in the Argyle community vary greatly, ranging from traditional small retail stores as noted in Light's study (1972) to modern enterprises, such as law firms, travel agencies, and CPA firms. Different types of businesses have different ways of operations and different customers, and relate to the local community, cultural groups, and the outside in different ways. The variety of businesses not only serves the different needs of the community but also enables the community to become sustainable and to further attract new businesses to the community.

### Major Types of Businesses

There are three major categories of Asian self-employed businesses in the Argyle community: traditional businesses, modern enterprises and firms, and intermediate businesses which are the businesses between the traditional and modern categories. These are mainly entertainment, communication, and education-related businesses. These three major categories include over 30 types of businesses (Table 2).[6]

Traditional businesses include laundries, groceries, gifts, garments, dry cleaners, beauty shops, and jewelry. Most of these businesses are operated by Southeast Asian refugees and do not require professional training. These stores attract Asian and non-Asian residents from the community, as well as customers from the outside. Although there is an increasing number of non-Asian customers from outside the community coming to shop in these stores, most customers are still Asians.

***Table 2.***   Type of Business

| Type of Business | Number of Businesses |
| --- | --- |
| ***Traditional*** | |
| Restaurant and fast food | 32 |
| Grocery | 18 |
| Furniture | 1 |
| Beauty saloon and beauty supply | 13 |
| Food and drink | 8 |
| Cleaner | 4 |
| Repair (auto, electronics, watch) | 4 |
| Gift | 8 |
| Photo | 2 |
| Clothing | 5 |
| General merchandise | 3 |
| Jewelry | 6 |
| Herb | 2 |
| Watch | 1 |
| ***Modern*** | |
| Dentist | 5 |
| Travel/tour | 4 |
| Insurance | 4 |
| Bank | 2 |
| Medical clinic | 4 |
| Auto sale | 1 |
| Realtor | 2 |
| Attorney | 1 |
| Architect and interior design | 1 |
| CPA | 2 |
| Pharmacy | 2 |
| ***Intermediate (communication, education, and entertainment)*** | |
| Electronic products (incl. pagers) | 2 |
| Videos | 6 |
| Printing | 1 |
| Books | 1 |

***Sources:***  Zhang (1989,1992,1996).

Modern service enterprises and firms require more professional training. They include banks, law firms, accounting firms, travel agencies, realtors, architecture and interior design, and health-related businesses such as pharmacies, dental clinics, and medical clinics. Health-related businesses in Argyle include both traditional and modern medical treatments. The health-related businesses mainly serve the local residents of all racial and cultural backgrounds, but traditional medical services such as herbiology, acupuncture, and massage may have patients both from the community and the outside. These traditional doctors treat patients with chronic health problems, such as arthritis, and also provide treatment for weight problems, smoking, and alcohol addiction. However, because traditional treatments are not covered by health insurance, the doctors often worry about how to reach and keep their patients.

Dr. Marcos, the medical doctor from the Philippines, exemplifies the patient profile. Most of his patients are Vietnamese, but he also serves Cambodians, Americans, and some Latinos. About 40% of all his patients do not understand English or Vietnamese. Most of time, they bring interpreters when they come to see him.

Earlier research indicating a dearth of Asian businesses in areas such as lumber, auto, and furniture was interpreted as reflecting a pull toward meeting the exotic taste of Asian population and a push away from other areas to avoid to compete with already established American stores (Light 1972). In the Argyle community, there used to be two car dealerships, one operated by Koreans and the other by Vietnamese. The Korean one has closed, but the Vietnamese one is still operating. The Vietnamese car dealer owned a car dealership in Vietnam before coming to Chicago. Therefore, this is not a new business experience. According to him, he is able to sell 400 to 500 cars a year. In Vietnam, people had to pay cash but here people can pay by credit, so his "business is quite good."

Business in the intermediate categories include entertainment, communication, and education-related services, such as bookstores, electronic products, video shops, and photo shops. These businesses attract mainly local residents. Occasionally, stores selling electronic products on Sheridan Road and Broadway also have customers who just drive through the community. Each Asian group has several video shops of its own. These video shops are important entertainment sources for the non-English speaking populations. They also have ethnic customers from outside the community. In addition to these three major categories, a manufacturing company, Phoenix Soy Bean Products Inc. on Broadway, produces soy bean products for different restaurants and stores in Chicago. It is just beyond the boundary of the Argyle community.

In addition to the Oriental stores, community residents have access to many non-Asian stores, including a Dominick's grocery store on Sheridan and Foster and the Goldblatt's Department Store on Broadway and Racine. Several thrift stores serve the low income residents in the community. Some "old" non-Asian businesses, which moved to the community a long time ago, serve the non-Asian residents or provide goods that most Asian stores do not carry. These non-Asian businesses include food marts, liquor stores, bars, and fast food restaurants. As more Asians moved to the community, the effective survival strategies for the remaining non-Asian business are to provide goods that most Asian stores do not carry. John, a liquor store owner on Argyle street whose father started this business in early 1970s, explained:

> For certain items they asked me, because they (Asians) do not carry up and down the street. Some of the items I didn't carry before—non-liquor items. Some customers

ask me about paper towel which I didn't have before, and bathroom tissue. The only thing they asked for many times that I don't want to carry is milk.

Since Asians moved to the community, most bars have closed. According to a local business leader and police officers, there used to be 30 or so bars and tarvens in the area. Now only seven are left. There are also several liquor stores. The decline of taverns and bars reflects in part the change in population, since many Asians, especially Buddhists and Muslims, do not consume alcohol.[7] However, increased property values in the area have driven out many bars that could not afford increased rents. An overall change in attitude toward alcohol and tobacco has also contributed to the decline. As John, the liquor store owner, indicates:

> The business [his liquor business] is a kind of going to hell because of different reasons. The customers and demographic change is one reason. But the biggest problem, I think, is the industry. People are drinking less. As far as the changes in the demography, there are more different customers. The differences as I see it is less demand for variety than before. Customers buy only certain brands. It has to do with the price. The customers are trying to adjust themselves to the policy change. It [opening a liquor store] was a better idea back in the eighties. It is not a viable idea to open a liquor store now. Nobody would have expected to see what has actually happened in this business. This business used to be called recession proof, but not any more.

The Argyle business area has its daily cycles. It attracts different people at different times of the day. On weekday mornings, Argyle street is relatively quiet with few customers, but there are still pedestrians and visitors, mostly residents of the community, including mental patients and welfare recipients and people who come to visit friends. Between 4:00 p.m. and 6:30 p.m., the streets become busy again with residents coming back from work and shoppers stopping at Argyle on their way home from work. Residents from the east side of Sheridan walk through Argyle street quickly from the El train station. In the evenings and on weekends, restaurants are visited by Asian families and Americans. Most of these restaurants close at 10:30-11:00 at night. During the weekdays, most of the customers are residents. On weekends, customers come from all over the city and even other states such as Indiana, Wisconsin, and Michigan.

The Asian Small Business Association has been working closely with the CTA and the 20th police district foot patrol officers to crack down on criminal incidents, such as graffiti, drug use, and alcohol-related crimes, in order to sustain economic growth in the community. Nevertheless, after 9:00 p.m., Argyle street—the main business street—is almost deserted, with only drunks swearing and fighting with each other outside the bars, or wandering on the street. Business owners on the main streets like Sheridan Road and Broadway Avenue feel safer than those on other streets.

## Business Expansion

Some Asian business owners have begun to expand their businesses in recent years. There are three primary ways to expand businesses for most Asians. They remodel the current store or purchase a larger space in the community and move there; open another type of business in the community; or open a similar business in the community or outside.

For example, some retail business owners have started wholesale businesses because it reduces the costs of their own retail business. Some business owners became travel agents. Phan, the Vietnamese pharmacist, became a travel agent "because it won't cost anything. The only thing needed is to install a computer and build connections with the airlines." He installed the computer in his drugstore. With the relaxation of diplomatic relationships between the United States and Vietnam and other Southeast Asian countries, refugees can more easily visit their families and start trading with their home countries. The demand for airline tickets has been increasing. For another example, the Thai grocery store on Lawrence also sells roundtrip air line tickets from the United States to Bangkok.

Dr. Chen, the dentist from Taiwan, provides an example of expanding in business. He has offices at three different locations, one on Argyle, another on Sheridan, and a third one in the northern suburbs of Chicago. In addition, he has five or six associates treating different dental problems at these different offices. A CPA opened his accounting office in 1986 and in 1994 he began to represent All State Insurance company. Dr. Marcos, the Filipino doctor, started working in another Filipino doctor's office on Lawrence. In 1993, he opened his own clinic while continuing to work part-time at the old business.

## Traditional Ways of Payment and Credit Deferment Criteria

In traditional Asian countries, ways of doing business and handling credit were much more informal than in the United States. In some stores, the number of business transactions are small, and therefore many stores do not accept credit cards or private checks.[8] However, if frequent customers or friends do not bring enough cash, business owners allow them to buy on credit, just writing down the name of the customer and the amount of money owed. At the next transaction, the customer pays the store all that is owed. Like stall-holders and peddlers, some Asian business owners are also willing to let customers negotiate the price. These types of payment practices are based on a relationship of trust between the business owner and the customer, and take place mainly if the customers speak the same dialect as the owner and the owner has known them for a long time, perhaps even before they came to the United States. If the customer forgets or neglects to pay, the owner bears the loss and refuses to extend credit again. Friendship and loyalty are very important to each cultural group. Some business owners would rather keep close ties with their old friends than stick to their business rules. Even officers at the International Bank and the New Asian Bank acknowledge that they have relative flexible lending standards for Asians.

## Informal Ties and Public Involvements

Business owners in the community want to expand, and several even asked me to help them bring in new customers from the outside. However, very few, including two newly developed banks, think that public relations or advertising are important. Although several of the modern service businesses have more connections with groups outside the community, especially formal professional organizations, many of the traditional businesses do not have any connections with other business organizations. They often do not even know about groups outside community, nor do they think it is necessary to know about business organizations. The major reason that most are not interested in public relations is that their businesses are either small-scale or new. They have to consider the economic benefits of developing public relations.

There are local business organizations, organized along ethnic lines. They function more like traditional social groups than like chambers of commerce. Sometimes, business owners gather together to discuss community issues and problems, such as community safety or lack of parking space, although they recognize that they can not solve these problems single-handedly. The best they can offer to the business community is education—providing information on business regulations.

Most business organizations lack the financial resources to do the things that they planed to do. Liang is a member of Chicago Chinese Council. "The council is to help different countrymen to find house, to start business. No matter where they are from, if they have financial problems, we would help them." However, this is only what they intend to do. "We don't have money to start now. Money is the most important and the second is the personnel." The members of the Council "get together during holidays. You can buy a ticket or more. There will be a lot of food. That's it. You can't do anything without money." He continued:

> Chinese Mutual Aid Association has money. They get money from the government, $400,000 a year. They also have personnel. More than 10. They hire people to teach computer skills and they also hire the Americans to teach English. That makes a lot of difference. They are the biggest and most influential agency in the community. Most work I talked about are done by them.

Many problems of individual businesses are solved with the help of Charlie Soo, director of the Asian Small Business Association, informally known as the "Mayor of New Chinatown" by different groups in the community. He has an office on Argyle and Broadway. The Association is a one-man office. Soo used to serve as commissioner of the Illinois Economic Development Commission, as project director with Chicago Economic Development Corporation, and as president of an international management and consulting firms with overseas offices. He is also active in the U.S. Chamber of Commerce, Chicago Association of Commerce and Industry, and Chicago Press Club. Being experienced in economic development, Soo has tried to bring resources to the community, for example, mobilizing resources from the city and private businesses to pave the sidewalk, and bring in money to compensate owners for remodeling business facades. Sometimes, he also takes community business owners to business exhibitions and events in the city.

Some business owners are not familiar with, and therefore do not trust, formal procedures. They rely on informal ways and personal trust. Induni was an Indian business owner. Her family had a laundromat on Argyle for seven years and another one outside the community for two years. When I first met her, she was discussing the Bible and selling jewelry to other customers with her friend Shanonn in the laundromat. Shanonn is an Iranian woman living on Hollywood Avenue. Both of them are Muslim. When I met Induni, her husband had just found a job in Florida and moved out of Chicago. She and the children were left-behind and she was anxious to sell her business and join her husband. She did not tell me about this initially, but did volunteer the information when I met her for third time. I was sitting in laundromat, chatting with her, like everybody else from the community, then she asked me to help her find buyers. She said, "I have friends, I like friends. I like you, you help me." I asked her why she did not put up a "for sale" sign outside the window. She said, "My husband does not like it. Before, I sell it at good price, but now I want to sell it at the lowest price. Tell people. The sooner the better." She let me help her because she trusted me, and also because I was a woman.

Afterwards, I went to Charlie Soo, Mayor of New Chinatown. He would not believe Induni was going to sell her business. However, after about two weeks, Induni told me people began to call her on the phone and visit her business establishment. Several weeks later, I noticed that the laundromat was gone. A few weeks later, a grocery store opened its business in the laundromat's old location. Induni also told me that several years ago, a journalist had tried to interview her, but she did not know why this journalist wanted to ask her those questions and she did not understand the questions, so she turned down the interview.

Most businesses are concerned about community safety because some business owners and employees have been victims of robbery, burglary and theft. Ms. Chou, the World Journal Book Store manager, was threatened by a gunman during the day and forced to open her cash drawer. A number of store owners I talked with mentioned that their stores had been robbed at night, but not all of them were willing to report the robberies to the police or talk about them in public. Some appeared to be indifferent to problems in the community. Yuan, a Cambodian Chinese gift store owner, shook her head after a community safety meeting with

Alderman Mary Ann Smith and said, "they did not address anything," but she did not raise any of her concerns at the meeting.

There may be several reasons why so few business owners are willing to report problems to the police. First, some business owners think things will not change even if they do report crimes to the police. Some do not want to be involved in the procedures of filing incident reports. Also, Asian culture and the philosophies of Buddhism, Confucianism, Taoism, and Islam teach Asians to avoid confrontation and unhappiness. As Phan explains:

> Normally, I forget things that I don't like. I don't like to remember what I don't like. I like to accept things that I can't change and I like to change things that I can change, because it makes you happy if you don't think other points. The happiness in my life is that I feel happy, therefore I am happy, but [if] I don't feel happy, I am not happy. Happiness of your life is what you think from what you get. The happiness of life is what inside you, not outside. Happiness is not just satisfying the instinct—what you want. You want to do this and to do that, because everybody likes that. You are happy because you think you are satisfied, not when you want to have something and get it. Sometimes if you don't think of it, and you are satisfied. That's why I don't want to remember and confront things I don't like.
>
> You only have 24 hours a day to live: eight hours to sleep, a couple of hours to eat, several hours to pick up your laundry, to wash, then you have about seven to eight hours to live your life. If you hate, if you get angry, if you are not happy because you are angry and you don't like certain things, it means you spent your time making your[self] unhappy. Then what do you live for? I don't look at the problems. I just keep working.

Lee, the Cambodian grocery store owner, a Buddhist, also reported several incidents of burglary or theft in his store, but he said, "we have time to make money and time to loose money."

Some service organization leaders complained about the lack of participation of the business community in community problem solving and financial contributions. "We helped them (businesses) get settled down and started their businesses. Now they are doing well, but they don't want to involve (themselves) in community development," a staff member from the Chinese Mutual Aid Association complained. However, some business owners have been generous donors to and supporters of ethnic cultural organizations in the community. For example, they provide financial support to different programs at the Chinese Elderly Association, Buddhist temples, and Teo Chow Mutual Assistance Association. Therefore, their investments contribute to ethnic cultural development in the community.

In contrast to earlier observations of ethnic business, I found that businesses in the Argyle community range from traditional "exotic Oriental" retail stores to modern service firms, banks, and health care providers. This variety of businesses on one hand has attracted customers from different residential areas and

different racial and ethnic backgrounds, and on the other hand has laid a foundation for further business development in the area. These businesses are
mostly family operations and rely on informal and traditional connections with
the community. Business owners work flexible hours and sometimes use their
traditional credit criteria based on the relationship of trust. Business expansion
for many business owners means providing more revenue for their family. They
expand in ways that allow them to best use their resources and reduce costs.
For example, working at different locations, developing sideline businesses in
the same store, or operating a wholesale business through a retail store are all
low-cost ways of increasing revenues. Business owners are unwilling to get
involved in public life for a variety of reasons. For example, many operate on a
small scale that is often barely enough to take care of their extended family,
and they have neither the time nor the resources to become more publicly
active. In addition, many business owners are not familiar or do not feel comfortable with the formal procedures required of public involvement. They feel
more comfortable with relationships of friendship and trust. Others avoid public action because they do not like the confrontation and unhappiness it
reminds them of. They take the philosophical stance of eastern religions, which
counsel them to ignore the unpleasant and focus on the good.

Because of changes in the larger society since earlier waves of Asian immigrants, the range of Asian businesses is wider than previously. The modern service sector developed later than the traditional sectors and provides services to
other businesses in the Asian business community, and the rest of the community. Different businesses serve different needs of customers from varied racial
and ethnic backgrounds and places of residence. The variety of businesses form
the foundation for a sustainable community. Most of the businesses operate in a
traditional way and rely on internal resources. Although the Asian Small Business Association (ASBA) is an important agency that brings in outside
resources, the majority of Asian business owners are reluctant to get involved
in formal structures because of a lack of understanding or trust and their cultural traditions.

## EMOTIONAL BONDS AND ECONOMIC SIGNIFICANCE:
## THE ROLE OF THE BUSINESS

Ethnic businesses have played a crucial role in the development of the community. They provide Asians both inside and outside community with exotic products and a cultural base. They provide the Argyle community with economic
means as well as cultural and emotional support. Businesses open windows for
outsiders to experience and learn about different cultures. Businesses also play
an important role in stabilizing the neighborhood and in generating local economic growth.

## Emotional Bonds

I have argued that businesses in the community not only create jobs for families, relatives, and people in the community at large, but also generate local economic growth. In addition, the business establishments in the community are also places for families to get together and to take care of each other. Many family activities take place in the stores. The businesses provide emotional bonds for the rest of the community—residents meet each other in the restaurants and bakeries, or chat with store owners. Each store also functions as a small information center for the community.

The importance of business for family life is shown by the example of Mr. Lee, the Cambodian Chinese store owner, who works in his store along with his wife and three children. When the children finish school, they come to the store to play, do their homework, and help their parents until the store closes at 7:00 in the evening. The family then drives home together to Roger's Park. I interviewed Lee and his family on a Friday afternoon before 5:00 p.m. Lee had told me to come then because after that time they would be very busy, especially on Friday. The eldest son was not in the store, but the second son, a high school student, was watching TV behind the counter in the front. He explained, "I don't have anything to do after school anyway. So I just come to the store to work."

The store is arranged to accommodate the children. The cashier's counter is on one side of the store entrance and the children's counter on the other side. The children play inside. They have a 12″ TV and game boards and puzzles. At about 4:00 p.m., Lee's son left the store on a bike. About 10 or 15 minutes later, he came back with his younger sister. His sister goes to a kindergarten in the neighborhood. Then brother and sister worked on putting a puzzle together. When the carrier arrived, Lee's son left the counter to show him where to shelve the goods.

I found a similar pattern when I was waiting to interview Dr. Marcos. Phan's eight-year-old daughter was writing in her father's real estate business office next to the doctor's office, and her four-year-old brother was leaning on her fast asleep. After a while, she ran out the office and practiced on a piano behind a screen at a corner of the lobby. The doctor's office is inside the lobby. Then she came over to me and the receptionist to talk and asked us questions about her spelling homework.

This pattern of involving children in the business early helps explain why many children stay in the business when they grow up. In fact, when I asked storewoners' children what they would like to do when they grew up, they would say "the family business." Lee's second son said:

> I like to work here when I grow up, because I have experience in this kind of business. I have some experience in grocery store. If I start something else it might be difficult. If I go to college, I will study business.

However, conflicts do arise. When I first met Liang's family in 1989, his wife was working in the store, while her mother was sitting in the store taking care of their youngest son. When the elder sons came "home" from school, they would play with their younger brother, and the grandmother would sweep the floor. The two elder sons also helped give haircuts. When Liang allowed me to talk to his oldest son who was attending University of Illinois at Chicago, the son said:

> Sometimes I want to read, but the customers come in, we will attend to the customers. Although I want to keep reading, I can read afterwards. Sometimes when customers come in, I can't help because I have exams the next day. When I tell my parents, they will let me go upstairs to study.

When I asked him what he would like to do when he finishes school, he said:

> As I observe, I would like to do family business. But I have not thought well yet. You see in America, all the big businesses are family businesses. I may work for a while and then move up [go back to school—he majored in public health and would like to become a lawyer]. But no matter what I do I would be helping my parents and work for them, help them with their business and work for them after my work.

In 1993, after Liang opened his second beauty shop on Argyle, I visited the family again. The oldest son had graduated from the university and was working in the store. He expressed the dilemma of working with his parents. His parents wanted him to contribute and help their business but he tried to avoid conflicts with his parents, as the two generations often disagrees on how to run business. "After all, this is their business," he said.

Although some children mentioned that they wanted to help their parents, not all parents want their children to work in a small self-employed store like themselves. Ms. Chou, the World Journal bookstore manager, who lives in Chinatown on the south side, said:

> Young people don't have a chance to develop themselves in this kind of job. They should do a job that they like. This job fits middle-aged people because they understand the mind of customers, and the money is not good. It is relatively stable. We don't want to suffocate their potential. Some part-time students go to school, in the afternoon they want to take advantage of the time and make some money to spend. That's OK. Young people have their own interest. They don't have the patience to sort and classify books even if they may like to read.

In addition to serving family needs, businesses also play a role in connecting residents to one another. For example, Dr. Cheng, from mainland China, used to work in a Vietnamese Chinese drugstore but recently opened his own clinic on West Winnemac. He visited the book store to buy a Chinese newspaper and also to get messages. He had left his business cards with Ms. Chou, the bookstore manager, so that his old customers could get in touch with him again at his new location. Some community organizations or families sent announcements of programs and events to stores as a way to reach customers and the rest of the community. Families also sent their announcements of family events such as weddings to the stores.

For non-Asian residents, the stores provide an opportunity to experience another culture, although most cannot afford to visit the stores regularly. Katie, a white welfare mother with two children, said she felt she was visiting some Asian country, "Hong Kong or somewhere," when she walked on Argyle or visited a store. In the course of my observations in the stores, I would hear sometimes non-Asian residents talk to storeowners about personal problems. The owners would give them advice about what to do. An African American woman I met during a New Year's parade said:

> They [Asians] are smart people because they tell proverbs. For instance, I once was not happy because I thought I could not get the things I expected. You know, when you want to have something but you can never have it. I was disappointed. They told me "grass may be looked greener on the other side, but beware." You see, if I always complain, and complain, because I could not have what I wish to have I am not happy. Just take it as it is. If you see somebody has a fabulous car, you could not have but you want to. People who own the car may have their problems. I am 47 years old, I used to use a lot of make-up hoping to look better, but it is not me. I used to take pills when I was under some tension, but I don't do it anymore. If I have emotions inside of me I just let them out. I feel much better. They have good doctors, they took care of people who are on public aid.

Many residents seem to be well acquainted with the business community. Dianne is a Native American who has lived in the neighborhood for several years. I met her at a New Year's celebration on Argyle and she became one of my earliest informants, taking me along to introduce me to different stores, telling me who had owned the store, at what period of time, and what they sold. She would point to the arts and crafts displayed in the store windows and said some of them were similar to Native American arts and crafts. "It seemed I found my long-lost cousins," she said. She was also self-employed at that time, making and selling jewelry, typing and printing documents or papers for community organizations or individuals, and providing counselling to people who have been abused.

John, the liquor store owner, describes his feeling of being part of the community, even though his family lives elsewhere:

> Over a period of time, you get to know your customers, not just what they buy, but you get to know them. Maybe a man would come over five, six times, and one day you will see him with his wife, so you will get to know his family and know the kids.

It's really a personal business. You have happy moments in their lives and also have sad moments. You become part of the community. . . . Some of the customers we have not seen for years because they either quit drinking or moved away. As they passed by the window they waved and some of them came in and just shook hands and said "hello." That's a great feeling. As a small business, it is much more of the personal touch. It is a neighborly type of thing, more interwoven, you do become part of the community.

## Cultural Base and Cultural Experience

Because the businesses in the community are predominantly Asian, the Argyle community attracts Asians from both the city and the surrounding suburban area. They come to Argyle to purchase oriental goods and experience the cultural atmosphere. The ASBA conducted a survey of shoppers on two weekdays and one Saturday when no special event was occurring. They found shoppers who came from all over the Chicago metropolitan area and from three other Midwestern states—Wisconsin, Indiana, and Michigan. Figure 1, a map of the Argyle Street Market Areas, shows the location of the customer base in the Chicago Metropolitan area. On the Saturday when the survey was conducted, there were more cars without Chicago stickers than with them (Nelis and Castillo Inc. 1992).[9]

Although the most frequent customers are local residents, customers from outside the community spend more. The bookstore owner explained that customers from outside are better educated and well-to-do, some come from further away and purchase greater quantities of books. The residents in the community spend less on buying books. In some cases, retail business owners and restaurant owners from surrounding states visit Argyle every a couple of weeks to get supplies. I met a saleswoman from Indiana at Kim's Jewelry. She bought more than 20 silk blouses from the owner. These blouses sell well among "her own people" in Indiana. She and her husband usually came to Argyle every week or two to get supplies for their own business. The competitive prices also attract customers from outside the community. According to a local business leader, grocery prices are 20%-30% lower on Argyle than in Chinatown on the south side of Chicago. Therefore, people from south Chinatown also come to Argyle. Two young women worshiping in Chua Turc Lam, a Vietnamese Buddhist temple, on a Saturday morning, said they came to Argyle once every two weeks when they needed to return video tapes. When they came, they also went shopping and worshiping at the temple. "It is only one hour drive. It's a lot of fun to be here," they said.

Businesses represent an authentic aspect of another culture for outsiders to experience. So they attract customers from outside the community. Indeed, the Art Institute of Chicago organizes cultural tours to Argyle, thereby contributing to its reputation as a cultural enclave. Tour participants come from different

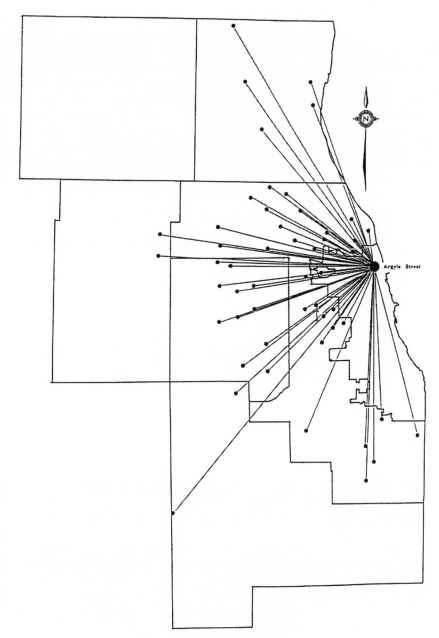

**Figure 1.** Argyle Market Area

parts of the city and suburbs. Several universities and schools have organized student field trips to visit the community to enrich their educational and cultural experience.

## Economic Opportunities

Argyle is a regional Asian shopping center which provides revenue not only to business owners but also to the city of Chicago through the sales taxes it generates. It has also produced a substantial increase in property tax revenues, as a result of which the area's development has caught the attention of city politicians. In fact, both city and state politicians have frequently visited in the community on different occasions. Formal recognition came when Charlie Soo, the ASBA director, won the Governor's Hometown award for excellence in economic development and community improvement. Local politicians have sought to capitalize on the economic vitality. For example, 48th ward alderman Mary Ann Smith remarked at a community safety meeting:

> As we know, over $70 million are earned from this community every year, therefore, it deserves more safety. We are negotiating with the city, tremendous monies are coming to your area to secure the worst buildings in the area. The community deserves police service. This is a tremendously diversified area. We have police patrols in the area. We want the merchants to feel safe. The community has undergone the first transition. We will do nothing else but improve it.

Several community institutions and older business and property owners, such as the manager of the Summerset, a nursing home for the mentally disabled on Sheridan Road, hope further improvement of the commercial area in Argyle will benefit them too. Business development in the area has helped improve the safety, sanitation, and physical make-up of the community.

Indeed, Argyle has become an example of revitalization in a low-income and high crime area. For example, a business delegation of the Milwaukee Department of City Development sent ASBA a letter after its members had returned from a trip to Argyle. The letter read:

> The Milwaukee representatives saw that commercial success is possible even in the mist of a low-income neighborhood which has problems with crime, taverns and so forth.

In May of 1994, Mayor Richard M. Daley announced that a $2.1 million grant from the Strategic Neighborhood Action Program (SNAP) would be used to improve residential and commercial property, storefronts, schools, streets, sidewalks, and other parts of the infrastructure in the community. The grant covered the area from Montrose to Foster, and from Sheridan to Broadway. Uptown community leaders believed that the program would set the stage for attracting more economic development and usher in an era of new vitality for

Uptown. Alderman Mary Ann Smith predicted that someday "people will say, 'I used to live there,' and wish they had never left" (Bess 1994).[10]

# CONCLUSION

Asians are often portrayed as successful or as a "model minority," especially when looking at their family income and their educational and occupational achievements, but a closer look tells a more varied story. In their communities are different Asian groups and individuals. Some are well educated and relatively wealthy, and some are disadvantaged with little education. Some are well established and some are still struggling for daily survival. Average family income is apparently higher among Asians than any other minority group, but the higher family income and business success are often the result of using family resources, which include pooling the savings of family and close relatives to start a business, using family members as workers to work long hours, and holding several jobs with income shared among more family members. National data also suggest that median household income for Asians is higher than non-Hispanic whites, but median per person income for Asians is lower than that for non-Hispanic whites (Table 3). In four metropolitan areas (Los Angeles, San Francisco, Oakland, and New York), which together house approximately 30% of all Asians, household income for Asians is also lower than for non-Hispanic whites ($37,200 versus $40,000). A 1990 study revealed that 18% of Asian families have three or more workers, while 14% of non-Hispanic white families had that number of workers in the family (AAPIP 1992).

Some Asians had experience in running self-employed businesses in their homelands before they came to Chicago. For these business owners, adjusting to a new business environment is the major task. For Chinese immigrants who came before the refugees, self-employed businesses provided a niche in an

***Table 3.***   Median Household Income and Per Person Income

|  | Non-Hispanic White | Asian Pacific American | African American | Latino |
|---|---|---|---|---|
| National |  |  |  |  |
| Median household income | $31,100 | $36,000 | $19,000 | $24,000 |
| Median per person income | $12,000 | $10,500 | $6,200 | $6,200 |
| Four Metro Areas |  |  |  |  |
| Median household income | $40,000 | $37,200 | $24,000 | $25,600 |
| Median per person income | $17,600 | $10,800 | $8,600 | $6,300 |

**Sources:**  Estimates based on observations drawn from the U.S. Bureau of the Census (1990), one percent Public Use Microdata Sample. Non-Hispanic whites were sampled at a rate of 1 in 10, and African Americans and Latinos were sampled at a rate of 1 in 2 (Ong and Hee1994).

urban area. Professionals with higher educations have experienced more cultural and structural barriers in finding jobs outside the community and moving up career ladders. However, even those who had business experience before coming to the United States found that self-employment or family business here is very different than in their home countries in terms of social status and the complexity of doing business. Compared to findings from earlier studies, Asians nowadays are operating a much wider range of businesses, although statistical data suggest these businesses are still relatively marginal compared to those of other groups. Different types of businesses are operated by ethnic groups who came from different countries and places at different times. These businesses link the Argyle community, the larger community, and the different ethnic populations in different ways.

Asian businesses in the community are also more likely to rely on a traditional and informal style of operation based on relationships of personal trust and friendship. Because most of them operate on a small scale, using family labor, they are unwilling to involve themselves in formal public life. Some business owners and organization leaders attribute the lack of involvement by the business community to cultural traditions or religion, but unfamiliarity with formal procedures also plays a part in their reluctance to take a more active role in community affairs.

Self-employment provides a way of making a living for many new Asian refugee and immigrant families. Because many family activities occur in the stores, the businesses help strengthen family ties. The stores in the community serve as an anchor place for people from different backgrounds to interact with one another and search for emotional support and cultural understanding. For Asians, ethnic stores supply exotic foods and cultural products that they can not find in mainstream stores. As symbolic ethnicity theory suggests, many Asian customers from all over the Chicago metropolitan area and several Midwest states (e.g., Indiana, Wisconsin and Michigan) frequently visit the stores, and they voluntarily identify themselves with this aspect of ethnic cultural life. In these stores, many can also speak their own languages and share their life chances.

Tensions among and within different groups were often observed—for example, between "new" and "old" Asian business groups and among major Asian business groups. However, the old businesses welcome the new (i.e., New Chinatown) because these new businesses and the new image of the community help improve business environment and therefore attract customers from outside. They also generate economic opportunities that benefit both the Argyle community and the larger community area. Therefore, both local business community and government agencies welcome the change and use the new name of the community—"New Chinatown"—in order to attract customers and mobilize resources for further development.

# ACKNOWLEDGMENTS

This research paper is based on one of the chapters from the author's Ph.D. dissertation. I am grateful to the residents, business owners, and people from not-for-profit organizations, business organizations, churches, and temples in the Argyle community. They provided me with the inspiration and insights for this research. I am grateful to Dr. Helena Lopata for her editorial suggestions. I am also grateful for the Graduate School of Loyola University Chicago, the United Way of Chicago/Crusade of Mercy, and the Hilda B. and Maurice L. Rothchild Foundation for sponsoring my dissertation research.

# NOTES

1. The Hip Sing Association is a nationwide fraternal Chinese merchant and cultural organization which was founded by early Chinese immigrants, mainly from Taishan (or Toishan) of Guangzhou Province. According to the current president of the Association in Chicago, there are about 17 associations nationwide.

2. Siu's study of Chinese laundrymen in 1987 indicates that since 1872 when the first Chinese laundry opened its business at the rear of 167 West Madison Street, there have been 591 Chinese laundries in Chicago, besides those in the suburbs. The pattern of the movement seems to be a stretching out from the center of the city toward its periphery. By 1940, Chinese laundries had spread to over 50 of Chicago's community areas.

3. According to Tsia (1986, p. 45), Chinese immigrants in the United States were not only socially and economically divided, but they also represented a variety of regions, cultures, and languages. The rich and more respectable merchants were generally the San-yi (from the districts of Nanhai, Panyu, and Shunde); the petty merchants, craftsmen, and agriculturalists were mainly among the Si-yi (from the districts of Enping, Kaiping, Taishan (Toishan) or Xinning, and Xinhui); and the laboring class came from a variety of regions. For example, the San-yi people at times controlled wholesale merchandising, the garment industry, and overall manufacturing. The Hakkas (Guest Settlers) dominated the barber business; and Zhongshan immigrants were the tenant farmers engaged in fruit growing in the Sacramento-San Joaquin Delta. The people from Guangzhou (Canton) and the San-yi spoke Cantonese, which came to be considered standard. Most people from Zhongshan district, about 30 miles south of Guangzhou, spoke a dialect closely resembling the standard Cantonese, but the surrounding countryside spoke a dialect akin to Amoy. The Si-yi people spoke a dialect almost incomprehensible to the city dwellers. The Hakkas, originally migrating from North China, were quite scattered, with strong concentrations in Jiaying and Chaozhou (Teo Chow) and other districts of Guangdong Province and Fujinan Province. They spoke a dialect more akin to Mandarin. Among these dialect groups, there was a long history of rivalry. The Hakkas and the Cantonese had long felt hostile to each other in China.

4. According to the studies of Strand and Jones, Jr. (1985) and Ong and Blumenberg (1994, p. 116), between 1975 and 1991 over a million Southeast Asian refugees were resettled in the United States, arriving in two waves. From 1975 to 1978, 178,000 refugees came to the United States. Of these, 83% were Vietnamese and the remainder were largely Laotians. Shortly thereafter (1985-1990), a protracted famine forced an additional 150,000 people into Thai refugee camps. Ongoing political turmoil in Vietnam motivated over 85,000 people, mainly ethnic Chinese, to risk travel in small crafts never meant for the open sea. Finally, refugees from Laos fled to Thailand as communists drove Hmong from their highland homes and seized businesses largely owned by ethnic Chinese. Rising antagonisms toward refugees in countries of first asylum such as Thailand combined with the continued massive exodus from the region prompted the U.S. government to admit additional refugees. After

these two major waves, a small number of refugees came to join their families or their American children.

5.  Most Asian business are small self-employed businesses. Some are family businesses, some are individual practitioners. A few businesses, although small, are franchises of large companies that have their main offices in Chicago or other large cities such as New York.

6.  I categorized the businesses on the basis of previous sociological research. The store that carries different products under the same name is counted as one business. The stores that are owned by the same person under different names are counted as separate businesses.

7.  Avoiding alcohol and other intoxicating drugs is one of five major moral principles in Buddhism. According to Buddhist teaching, even a small amount alcohol distorts consciousness and disrupts self-awareness (Ven. Dhammika 1991).

8.  Most of the stores with fewer customers only accept cash, although increasing numbers of stores accept credit cards if the customer purchases a certain amount or more, especially the stores that have more customers, such as grocery stores.

9.  This survey was conducted by identifying and recording the origin of out-of-state license plates and windshield stickers of cars parked on city streets and public and private parking lots within one half block of Argyle Street (Nelis and Castillo Inc. 1992).

10.  Under SNAP, which began in 1992, the city channels resources from several departments into a targeted area over a 12-month period. Other SNAP grants went to Austin, Southeast Chicago and the Near West Side (Bess 1994).

# REFERENCES

Asian Americans and Pacific Islanders in Philanthropy (AAPIP). 1992. *Invisible and in Need: Philanthropic Giving to Asian Americans and Pacific Islanders.* San Francisco, CA: Asian Americans and Pacific Islanders in Philanthrophy.
Bess, J. 1993. "Low-Cost Housing has Uptown Split." *News Star* (June 2).
Bess, J. 1994. "It's a SNAP for Uptown." *News Star* (May 4).
Boggs, V., G. Handel, and S. Fava, eds. 1984. *The Apple Sliced: Sociological Studies of New York City.* Prospect Heights, IL: Waveland Press.
Boswell, T.E. 1986. "A Split Labor Market Chinese Immigrants." *American Sociological Review* 51(June): 351-371.
Dennis, W.J., Jr., B.D. Philips, and E. Starr. 1994. "Small Business Job Creation: The Findings and their Critiques." *Business Economics* 29: 23-30.
Doeringer, P., and M. Piore. 1971. *Internal Labor Markets and Manpower Analysis.* Lexington, MA: Health.
Feyder, S. 1977. "Chicago's Chinatown Has No Room to Grow, So Property Values Going up." *Chicago Tribune* (September 4): sect. 12, p. 1.
Freedman, M. 1985. "Urban Labor Markets and Ethnicity: Segments and Shelters Reexamined," Pp. 145-166 in *Urban Ethnicity in the United States: New Immigrants and Old Minorities,* edited by M. Loinel and J. Moore. Beverly Hills, CA: Sage.
Gleen, E.N. 1983. "Split Household, Small Producer and Dual Wage Earner: An Analysis of Chinese-American Family Strategies." *Journal of Marriage and the Family* 45: 35-46.
Hilton, M. 1979. "The Split Labor Market and Chinese Immigration, 1848-1882." *Journal of Ethnic Studies* 6(Winter): 99-108.
Hunter, A. 1988. "Persistence of Local Sentiments in Mass Society." Pp. 174-191 in *New Perspectives on the American Community,* 5th edition, edited by Roland Warren and Larry Lyon. Chicago, IL: Dorsey Press.
Kitano, H.H.L., and R. Daniel. 1988. *Asian Americans: Emerging Minorities.* Englewood Cliffs, NJ: Prentice Hall.

Knoll, T. 1982. *Becoming Americans: Asian Sojourners, Immigrants, and Refugees in the Western United States*. Portland, OR: Coast to Coast Books.

Lau, Y.M. 1988. "Alternative Career Strategies Among Asian American Professionals: The Second Rice Bowl." Ph.D. dissertation, Northwestern University.

Le, N. 1993, "The Case of Southeast Asian Refugees: Policy for a Community 'at Risk'." Pp. 167-188 in *The State of Asian Pacific America: Policy Issues to the Year 2020*. Los Angeles, CA: LEAP Asian Pacific American Public Policy Institute and UCLA Asian American Studies Center.

Le, N. 1995. "The Case of the Southeast Asian Refugees in Chicago: Policy for a Community 'at Risk'." Paper presented at a Consultation conducted by the Illinois Advisory Committee to the U.S. Commission on Civil Right.

Lee, R.H. 1960. *The Chinese in the United States*. Hong Kong: Hong Kong University Press.

Li, P.S. 1988.*Ethnic Inequality in a Class Society*. Toronto, Canada: Wall & Thompson.

Light, I.H. 1972. *Ethnic Enterprise in America: Business and Welfare among Chinese, Japanese, and Blacks*. Berkeley, CA: University of California Press.

Lopata, H.Z. 1994. *Polish Americans*, 2nd revised edition. New Brunswick, NJ: Transaction.

*Local Community Fact Book, Metropolitan Area*. 1980, 1990. The Chicago Fact Book Consortium.

Loewen, J.W. 1988. *The Mississippi Chinese: Between Black and White*. Prospect Heights, IL: Waveland Press.

Marciniak, E. 1981. *Reversing Urban Decline: The Winthrop-Kenmore Corridor in the Edgewater and Uptown Communities of Chicago*. Washington, DC: National Center for Urban Ethnic Affairs.

Marciniak, E. 1986. *Reclaiming the Inner City: Chicago's Near North Revitalization Confronts Cabrini-Green*. Washington, DC: National Center for Urban Ethnic Affairs.

Nishi, S.M. 1979. "Asian American Employment Issues: Myths and Realities." Pp. 495-506 in *United States Commission on Civil Rights, Civil Rights Issues of Asian and Pacific Americans: Myths and Realities*. Washington, DC: U.S. Government Printing Office.

Nelis and Castillo, Inc., Urban Planners. 1992. *Gateway to the Future: A Community Development Plan for Argyle Street*. Report prepared for the Asian Small Business Association. Chicago, IL: Nelis and Castillo, Inc.

Ong, P., and E. Blumenberg. 1994. "Welfare and Work among Southeast Asians." Pp. 113-138 in *Economic Diversity, Issues and Policies*. Los Angles, CA: Leap Asian Pacific American Public Policy Institute and UCLA Asian American Studies Center.

Ong, P., and S.J. Hee. 1994. "Economic Diversity," Pp. 31-56 in *Economic Diversity, Issues and Policies*. Los Angles, CA: Leap Asian Pacific American Public Policy Institute and UCLA Asian American Studies Center.

Roger, W., H. Aldrich, and R. Ward. 1990. *Ethnic Entrepreneurs: Immigrant Business in Industrial Societies*. Newbury Park, CA: Sage.

Siu, P.C.P. 1987. *The Chinese Laundryman: A Study of Social Isolation*. New York: New York University Press.

Strand, P., and W. Jones, Jr. 1985. *Indochinese Refugees in America: Problems of Adaptation and Assimilation*. Durham, NC: Duke University Press.

Tang, J. 1996. "To Be or Not to Be Your Own Boss? A Comparison of White, Black and Asian Scientist and Engineers." Pp. 129-166 in *Current Research on Occupation and Professions*, Volume 9, edited by H.Z. Lopata and A.E. Figert. Greenwich, CT: JAI Press.

Tsai, S-s. 1986. *The Chinese Experience in America*. Bloomington, IN: Indiana University Press.

United Way of Chicago. 1995. *Assessing Chicago's Human Needs: Community Development Series III*. Chicago, IL: United Way of Chicago.

U.S. Small Business Administration. 1994. "New Data Shows Smallest Firms are Nations Greatest Job Creators." News Release (September 30).

Ven. Dhammika, S. 1991. *Good Questions, Good Answers*. Taipei, Taiwan: The Corporate of the Buddha Educational Foundation.

Wong, B. 1988. *Patronage, Brokerage, Entrepreneurship and the Chinese Community of New York.* New York: AMS Press.

Wong, M.G. 1982. "The Cost of Being Chinese, Japanese, and Filipino in the United States: 1960, 1970, and 1976." *Pacific Sociological Review* (January): 58-78.

Woo, D. 1992. "The Gap between Striving and Achieving: The Case of Asian-American Women." Pp. 191-201 in *Race, Class, and Gender*, edited by M.L. Andersen and P.H. Collins. Belmont, CA: Wadsworth Publishing.

Yoon, I-J. 1990. "Self-employment in Business: Chinese, Japanese, and Korean Americans." Paper presented at the Annual Meeting of the American Sociological Association, Washington, D.C.

Zhang, J. 1989. "Chinese Immigrants and Small Family Business." Unpublished manuscript, November.

Zhang, J. 1992. "A New Community in an Old Neighborhood: Cultural and Economic Dimensions." Paper presented at the Mid-West Sociological Society Annual Meeting, Kansas City.

# INDIAN-AMERICAN SUCCESS STORY OF "POTEL"–MOTELS:
## IMMIGRATION, TRADITION, COMMUNITY, AND GENDER

Nandini Narain Assar

## ABSTRACT

Immigration policy and tradition dovetail in their impact on the social organization of immigrant communities. Here, the focus is on Asian-Indian Patels, who are concentrated in the budget motel business in the United States. Family reunification policy does not recognize immigrant families as labor, so a majority of documented immigrants are exempt from labor certification. Most Patels enter the United States under these rules and are part of a chain migration. Biased labor and financial markets in the United States herd immigrants into family businesses with a heavy concentration of family labor. Difficulty in finding commercial credit has led to the practice of community financing. Traditions define Patel women as housewives, even when they work in motels, and this definition is in opposition to that of worker/male breadwinner. So, a housewife is not a worker. Patels are successful in budget motels both due to community financing and because of the use of family labor. Both these advantages are escapes from the market economy.

Current Research on Occupations and Professions, Volume 10, pages 67-86.
Copyright © 1998 by JAI Press Inc.
All rights of reproduction in any form reserved.
ISBN: 0-7623-0034-5

When there is a direct link between contributions of labor and decision making, then traditional gender hierarchy is challenged. When there is no such link, then gender hierarchy is exacerbated. In families that sponsor subsequent links in a chain migration, gender hierarchy is exacerbated. In families which are the last link in the chain, gender hierarchy is challenged.

# INTRODUCTION

In the mornings, we are *bhangis*[1] whose work is to sweep and clean the rooms. In the afternoons, we are *bawarchis*:[2] we cook and feed our families. And in the evening, we are *sethanis*:[3] we dress up nicely, sit in the office, and collect money from the customers. That is our life in the motel business.

When asked to describe a typical day in the motel business, Bhavna[4] gave this succinct response. Although I found it amusing, there is a great deal of insight here.

Recent estimates are that Asian Indians control more than 65% of budget motels *nationwide* (Lister 1996). They have achieved the "American Dream" of economic prosperity. For Indians, affiliation with a linguistic or regional subgroup is more relevant and stronger than identification with the national group (Sheth 1995). Most Indian motel-owners trace their origins to the state of Gujarat[5] in western India. Of Indian immigrants to the United States, about one-third are Gujarati (Weiner 1990). Although Patidars[6] are only about 5%-8% of Indian immigrants to the United States, they comprise 90%-95% of Indian motel owners (Mehra 1993). Their most common last name is Patel; consequently, the motels owned by Patels are popularly referred to as "Potels." Within the community, hopes of owning a motel have become a popular jest, particularly since a singer named Bali composed the "Patel Rap" in 1989. The lyrics describe a man who keeps his shop open late to save enough to buy a motel (Woodyard 1995). There is a grain of truth in this characterization. Gujarati motel-owners were the focus of the movie "Mississippi Masala" as well, also in the 1980s.

Patidars are successful in making money. It is now well-known that Asian Indian Americans are at the very top of the economic ladder, with the highest household income of all groups, including whites. Gujaratis are at the top of the heap of Indians. Twelve percent of Gujaratis report household incomes in excess of $75,000, and less than 6% reported household incomes less than $15,000 (Mehra 1993).

In this chapter, I examine the work involved in running budget motels in the United States—who does it and under what conditions. Most Indian motel-owners trace their origins to the state of Gujarat in western India. First, I lay out the theoretical framework, including a discussion of immigration and tradition as they interact with family relationships and family businesses. Then, I describe the methodology and the participants. Next, I discuss the nature of motel work, and present the findings. Finally, I evaluate whether the data supports my theoretical contention that the gender division of labor is linked with the international division of labor and has direct impacts on economic outcomes and family relationships.

## THEORETICAL FRAMEWORK

The international division of labor is a central feature of a global structural analysis. I contend that the global economy is based on the gender division of labor as well, in line with Mies' (1986) analysis. Linking the international division of labor (at the macro level) and the gender division of labor (at the micro level) explains the working of the global economy more accurately and completely than a focus on either one alone. Further, the gender division of reproductive labor in home interacts with and reinforces gender division in the labor market (Glenn 1992). Gujaratis, for example, immigrate to the United States under the influence of global forces that organize movements from the Third World to industrialized nations. The international division of labor explains this part. The focus on Gujarati motel owners is theoretically driven: access to family labor is a key competitive advantage for immigrant businesses, and it is linked to the gender division of labor.

The most widespread conceptual dichotomy about work is between the productive/waged/public sphere and the reproductive/unwaged/private sphere. However, motel work conflates the dichotomy: it is productive/reproductive/unwaged/public/private. It cannot be easily categorized in either domain. Glenn (1992) has argued that reproductive labor is divided along racial and gender lines, and that the specific characteristics of the division vary regionally and change over time as capitalism reorganizes reproductive labor, shifting parts of it from the household to the market. The relationships of men and women who work in budget motels also span the public and private. However, the work is disconnected from the labor market and the conditions of work are determined by family, kin, and community relationships, characterized clearly in the private/domestic domain.

### Immigration

Most Gujaratis who apply for entry into the United States under the family reunification laws report a smooth and hassle-free immigration experience. Typically, the entire family comes to the United States together, with full immigrant status. Immigrants who experience a smooth entry with documented status either are people whom the United States state wants or have devised a strategy to subvert the legislation to restrict entry into the United States. Both processes are at work in this context.

One explanation for the relative ease of immigration of Patidars rests on the centerpiece of United States immigration law—the family reunification policy—which encourages immigration in a family context. Since the reform of 1965 and through the changes to date, anyone who applies for immigrant status under the family reunification rules is not required to obtain labor certification (Heer 1996, p. 55). Family reunification leads to the admission of people without regard to their impact on United States productivity and standard of living (Isbister 1996, p. 69). In other words, an immigrant who has family (defined in particular, prescribed ways) in the United States is not considered in relation to the labor market. This is centrally important since the vast majority of documented immigrants enter the United States under the family reunification policy, and as such are exempt from labor certification (Jasso and Rosenzweig 1990, p. 188). This means that the state does not recognize them as labor. The category of immigrant "family" is defined as non-workers.

The pattern of family reunion is closely connected with the rate of business startups in labor-intensive industries. In countries where family reunion and self-employment was not encouraged, there is no evidence of immigrants' involvement in entrepreneurship (Phizacklea 1988). The connection between immigration policy and family enterprises pertains to Gujaratis in the United States: motel owners live on the motel premises, saving substantially on rent, utilities, childcare, and transport costs. Family members work there, thus reducing labor costs. In the pilot study,[7] about 85% of Gujaratis who came in under family reunification rules enter family businesses, mainly motels (Assar 1990).

Ostensibly, then, family reunification immigration policy is based on social and not economic concerns. The paradox is that a majority of immigrants (who are not considered to be workers in the immigration policy) are herded into family businesses by the structure of the labor market, with a very high component of family labor. This formulation allows the economy to benefit from their labor without bearing the costs of migration or recognition of their status as workers. So, the vast majority of documented immigrants with green cards are also denied all access to social services for five years, while they are liable to pay all taxes. Contrast this with the common perception of immigrants as a drain on national resources.

## Tradition

Recent research about immigrants shows that there is a direct relationship between economic contributions and the sharing of decision-making power. There is change from traditional to more egalitarian patterns when women make substantial economic contributions after immigration (Hondagneu-Sotelo 1994; Kibria 1993; Lamphere 1987; Pessar 1986). My contention is that "traditional" gender hierarchy is contested when there is a direct relation between economic contributions and decision-making power. When the two dimensions are disconnected, it results in exacerbation of gender hierarchy. Therefore, I examine the link between economic contributions and decision-making power among Gujaratis in the motel business.

Most of the men were either farmers or had small businesses prior to immigrating. Most of the women were housewives. Post-migration, both men and women share the work in motels, but the women remain housewives. The category of housewife defines women as non-workers and creates the space for the category of male "breadwinner" (Mohanty 1997, p. 13). There is a reformulation of gender relations, reflected in the jobs each is expected/likely to perform.

Motels are clearly an example of the blurring of public and private domains. Even the physical separation of work and home is challenged because most owners reside on the premises of the motels. Further, the nature of the work is identical to housework. It is only the context in which the work is performed that distinguishes it from housework. However, when women engage in this devalued "economic" activity, they remain rhetorically within the domestic domain (they do not "go out to work"), so there is no change in the "traditional" status of housewife for women. The designation "housewife" is in opposition to the status of "worker." That is, it defines women as non-workers (Mohanty 1997, p.13). The conditions of work are determined by "traditional" family and kin relationships, so women's labor can be integrated without a corresponding change in their status as "housewives."

Typically, budget motels involve relationships that span production and reproduction. That is, the men and women work together and also live together:

> Owners can put their families to work, reducing labor costs to practically nothing. The wife did the housekeeping. The husband did the front desk, and both of them together at night did the laundry and folded the sheets. The husband and wife, as a team, worked almost 16 hours a day. Doing that, [they] saved money from the payroll, and could turn around and pay off the debt. . . . It hardly took two or three years to return the money to the lenders (Woodyard 1995).

## METHODOLOGY AND SAMPLE

Most people are wary of revealing family relationships. Among immigrants from the Third World, this becomes more important because negative connotations connected to "foreign" cultural norms are prevalent stereotypes in the United States, linked to low status of women in these groups. I needed information about family relationships among immigrants. Therefore, I expected that the information I was after was likely to be difficult to elicit.

Miles and Huberman (1994) argue that field research can be considered an act of betrayal, no matter how well-intentioned or integrated the researcher. They mean that the researcher usually gains more than participants from the research, although participants usually expose more about their lives (Behar 1993). Thus, the research relationship is inherently hierarchical.

Attention to both these issues is critical to research. My location as an Indian immigrant woman is a decided asset in gaining access to Patels[8] from Gujarat who operate budget motels. I am familiar with the general cultural codes, but I am not Gujarati. In other words, although as an Indian immigrant woman I am an insider in relation to the wider society, I am not part of the Gujarati community, so I am an outsider as well. I had both statuses simultaneously. Since I am Indian, I shared Indian-ness—food, language, clothing, and general ethos—with participants. Since I am not Gujarati, I did not have preconceptions about particular norms and could hear them without the filter of Gujarati-ness.

I conducted semi-structured, tape-recorded interviews with Gujarati motel-owners and families. They varied in length from 30 minutes to four hours, depending on the participant's interest and availability. Participants ranged in age from teenage to in their seventies. They were located in three southeastern states. The selection criteria were that they were Gujarati, they operated a budget motel or had done so in the recent past, and they were within driving range. The geographic locations are not significant for this study. I could have interviewed anywhere in the continental United States since Patels own budget motels nationwide.

First, I approached prominent Gujaratis locally, explained the project, and requested referrals to motel-owners. These were my initial "key contacts." They provided me with a directory of regional Gujarati families, and pointed out four families whom they thought would be receptive to taking part in the project. They agreed to call these families and tell them about me and my project. Next, I called the four families and arranged to meet them as potential participants. Since I wanted access to the entire family, including teenagers, it was crucial that I approach participants with a reference from someone they knew and trusted.

I went out to meet them and observe as they went about their work. When possible, I stayed overnight in the motel to document their daily routine. Most participants invited me to eat with them and this was significant in establishing rapport. I went out a second time to interview each member of the family. My attempt was to talk with each person without the presence of family members. In a few cases, it was not possible to do the interviews in private. I attempted to interview women first, but this was not always possible. I did some interviews while participants were attending to the front desk or doing cleaning, laundry, or cooking. I worked with them when possible, for example, making beds, folding laundry, or cooking. I also attended community activities, festivals, celebrations, and prayer meetings. After I had interviewed each family, I left it open that I might need to contact them again for clarification or follow-up. I requested one referral from each family to continue the interviews. I was after a snowball sample.

The referrals were participants' only assurance that the research would not have negative impacts on their lives. Had I approached them cold, I knew from past experience in the pilot study, I would have been hard pressed to establish credible field relationships with the necessary levels of trust and rapport. As an added measure, I asked for feedback from Gujarati immigrants before making the findings public. This effort was not to ensure that I had "got it right" but rather that I had not missed the point (Wolcott 1995).

## ABOUT POTELS

The first Patel to own a motel in the United States came from Gujarat in 1949. He picked apricots, cotton, and grapes in Northern California for six months, and used his savings to buy a residential hotel in San Francisco. In the late 1970s, Gujaratis bought up dying family-run motels that were afflicted by reduced travel due to the oil price hike. They gained access to the United States through the immigration reforms in which family reunification was the central motif. Most arrived with no capital and few skills or experience that translated well in the United States labor market. Financial institutions were skeptical about extending loans to them (Lister 1996). They proceeded with interest-free loans from friends and family, fixed up dilapidated properties, and turned them into franchised units of national chains. Community financing at no interest is widespread among Gujaratis, especially in the motel business. They do not look to make a profit from each other.

Motels are attractive because they provide a stable income. Today, Indians own an estimated 27.2% of all hotels/motels in the United States. In 1989, they formed the Asian American Hotel Owners Association (AAHOA) in response to experiences of financial and industry biases and discrimination. Their current membership is more than 4,500 (Lister 1996).

### Occupational and Gender Hierarchies

"Gender is fundamental to the way in which work is organized, and work is central to the social construction of gender" (Game and Pringle 1983). Since gender division of labor is central to maintaining "traditional" gender hierarchy in families *and* motel work is for the most part composed of "women's work" also done by men, logically, it ought to lead to dramatic shifts in gender hierarchy. Instead, gender relations remain largely unchanged.

Occupational and gender hierarchies are inextricably linked in the status and evaluation of motel work. We can observe both hierarchies in the allocation of work among family members, and its status among Gujaratis in general. Tension and contradiction also serve to highlight them. One major contradiction is that though Patidars prosper in this line of work, the nature of the work is devalued, since it is defined as domestic work:

> They must confront, acknowledge, and convey the undesirable nature of the work they do to their children, as an object lesson and an admonition, and at the same time maintain their children's respect and their own sense of personal worth and dignity (Dill 1980).

The following experience highlights the devaluation of skills that are necessary in the motel business, indicative of the negative status of the work. One potential participant (male) at first agreed to talk with me. Further into the conversation, when he learned that I wanted to observe him at work, he angrily declined to take part, saying: "What? You want to see me working? Then I am not interested. You listen [to] me, I am not interested in your project." Significantly, his wife and daughter-in-law, both of whom I had spoken to earlier, could not make the decision as to whether they could take part in the study without conferring with him first.

Another participant, Dev, is in his forties, the father of two, and arrived with his family from India three years ago. He has a BA and had a flourishing business prior to migrating. He articulated the contradictions in motel work in this way:

> I:   Do you think your children will continue in the business?

> Dev:   If they [the children] study well, they should do some management courses, do something in the hospitality industry, and then—I mean, I wouldn't like them to continue at this place. I'm sure we'll have some better place by the time they grow up and if they want to continue, it'll be a good thing for them—it'll be a better start for them than what we had. But then, I don't know—they study something different and then they want to pursue that—it's up to them.

> I:   Would you consider going back to school to improve your own employment prospects?

> Dev:   Well . . . I don't want to study anything from the job point of view. Because I hope everything goes all right and I'll not have to work for somebody. So only if I have to do something for my business, like a basic knowledge of computers, or . . . I'm planning to get that CHA degree, not a degree, it's kind of an examination you have to take—Certified Hotel Administrator. It's considered to be a good thing. Otherwise going to school and all, for getting a job . . . no. I am satisfied with what I'm doing and the income I'm getting. I don't think that a job can give me more income than this gives me.

Clearly, Dev does not want his children to do what he is doing when they grow up, and at the same time, he does not intend to change to higher status work because the income is good.

Gender hierarchy is evident in a similar devaluation of "women's work" regardless of who does it. In other words, when men do this work, it remains devalued. Another man I interviewed, Das, picks his grandchildren up after preschool and takes care of them in the afternoons. I met him for the first time during this time of the day. He was attending to the front desk and to the children. This is possible because the front desk/office usually adjoins the living room in motel accommodations. We noticed an advertisement for a baby doll on TV. Das asked his smaller grandson (age 3), tongue-in-cheek, whether he would like to have that toy. The child vehemently refused and said it was a toy for girls. Das was pleased and smiled.

Immediately after this exchange, Das changed his grandson's diaper, sang to him and tenderly rocked the child to sleep in his arms. Although he was involved in child-care on an ongoing basis, he was careful to devalue these skills to the boys and to me as "women's work."

Men explained that they shared the cleaning jobs, but women pointed out that the most devalued jobs remained firmly in the women's domain. Here is a typical man's account of the sharing of work:

> Dev:   If we think that the maid is not here, who will clean the bathrooms, it won't work. If we think, "No, I can't do the bathrooms"—Like the maid doesn't come on Wednesdays, we clean the bathrooms. We can't afford not to keep all the rooms ready, maid or no maid. . . . There is the better side of things too. There are good things here also . . . most of the people here send money to their family in India. People in India may think that doing such work is not good. But when the money goes, they feel good. You do get hurt when they say that you do such jobs here. . . . We did not take it as a hardship, because we got better opportunities.

However, women clearly expressed that it is beyond men's domain to clean bath-rooms. Dev's wife, Nisha says:

> Nisha:   Last time you came in earlier, so that you could watch them working. She (Nina, the maid) does the bathrooms and vacuum cleaning. We help her making beds in the rooms. During summer, when we are on leave, it is a great help to have her. She comes every day except Wednesdays, when she has to go somewhere for house-cleaning.
>
> I:   So on Wednesdays you have school, and Nina doesn't come. How do you do all the work in the motel?
>
> Nisha:   He (Dev) would do all the beds, etc. When I come back from my class, I clean the bathrooms.
>
> I:   How about at home? Who does all the cleaning at home?
>
> Nisha:   Me. (Laughter). Of course, I have lot of work at home. Every third day I have to do laundry for everybody. The children go to school and they are so busy. It is good to do all this work, as I was thinking what to do in vacation time. I like to do all the cooking, cleaning, laundry, etc. After June 15, my children have their vacation. That is why I took the Home Science class.

This is a contradiction that highlights gender hierarchy. The rhetoric remains that both partners share the work. However, it is either the wife or the maid who cleans the bathrooms in this motel. The gendered nature of the work is apparent. Domestic work remains devalued as long as men continue to refuse to take part, even though women are increasingly making economic contributions (Hochschild 1989).

However, women's relationship to domestic labor is not universal. Like other minority groups in the United States, ethnic relations are linked to domestic labor for Gujarati immigrants (Glenn 1992). Men in this group do perform "women's work" in motels and in families, as shown below. However, it remains devalued to maintain "traditional" gender hierarchy.

Among second generation[9] immigrants, work is gendered according to tradition when they remain in the motel business. They typically move up from budget motels to franchised enterprises with employees. Women no longer do the daily cleaning and laundry but are still in charge of day-to-day operations. They are educated and fluent in English, and may consider themselves removed from immigrant status. However, they are still the backup for the cleaning jobs when employees do not show up. Men in the second generation are engaged in making financial deals, overseeing construction and expanding the business. Although the rhetoric remains that both spouses share all the work, women reveal that they normally do not take part in business decisions and do not go against male relatives' ideas, generally going along with what men decide.

Although women contribute their labor in family motels, only when there are no more links in the chain migration is it possible for women to negotiate a more equitable gender hierarchy. I found that families who sponsored further links in the chain migration had more gender hierarchy than families who were the last link in the chain. This would be because each sponsored immigrant has costs of migration, which are met by increased contributions of labor by family members. That means everyone has to work more to support the next link in the chain. However, women's increased labor does not result in increased status because the new immigrants also reinforce "traditions."

## ADVANTAGES OF BUDGET MOTELS

First generation Patels typically buy a motel with financial backing from the community, and live on the premises. They save on living expenses and are available round-the-clock. A motel provides at once real estate, living quarters, a business, and employment for the entire family (Jain 1989). They make a profit, return investors' money, and look for bigger units. Now they can get commercial financing. They then identify another Gujarati family who is looking for a start, and sell the old property. Often, on-the-job training is part of the deal. It may involve sharing living space for some time. They typically do not look to make a profit, just to break even on this transaction, which works to the benefit of both parties. The seller gets a reasonable price, and the buyer gets a fair deal, plus training.

Participants reported the following reasons for choosing the budget motel business:

1.  They have access to Gujarati community experience, expertise, and financing on non-commercial terms. Gujaratis loan money to each other to help businesses start at no interest, like Korean immigrants' kyes or lending cir-

cles. For example, a 40-room building in a good location, bought for $500,000 can turn $100,000 profit a year. Owners can pay off the debt in a short time (Woodyard, 1995). As Das proudly outlined:

> Das: We have friends, we borrow money, paying them when they need it. Or if they don't need, [we can] hold it. . . . I can raise two million dollars within two weeks. Just call in today and give the guy one week, and they send the check. When we got the land, we didn't have money. Only $7000 in pocket, and they [the sellers] wanted $400,000 cash. I called in every one in New York, just borrowed and paid them [later].

2. They can meet family and childcare responsibilities:

> Sudha: My children were growing that time. So I want to take care of my kids, and [think of] some way I can do business. In a store, you cannot do business and take care of your kids. I can cook, I can take care of customers, I can take care of my kids, I can take care of my husband, I can cook for them. I can give food, home food, and same time I can take care of my kids too, you know—and run the business same time. . . . All in the same place. See, the kitchen is attached with office, so I can take care of both things. In another business, I could not take care of that same things. If I have a gift store, I cannot go home and cook and do these things, you know, because it is not a side-by-side—you run the restaurant, you cannot get both things together. Only this business, you get together—you are not apart from your children. I didn't like to give my children to a baby-sitter. I wanted to take care of them myself, so that later on, they wouldn't tell me that I put them the whole time with a baby-sitter and "we grow up by ourselves." What kind of values could I give if I the baby-sitter was their teacher, you know.

Sudha's concerns were right on target. Even though parents kept childcare and family responsibilities as a priority in selecting the business, second generation participants experienced "growing up by ourselves" because both parents worked in the motel:

> Jagdish: We did whatever we had to. It was always me and my sisters together. Sometimes [I feel like] we raised ourselves because Mom and Dad were busy with business—well, doing things for our benefit. That's why I want to have a house, 5 o'clock I want to be home, and I want to spend time with my kids.

They can also save on living expenses. First-generation immigrants expressed the benefit of these savings:

> Dev: I don't think that a job can give me more income than this gives me. Because you see very little cash coming in, but other things if I consider, it's O.K. If two of us go to work outside, we can't make [a] better living than this. Because here we are, with all these things . . . housing, utilities, everything's included. . . . For all these things, you would need one thousand dollars in a month.

However, second generation participants unanimously reported a sense of acute loss and deprivation because of the unusual and often cramped quarters:

> Deepa:   I would have liked our parents go ahead to have a home for us, a house home, where you have your living room, bedroom and kitchen—those are the things I was obsessed with, having grown up in a motel. I couldn't stand it that we couldn't stay in a house, we had to be in a motel. It just didn't feel like a home to me. I always thought, when I grow up I wanted to have a home of my own. That's what I wanted the most. If I had anything to do with it, I would say, "Buy a house."

3.   All members of the family can contribute their labor towards the motel business. Children pitch in after school, on weekends, holidays, and vacations. Girls are likely to stay involved until marriage, and boys until they leave home. Once married, daughters are only expected to contribute if their husbands are involved/interested in the enterprise. Sons are expected to contribute even when they are established on their own. When families sponsor subsequent members to immigrate, they share living quarters, and also work together. Families often also share living quarters during the transition periods when the business is changing owners. This is another point of contradiction to the common perception of immigrants being a drain on national resources. The key characteristic for immigration under family reunification rules is immigrants who will make minimal demands on the state and whose costs of migration are borne by families and communities.

4.   Patels who operate budget motels are buffered from the biased labor market: they need not compete for jobs or employees. Discrimination in the labor market was a common reason participants gave for going into a business. For instance, Bina's husband was laid off from three corporate jobs, even though he had a Ph.D. from an Ivy League university and extensive work experience in the United States. She identified racial discrimination as the cause for his repeated lay-offs. He would be hired due to his qualifications. After a few years, he would become eligible for increments in pay and benefits. Even though the companies were satisfied with his work, they were reluctant to move a dark-skinned foreigner into a senior position. So he would be laid off, and a junior person who could not claim higher pay would be hired to take over from him. Bina's husband would then be easily hired again on the basis of his qualifications and experience, but would not be eligible for a senior position in the new company, and would repeat the cycle. After the third lay-off, they decided to go into motels, because it provided more stability than the biased job market.

Parallel to the experience of racial bias is the well-known devaluing of training, skills, and experience gained abroad. Several participants held professional engineering degrees and had experience working as engineers prior to immigration. However, they were unable to translate their experience and training into comparable employment in the United States. Participants experienced it as a painful and disappointing realization. This downshift in occupational hierarchy is an accepted part of migration. Justification for this need not be elaborated in specifically racial/ethnic terms: instead it can be cast in terms of differences in skill, training, or education—to lack of "human capital" (Glenn 1992, p. 33). It is one of the mechanisms by which new immigrants are herded into family enterprises.

## NATURE OF MOTEL WORK

I am differentiating between franchised operations with employees and small budget motels with no employees, because there are important variations in who does the jobs between the two levels of operation. In small budget motels, the work is shared among family members, with small additions of employed labor.

The bulk of the work consists of cleaning and laundry: skills in which women are already trained and socialized, and which are easily incorporated as an extension of the domestic domain, facilitated by the fact that families reside on the motel premises. So, this work is done at home, interspersed with housework and childcare. Women require no additional training or language skills because there is no direct relationship to the labor market and minimal contact with customers. However, lack of fluency in English means that women must survive entirely within the community. On the other hand, most men do need to be trained to do this work because they are not used to cleaning and laundry. These devalued skills are transmitted smoothly within the family context from women to men when necessary.

Family businesses which depend on cultural and ideological resources within the family to transform immigrant women into workers committed to common goals are also anchored in their roles of daughters, wives, mothers, and keepers of the family honor. They are based on ideologies of domesticity and womanhood that pervade the domains of production and reproduction (Westwood 1988). Women and men's "traditional" statuses remain firmly in place.

The business demands round-the-clock attention. This fits in well with "traditional" ideas about women's place in the home and their domestic and childcare duties. The home now includes the business. Paradoxically, while Patels choose motels so that women can continue fulfilling their "traditional" duties, the situation also provides for the most radical changes in gender hierarchy through the work. That is, men who work in motels typically do "women's work." While there are many examples of women expanding their responsibilities to include more "traditional" men's work, the parallel move of men taking on "women's work" is relatively uncommon. Gender is socially constructed by devaluing "women's work," so as to not be a challenge to "traditional" gender hierarchy. This is clearly evident in the incident regarding the doll.

The term "women's work" does not mean the work that women do, or even the occupations that women are concentrated in. Rather, it refers to the ideological construction of jobs and tasks according to notions of appropriate femininity, domesticity, heterosexuality, and racial/cultural/ethnic stereotypes. It structures the work women do or precludes women from being defined as workers altogether (Mohanty 1997, p. 6). Patriarchal ideologies which fix women's role in the family are based on immigrants' "traditions." These very ideologies are reproduced and consolidated to provide the glue for profit in United States economy. The aspects of "tradition" that immigrants exploit for social or financial gain are unlikely to be those which provoke a desire for change (Shah 1996).

Although all participants claimed the work was equally shared among both spouses, none of the men I observed cleaned rooms. Only one man made beds and did laundry in my presence, and he was working with a maid. The other men I observed took the front desk, did yard work and maintenance, or cleaned the pool while I was around. The women cleaned, cooked, did childcare, laundry, beds, front desk, maintenance, pool, and yard work. I worked along with them when possible.

Men almost always are in charge of financial transactions and business decisions—whether to take a loan, expand the business, add property, make improvements, hire help, or buy a franchise; women are responsible for accounting, bookkeeping, and maintenance work. Although I could not observe these gendered divisions: this was a common allocation, according to the interviews.

## Decision Making

Family businesses are easily seen as being in line with the ideology of joint resources and inputs, if the family is the unit of analysis. However, examining whether there is a link between contributions and rewards within the family will allow us to see what is going on inside this unit and chart the impact of social structures on family relationships. Recent research about immigrants shows that there is a direct relationship between economic contributions to the success of the family and the sharing of decision-making power. There is a change from traditional to more egalitarian patterns when women make substantial economic contributions after immigration (Hondagneu-Sotelo 1994; Kibria 1993; Lamphere 1987; Pessar 1986). Their work shows that "traditional" gender hierarchy is contested when there is a direct relation between economic contributions and decision-making power. Building on this, my research shows that when the two dimensions are disconnected, it results in the exacerbation of gender hierarchy.

The nature of the work is identical to housework. It is only the context in which the work is performed that distinguishes it from housework. When women engage in this devalued "economic" activity, they remain rhetorically within the domestic domain, and so their decision-making also remains strictly within the domestic domain.

For example, Das said that business decisions were made jointly by the family. He is 59 years old and has been in the United States for 17 years. He has three daughters and one son, all married. The son (Jagdish), daughter-in-law (Champa), and two grandsons live with Das and Lekha in a two-bedroom apartment attached to a 20-unit motel. Das represented evidence of family involvement in decision making:

> Das:  I only take this decision [to buy motels] with the whole family. We discuss this openly that we are going to do it. We don't know [whether] we might succeed, or we might fail. "Are you all ready to? If we fail, we have to prepare how to live." After three years, we lost $120,000. I said, "What can we do? Don't worry if our luck is off." One of my daughters said—she is smart—she said, "I'll go to college, and I'll look after this business—one of the motels." I said, "O.K. I'll try to get a convenience store, try to get extra funds, and try to come back again [financially]."
>
> I:  So when you make a decision, then is it a family decision? You talked about it?
>
> Das:  Ya, family decision. We talk, discuss about it. Sometimes there is difference of opinion. Wife says, "No, we don't want to do it. We don't want to do this." Choice is sometimes hers, "No, I don't want to do this." We try to convince them [the women]. And the right to make the decision is based on who does the work. Like, they [the women] are making decisions about the house.

Clearly, then, the rhetoric is that the person who does the work has the right to make decisions in that arena. In this family, both parents had worked for 27 years to support the family. At first, both were employed in blue-collar jobs. Then they bought several grocery stores and worked them together. Next, they came to the United States and worked in motels together. However, the wife's decision making is recognized primarily in the domestic domain by the husband. So she is not considered to have "worked" in the business.

Her account is considerably different. Lekha is in her mid-fifties. There is no doubt in her mind that she shares both the responsibility and the work:

> I:  What did you do in the store?
>
> Lekha:  Everything. All round. We had to go everywhere, like here in the motel. If [the] maid won't come, I have to clean the rooms. When we go into business, we must learn everything. If you have got a restaurant, you [must] know how to cook, you [must] know how to set up a table. If you are in the motel, you should know how to be desk clerk, do night duty, how to do the maid's job, and how to serve the customer. You should know everything.

On the other hand, Lekha and Das's adult son, Jagdish, who is also in the motel business, said that all business decisions were entirely his father's:

I:    Do they usually have a joint decision?

Jagdish:    I think it's Dad 95 percent and Mom five percent. Dad pretty much tells Mom what he wants to do. I mean, my father—I admire him, I love him. He did some great things for us. I wouldn't want to be like him. I [would] love to have some [of his] skills but I wouldn't want to be like him all the way. I saw that he should treat my Mom better, maybe as an equal.

I:    So let me just say what I am hearing from you: that your mom and your dad worked together in the motel. You also said that most of the business decisions are your dad's.

Jagdish:    100% business decisions. Dad would explain it to Mom, and Mom, O.K.—I mean no disrespect to my mother, she has never been educated. She has been married ever since she was 13 years old. It happened back there in India, I can only imagine what she went through. O.K., you are married at 13, now you can now learn how to cook, now you can learn how to stay home.

This family's living arrangements are exactly in line with the ideal of the joint family in India. They are also a perfect fit with the demands of the labor market in the United States economy. Here is the linkage between the international division of labor and the gender division of labor. They work in tandem. Lekha describes the arrangements thus:

Oh! Yes. We have to be baby-sitting (Laughter). Because she (Champa, the daughter-in-law) works, she's coming 5 o'clock, Jagdish coming at 6.30 or 7.00. So me and my husband both have to. You can't go out, we just baby-sitting. When you got older people [in the family], they have to take such responsibility, otherwise won't go together. When we got small ones, it's very hard. If you want to do business, you have to go together, give up something you do together, then you grow together.

## CONCLUSIONS

The data support my contention that when economic contributions are not linked to decision-making power, then the outcome is an exacerbation of the "traditional" gender hierarchy rather than more equality. Participants in the study made this connection, too. For instance, Naresh, a middle-aged man with two teenage sons who has been in the motel business for eight years, said:

For Indian ladies, life in United States is harder than in India. In India, only job they had was to take care of the family—I mean, prepare nice dinner and take care of all the family chores. Here, their job is still there: they have to take care of cooking, cleaning and all the family—on top of that then, they have to work. She works whole day with me, taking care of motel, plus she has to take care of the family: cooking, cleaning and all that. I can help and I do help her, you know, that's just help. But she is the one who does most of the work.

The definition of work as distinct from housework is evident in his analysis. This highlights the paradox for Gujarati women working in family motels. They contribute more to their families' economic success after immigration. But because their labor is viewed as an extension of housework, it is disconnected from changes in status or in gender hierarchy. "After all, part of what needs to change is the very concept of work/labor" (Mohanty 1997, p. 11). We need to highlight the current fundamentally masculine definition of laborer/worker, which is constructed in opposition to the "housewife." Bhavna, Naresh's wife, presents her analysis of women's position in family motels:

> Bhavna:   This is a main business run by the woman, I can say woman is a most important factor in this industry. Everybody was saying so, it's not just me but everybody. You can say that all the women, the majority you can say 90-95% women are in this business. They can do anything in the motel. The men cannot do everything without them.
>
> I:   Usually, men have more exposure, more experience more education . . .
>
> Bhavna:   Ya, everything—they are totally dominating. In small motels, they don't have a desk clerk or anything—totally family-oriented business. When we go out and when we [women] come back, there is some work left for us. When they [men] go out, there is not a single thing left for them. Everything is completed. All accounting work, banking work, kitchen work, laundry work, motel work, everything. I have grown-up kids, but if you have small kids, then it's a. . . . If the woman is not doing anything in India, then in this country she has to go out and do a job or do a business, come back home, cook for man and children and then washing dishes, change the diaper for the kid and everything. I didn't see a single man who can help in the kitchen, or who could make food for the wife. No, not in our caste! They don't do, they have an ego, you know. These men have an ego. They came from India but they have never changed themselves. They have an ego. Oh, no! We cannot do that thing, not that.
>
> I:   So, the men don't change after they come here.
>
> Bhavna:   No, they don't change. They don't change, they have a certain insulting thing, they have an ego. My husband is 100% exceptional. He is real nice, he does every thing, he helps me in everything whatever I do in kitchen, washing dishes, doing my home laundry, everything you know. Sometimes he can sweep this floor too, vacuum and everything.

Post-immigration changes for Patidars are evidently positive for the family and the community as a whole, without changing the status of women within the family. Here is Bina's perspective as a non-Gujarati married to a Patel:

> Bina:   Women got more power [in the United States] than men. But Patel women don't have power. They should stand for their rights.

I:    But the thing is that they work so hard. This is the thing that really interests me. They work really hard . . . they work the motels, they clean the bathrooms, they do everything.

Bina:    Everything. . . . And their husbands . . . make tea for husband, make coffee for husband! Small motels are killing. They don't get nothing. These Patel women, they got 16-room motel and they possibly clean 10 rooms—two hours. Give $12, how much time you save? $12 is no big deal. They can get somebody, right? Their husbands say, do it and they do it. We had a 20-room motel, and somebody took four hours. For four hours, we paid $20. If you are going to clean 14 rooms, your back will hurt, your legs will hurt, your hands get red, you get so tired in the evening. . . . Why? But these people don't get any help.

The labor market operates to support this system of unpaid family labor. The work is low status and low waged, so there is a high turnover, even when there are employees to do it. Consequently, even when motel work is done by employees in the larger franchised businesses, family labor remains the mainstay. They are the backup when workers do not show up, which is common. Even though men do this work, they are careful to devalue it in order to maintain "traditional" family relationships. It is more acceptable for men to do "women's work" in the business than it is at home. For example, one Gujarati woman put her finger on this point:

Mira:    A typical Indian man is just demanding, you know. And my husband believes Indian men should not do household work. He should not do any work in house, you know. He should just sit on the sofa and have everything, just like macho men . . . (Laughter) . . . But he is very flexible, you know, in the business. He has cleaned the toilet. It's not like he has never done it. He has done it.

I:    Are you saying then that he separates that is his business and that is his home? Like he will do it in his business, but he won't do it here?

Mira:    Ya. I think because. . . . I think the way he was raised.

However, when women work in motels, they view it as an extension of the domestic role. In the following excerpt, one woman articulates how Patel women typically think about motel work:

Sudha:    See, Indian mentality is there, that "I am educated, why should I do cleaning job?" No, you have to do it because you want things, *you have to do like [a] 30-room house.* You have to . . . afterwards you [are] thinking that this is a 30-room house we are cleaning every day. [You cannot think] "I am not going to clean people's stuff." No! You have to . . . time comes, you have to do everything. This is a part of business.

Working in motels does not bring Patel women the status of workers: they remain housewives, as they were before immigration. Motel work is an extension of their domestic role. In contrast, for Patel men, motel work is separated from their domestic role—they may do it as part of business, but not at home. This separation is particularly striking because the business and the home are the same location. Immigration laws also do not recognize immigrant families as workers: they remain "dependents" even as their labor is incorporated into the economy.

The success of Patels in budget motels in the United States can be attributed to community financing and family labor. Both these factors operate outside the market economy. And the family reunification policy feeds this particular scenario, by allowing more family members relatively easy movement into the United States without recognizing the labor that they contribute to the economy. This is evidence of the interlocking nature of the gender-division of labor and the international division of labor.

## ACKNOWLEDGMENTS

The author would like to acknowledge Toni Calasanti for her support and encouragement in this work and Manisha Singal for her close reading and comments.

## NOTES

1. A lower caste.
2. Domestic cooks.
3. Business caste.
4. Names of people, places, and businesses have been changed to preserve their anonymity.
5. Gujarat is one of 25 states in India.
6. *Pati-dar* means landowner. In Gujarat, they comprise about 18% of the population. In this chapter, I have used Patel and Patidar interchangeably—although not all Patidars use the name Patel, all Patels are Patidars.
7. As part of my graduate training, I conducted a pilot study of Indian immigrants in small family enterprises in 1990: a convenience store, a motel, and a restaurant. I did semi-structured, taped-recorded interviews with six people and did 60 hours of observation at the business sites.
8. Patel is an agricultural caste group from Gujarat. It is also the most common last name they use.
9. I categorized as second generation those who graduated from high school in the United States, regardless of whether they were born in the United States or abroad. I believe that high school is a definitive socializing influence that determines how immigrants see themselves in the United States. Therefore, this category includes people who were born and raised in the United States and also those who were born abroad but came to the Unites States as children with their parents—basically, all those who "came of age" in the United States.

## REFERENCES

Assar, N. 1990. "The Mutual Impact of Family Structure and Business Practices Among Indian Immigrants: An Exploratory Field Study." Unpublished manuscript of a study conducted under the guidance of Prof. E.N. Glenn at SUNY—Binghamton.

Behar, R. 1993. *Translated Woman*. Boston: Beacon Press.

Dill, B.T. 1980. "The Means to Put my Children Through: Childrearing Goals and Strategies Among Black Female Domestic Servants." Pp. 107-123 in *The Black Woman*, edited by L.F. Rodgers-Rose. Beverly Hills, CA: Sage.

Game, A., and R. Pringle. 1983. *Gender at Work*. Sydney, Australia: George Allen and Unwin.

Glenn, E.N. 1992. "From Servitude to Service Work: Historical Continuities in the Racial Division of Paid Reproductive Labor." *SIGNS* 8(1): 1-43.

Heer, D. 1996. *Immigration in America's Future: Social Science Findings and the Policy Debate*. Boulder, CO: Westview.

Hochshild, A.R. 1989. *The Second Shift*. New York: Avon.

Hondagneu-Sotelo, P. 1994. *Gendered Transitions*. Berkeley: University of California Press.

Isbister, J. 1996. *The Immigration Debate: Remaking America*. West Hartford, CT: Kumarian Press.

Jain, U. 1989. *The Gujaratis of San Francisco*. New York: AMS Press.

Jasso, G., and M. Rosenzweig. 1990. *The New Chosen People*. New York: Russell Sage Foundation.

Kibria, N. 1993. *The Family Tightrope: The Changing Lives of Vietnamese Americans*. Princeton, NJ: Princeton University Press.

Lamphere, L. 1987. *From Working Daughters to Working Mothers*. Ithaca, NY: Cornell University Press.

Lister, H. 1996. "Accidental No More." *Lodging* [management magazine of the American Hotel and Motel Association] (December): 54.

Mehra, A. 1993. "Keeping Up with the Patels." *Little India* (December 31).

Mies, M. 1986. *Patriarchy and Accumulation on a World Scale*. London: Zed Books.

Miles, M.B., and A.M.Huberman. 1994. *Qualitative Data Analysis*. Thousand Oaks, CA: Sage.

Mohanty, C.T. 1997. "Women Workers and Capitalist Scripts: Ideologies of Domination, Common Interests, and the Politics of Solidarity." Pp. 3-29 in *Feminist Genealogies, Colonial Legacies, Democratic Futures*, edited by M.J. Alexander and C.T. Mohanty. New York: Routledge.

Pessar, P. 1986. "The Role of Gender in Dominican Settlement in the United States." Pp. 273-294 in *Women and Change in Latin America*, edited by J. Nash and H. Safa. South Hadley, MA: Bergin and Garvey.

Phizacklea, A. 1988. "Entrepreneurship, Ethnicity, and Gender." In *Enterprising Women: Ethnicity, Economy, and Gender Relations*, edited by S. Westwood and P. Bhachu. London: Routledge.

Shah, S. 1996. "The Co-optation of Asian-American Feminism." Pp. 242-246 in *Experiencing Race, Class, and Gender in the United States*, 2nd edition, edited by V. Cyrus. Mountain View, CA: Mayfield Publishing.

Sheth, M. 1995. "Asian Indian Americans." Pp. 169-197 in *Asian Americans: Contemporary Trends and Issues*, edited by P.G. Min. Thousand Oaks, CA: Sage.

Weiner, M. 1990. "The Indian Presence in America; What Difference Will It Make?" In *Conflicting Images*, edited by S.R. Glazer and N. Glazer. Glenn Dale, MD: The Riverdale Company.

Westwood, S. 1988. "Workers and Wives: Continuities and Discontinuities in the Lives of Gujarati Women." In *Enterprising Women*, edited by S. Westwood and P. Bhachu. New York: Routledge.

Wolcott, H. 1995. *The Art of Fieldwork*. Walnut Creek, CA: Altamira Press.

Woodyard, C. 1995. "Roadside Revival by the Patels." *Los Angles Times* [Home Edition] (July 14): A1.

# THE LOBSTERMEN OF NARROW RIVER:
## BECOMING ONE OF THE GUYS

Richard V. Travisano

## ABSTRACT

This paper grew out of the author's experience working at lobstering in Rhode Island waters. From participation, observation, conversation, and introspection, it addresses how part-time lobster fishermen comport themselves and deal with one another in a rather competitive, somewhat dangerous, and physically demanding activity. The rather exacting morality of lobstering, involving a balance between competition and cooperation as well as the careful respect for the territory and gear of others, is delineated. Further, the conditions under which one can justifiably side-step such rules are also detailed.

## INTRODUCTION

I live in Rhode Island on a tidal estuary called Narrow River. My dock and my boat are 45 feet from my door. The Atlantic Ocean is three miles away. In 1989, I began going out to pull lobster traps with a friend. In 1990, I did so again. In 1991, I got my own boat, my own commercial license, and my own lobster traps. Fishing for lobsters, you understand, became something that I do. I am one of the people who

Current Research on Occupations and Professions, Volume 10, pages 87-113.
Copyright © 1998 by JAI Press Inc.
All rights of reproduction in any form reserved.
ISBN: 0-7623-0034-5

lobsters out of Narrow River. In 1992 and 1993, there were twelve of us. Some fished as few as 20 traps, some as many as 150. We all sold lobsters, but none of us did so exclusively for a living. Joseph "Bill" Poirier, for example, was a social worker who spent his last 15 working years as Social Services Director of Woonsocket Rhode Island Hospital/Landmark Medical Center. He's now retired, and he lobsters. Arthur Johannis works fixing generators for the electric company and lobsters. Peter Brodeur builds and sells lobster traps and lobsters. I teach at the university, and write, and lobster.

For me, fishing for lobsters and becoming part of the network of fishermen came first. Thinking and writing about it came later. From the perspective of the discipline, I am an observant participant. I am in the network; I take part in the activity. From my perspective as a person, the network and the activity are part of me. For whatever reasons, it suits me. And I fit. (In fact, when I asked a fisherman I had known for a while if I could tape our conversation, he looked at me quizzically and asked what I did at the university anyway. So I told him and he said he had thought I was a janitor.)

My immersion in lobstering raises questions of theory and method. Many years ago I researched conversions to fundamental Christianity. In that arena, I was a spy—I knew who I was in the situation; my subjects did not (Travisano 1971). But, in this case, I was not visiting a social world, I was joining it. And the other fishermen knew who I was—I was the new guy. I knew precious little about what I was doing. What I did know was that like the people I grew up with (and the person I grew up as), these were mostly working-class guys. So if I went easy but worked hard, it should work out. My strategy (if we can call it that) was to dive in and become part of what was going on and see where I ended up later.[1]

Not surprisingly, given my bent, I find myself writing what Denzin calls a minimalist sociological text. Of such texts, he writes:

> They are written from the ground level, from the point of view of what situated individuals see, feel, hear, and say . . . . The world captured interprets itself, for its prose harbors the interpretive theories and everyday sociological imagination people in the situation summon to make sense of what is happening to them (Denzin 1990, p. 5).

This essay fits what Denzin is pointing to fairly well. It is written from the ground and tells of what I experienced, learned, and figured out in the situations of lobstering, as well as some of what I became. This essay, in part, could be what Van Maanen (1988, pp. 106-107) calls auto-ethnography, as I am attempting to convey at least some of the story in terms of lived experience. But only some of the story. The truth is once lobstering became what I was doing 30 hours a week in my third season at it, I was fully part of the network, and distinctions between the everyday sociological imagination that was imbedded in the network and my "professional" sociological imagination simply got lost. I had crossed over.

So. I have a story to tell. It might be ethnography, or even auto-ethnography. It might be a minimalist sociological text; it might be more than minimal. It might be science; it might be sociology; it might be journalism. What do I think about these issues? In this case, I don't. What I know is that I went out on the water to find out what it was about. And I found out—about the sea, lobsters, tides, currents, motors, boats, traps, bait, and more. And I found out about the guys, how they comport themselves, and how they relate to one another. I gathered a lot of information. It is solid, and it is interesting if you like the topic, so I am writing about it.

In an earlier essay on this subject (Travisano 1994), I focussed on the tasks involved in lobstering and on the relationship between the lobsterman and the sea. This essay is about lobstermen dealing with one another, about a network of lobstermen and its rules and its ways. But to present this in context, I must first summarize what lobstering entails—that is, what these lobstermen do, how they do it, and why they do it.

## WHAT LOBSTERING ENTAILS

Obviously, to go lobstering you need a boat. As Narrow River is very shallow in spots at low tide, and since a bridge one must pass under has a low clearance at high tide, the boat cannot draw too much water or extend too high above the water line. On the other hand, the mouth of the river is treacherous and often has 3- to 5-foot breaking waves (or bigger; but then, one does not go out). Further, you are going out on the ocean—usually not farther than two miles, but on the ocean nonetheless. Your boat, then, should not be too small either. Our boats vary from 12 to 26 feet, with the majority between 17 and 22. (The fellow with the 12-footer is very careful about when he goes out; the fellow with the 26-footer goes out further than the rest of us.) The smaller boats are powered by 35 to 45 horsepower outboards, and the range goes up to 225 horsepower on the 26-footer.[2]

You also need lobster traps (also called pots). These are essentially boxes, usually some three-feet long, two-feet wide, and one-foot deep. About half used today are made of the traditional oak lath, while the others are made of one-and-a-half to two-and-a-half inch wire mesh, coated with green, black, or red plastic. These traps typically have two openings with knitted nylon net funnels (or heads) fitted in them. The idea is that a lobster can find the way in but usually cannot find the way out. There is a door on top which opens to put in bait and take out lobsters. Bait (usually filleted fish carcasses or salted fish) can be hung in the traps on strings, or in bait bags (nylon mesh bags similar to onion bags but of a heavier gauge). Traps are weighted with bricks (which are secured) to keep them on the bottom. Lines with buoys (painted in individually chosen colors) are attached to traps. Traps are usually fished singly by the lobstermen under consideration herein, but lobstermen who fish larger numbers of traps fish them in trawls, with five or more traps attached to one line, with perhaps 100 feet of line between the traps. These are pulled up with electric haulers. Traps of the size and heft we use for the places where we fish cost $30 to $35. With ropes and buoys, it costs some

$40 to $45 to put a trap on the bottom of the sea.[3] (I mostly fish in anywhere from 15 to 35 feet of water, though I do fish a few traps at about 50 feet, so my lines are 50 feet and 75 feet accordingly, as one needs some slack because of tides and swells.) Of course, traps can be purchased used, or sometimes salvaged from the beaches after storms. Getting used traps is common among part-time lobstermen. Such traps usually need work, but all lobstermen are constantly fixing gear anyway, so no one makes much of such effort. In any case, anyone lobstering has some money tied up. The range for getting into the activity for the twelve in question (counting boats and traps) ranged from $1,600 up to $10,000.[4]

So, with a boat, and some traps, and a license, you can go lobstering. This means you get gasoline and bait, pile traps in your boat, and go out on the sea and drop the traps in. Then, every two days, when the lobstering is good and the weather and sea allow, you get more gasoline and more bait and go out to service the traps. When the lobstering is slower, we usually do three- or four-day sets. One man can service 80 to 100 traps a day in a good sea (that is, if he likes working hard and fast). Theoretically, then, one man can run up to 200 traps pulling them by hand, if he goes out every day. Of course, the weather and sea will not allow him to go out every day, so sometimes he is going to have to do more than 100 to catch up. This means that, realistically, about 160 traps is the maximum without trawls and a hauler.

Lobstering, then, means some measure of monetary investment and a great deal of work. Lobsters, of course, are a choice food and are costly to buy. But there are many middlemen between the lobsterman and the lobster dinner, so lobstering (despite the prevalent belief in Rhode Island popular imagination) is not easy money. Part-time lobstermen seldom make any significant profit (by which I mean, we do not make enough to pay for our boats). And full-time lobstermen seldom get rich. So numerous people who try their hand at lobstering give it up. Many are called by the lure of the sea, but not all are permanently hooked. While the original investments of the twelve from Narrow River varied from $1,600 to $10,000, and while getting deeper into the activity has raised the investments of two fellows to around $20,000, no one is set up to earn more in a year than they have put into it, and most are set up to earn less.

My own case is a typical example. I ran 50 traps for two summers. I cleared $4,200 after the cost of the licenses, some used traps and clips and other items to fix traps with, rubber bands for banding lobsters claws, name tags for my traps, ice, bait, gasoline, boat paint, and minor engine repair. But that does not figure the cost of the boat ($4,500) or the cost of the motor job ($1,400) which I needed at the end of the second season. And while I was waiting for the motor to be fixed, a storm destroyed 30 of my traps. (I found seven which I could repair.) Those 30 traps seemed a lot to me, but think about the full-time lobsterman from Maine whom I met. He fishes over 900 traps and lost over 500 to a storm, a northeaster that blew for five or six days. That's $25,000 worth of gear. "That's really tough," I said to him. In classic droll Yankee manner, he answered, "Yep, set me off my stride for two/three days." I once asked a clam digger if he ever thought about lob-stering for a living. He said "No," and then added, "It takes a special kind of man to kick $30,000 off the back of a boat every spring."

Why, then, do people do it? Essentially, because they are drawn to it. They like running a boat. They like working lobster traps. They like being on, and transact-ing with, the ocean. They like the experiences which lobstering affords and at least helps to pay for (Travisano 1994). These lobstermen are practicing existentialists, engaging in what Weigert (1991) has dubbed *transverse interaction* with a *gener-alized environmental other*. In fact, this varies for the full-time and part-time lob-stermen in some ways. The full-time fellow is apt to be ranging over more miles of water much farther from shore. So he's going to see bigger seas and experience the massiveness of the sea more. Of course, he has a bigger boat, so it will take bigger seas to push him around or sink him (though it happens often enough). Fur-ther, he is using trawls and a hauler, so he is not going to be directly connected with the sea in the same fashion as the part-time guy without a hauler. A part-time guy can find a particular underwater rock or two and place his traps on one side or the other of them depending on how he wants to play the tide. Further, you can feel a current, the tide, a swell tugging on your trap as you haul it through the water by hand. You learn quickly to work with the swells to save energy when you need to. Let me give two detailed examples of being out there.

I am out hauling traps in a tough sea. The sky is a dead grey, and the water looks like steel. It is going to storm, and the sea is coming up. I kick my motor into neu-tral and grab a buoy in swells of three feet and more. I am hauling a trap on a 75-foot rope from 50 feet down. It is a heavy trap with extra bricks in it because there is a strong tidal current in this spot. So it is tough coming up. I am a little arm weary, because I have already pulled 40 traps in a tough sea, so I am holding the line firmly but still as my boat rises up, so that the swells are pulling the trap up. Then, as the boat slides into a trough, I quickly pull in the few feet of slack. So I am pitching up and down in a tough sea in an open boat 17 feet long and 5 feet wide. Furthermore, the tidal current and the wind are pushing me about 100 to 200 feet per minute when I am not under power. I am not holding onto the boat as I have to hold on to the line. Sometimes I brace my bent knees against the 20-inch high gunnel. I feel very little, even though the spot I am at is only a half mile from shore. I finally leave without hauling five of the ten traps I have in the area because I am worried about getting in the mouth, and because I am too tired—and if you are too tired, it is dangerous.

Now, for a more positive example, here is Arthur beautifully expressing the lure of the sea:

> You know, Dick, once you've got pots out there, you've got to take care of them. It
> gets you out there, you know? I mean *any* reason to go out there is a *good* reason.
> Like, say you wake up and its a real shitty morning; it's raining and the wind is blow-
> ing. So you don't really want to go. You want to stay in bed. But you haven't been out

for three days because of the weather, and you've got to go because you know you've got lobsters out there that are going to eat each other up. So you go out. You get up and get all your shit, and get into your rain gear, and you go. And it's shitty. But maybe, once you're out there, the rain slows down. And maybe, just maybe, a window opens in the clouds and the sun comes through. *And all of a sudden you're in the middle of a fucking rainbow!* And you say, "Holy shit, I could have missed this!"

Now, which situation do we prefer? The clearing skies and the rainbow, of course. But the tough sea experience is also part of what the ocean is about, so we value it as much. If we didn't, we would stop lobstering and just go out when we thought we might catch a rainbow. (But we do know that there is not much gain in chasing rainbows, don't we.)

As to the relative importance of profit and the experience of working the sea, consider Ben, a man of 71 years who fishes about 120 traps. His boat can only carry 20 traps at a time, so he got caught with a lot of traps still in the water when a hurricane hit. He lost a lot of gear. His profit for the season was $300. His comment? "I wouldn't keep going if it cost me money, but as long as I don't lose money it's worth going out there."

A second story about Ben's devotion to going out involves a fishhook. When sport fishermen hook a lobster trap line, they have to break their line, leaving either a fishing plug or a fishhook imbedded in the lobsterman's line. Such was the case when Ben pulled on his line to hurry his power hauler along and sunk a good-sized fish hook deep into his hand between his thumb and palm. Worse, while he managed to shut the hauler off, a current was carrying his boat away from the four traps of the trawl left in the water. He was in imminent danger of going over the side and under. Realizing he had no other option, he ripped his hand away from the line, tearing an inch and a half gash through his flesh with the fish hook. But he went back out.

## HOW MANY LOBSTERMEN? HOW MANY TRAPS?

As it happens, there were 1,245 people (1,235 men and 10 women) holding commercial lobstering licenses in Rhode Island in 1992, the year during which most of the information in this study was gathered. (This breaks down to 151 $200 "lobsters only" licenses," 940 $300 "all species" licenses, and 154 $160 "all species and gill net" licenses.) However, not all of these people lobstered. Some with multipurpose licenses were fishing for other species. Some were keeping their licenses current even though they no longer fished. What is working in both cases is that there is always talk of limiting the number of licenses allowed, so people want to protect their future options. So, an informed guess would be that some 1,000 people were fishing lobsters commercially in Rhode Island waters in 1992. Those who do use their commercial licenses of the $200 or $300 variety to lobster, can drop any number of traps they can manage into the ocean—be it 50 or 5,000.

The Rhode Island Lobsterman Association estimated that some 300 people were fishing lobsters full-time during the time of this study. It is hard to say what full-time means. Some full-timers fish only 400 or 500 traps and are willing to live a more modest life than the credit card moguls like to see. To make $30,000 annually, a modest full-time income in this place and time, a lobsterman would have to fish some 600 traps, keep his expenses down, and really know what he was doing (which means he has to be experienced and smart). In my network, only Peter Brodeur has had experience (and success) at such a level. If you are fishing 500 or 600 traps, you are most likely to have help—a deckhand. And if you are fishing 700 or more, you certainly will have help. People do fish 1,000 to 3,000 or even more traps. Quite obviously, the more traps, the more help. In some cases, the help is hired, while in others there are partners in the operation. If we estimate (conservatively) that 200 full-time lobstermen average 700 traps each, we have 140,000 traps, and if we estimate (conservatively again) that the other 100 full-timers average 1,000 traps each, we have 100,000 more traps. Even if we subtract 20,000 traps because we hypothesize that 20 pairs of partners in these 100 lobstermen carry double licenses in case they come to a parting of the ways, we still have 80,000 traps.

Then there are part-time lobstermen, whose number we may estimate at 700. If they average 65 traps each, it adds another 45,500 to our total. And finally, 502 people (471 men and 31 women) held noncommercial licenses ($50 to fish five traps). This amounts to another 2,500 traps (not allowing for the fact that some of these people illegally fish a few more than five traps). Adding all these, we can reasonably estimate some 268,000 lobster traps in Rhode Island waters. This sounds like a lot, and it is.

There are areas where there are lobster trap buoys pretty much all over. One has to motor carefully through such areas, as a line wrapped around a propeller numerous times means hanging over the transom or getting into the water to work it free. So there is great pressure on the supply of lobsters. It is estimated that we have been annually taking some 90% of the legal-sized lobsters within five miles of shore for the last number of years. Up to now, the animals have replenished themselves. A factor in this is that populations of several species of "ground feeding" fish (such as cod, haddock, and flounder) have been much reduced through overfishing. Thus, fewer little lobsters are being eaten, and so more are growing to legal size. Also, lobsters in Rhode Island offshore waters migrate to inshore waters each June. As the destruction of many major fisheries around the world has sadly demonstrated, our knowledge of what goes on in the seas is obviously limited. So no one knows if Rhode Island lobsters will continue in the same numbers, or if the catch will drop off slowly, or whether it will crash (Ballenger 1988, pp. 89-103).

Given this basic information about lobstering, I now turn to the focus of this paper—the networks of communication and joint action among these fishermen. To be addressed are the nature of the networks themselves, the issue of maintaining self-respect and the respect of others, the issue of helping one another, and the rules (formal and informal) governing behavior toward the resource, as well as toward one another's territory and gear.

## THE NETWORK: BEING ONE OF THE GUYS

With close to 2,000 people fishing for lobsters commercially, we obviously do not all know one another. There is a formal organization, the Rhode Island Lobsterman's Association, though in 1992 this organization had only 65 active dues-paying members, and perhaps another 30 or 40 interested hangers-on. (The romance of the independent man is alive and well among Rhode Island lobstermen.) So most contact between lobstermen goes on informally—on the water, on the beach, in wholesale and retail fish outlets, in marine and hardware stores, in breakfast joints, in bars (though less often than many Rhode Islanders assume), and in the garages, cellars, and backyard workshops where they repair and/or make their traps.

Networks tend to be local. Guys who live in the same area, or fish in the same area (as in the case with those of us out of Narrow River) or dock at the same marinas, tend to develop friendly reciprocal helping relationships. Proximity is the first basis for networks. This is mitigated, or enhanced, by other factors. First, there is the obvious issue of whether people like one another. Second, a guy from one area may fall into a network based in another area because he particularly likes the guys there and they, him. Some networks have key figures, individuals whom the networks center around. Not that they are leaders—there are no leaders or followers in this activity.[5] Rather, such key figures as usually guys who are in positions to be centers of communication and information, and may also have more experience and hence more knowledge. A guy might live at a location which is handy for others to drop by. A guy might spend a lot of time on his boat at the state dock. (I met one fellow who lives on his boat from spring to fall.) In the case of my own network, the key figure is Peter Brodeur. Peter is very knowledgeable, and when he is not on the ocean, he is very often working in his backyard workshop. His location is handy to us who fish out of Narrow River. It is a small neighborhood. Further, guys who know Peter from when he fished a big boat out of the state docks come around. Guys who want him to build traps for them, or who want to get materials from him to build or repair traps themselves, also show up. Add his prominence in the lobstermen's association to this, and it is clear that he is in contact with a lot of people and a number of networks. So you can meet a lot of different people and pick up a lot of information hanging around Brodeur's. I met the State of Rhode Island biologist who is responsible for overseeing the lobster fishery at Peter's. He runs a few traps himself, and buys his traps from Peter. (Peter introduced me as a lobsterman, not as a sociologist: "Trav here is running some traps out of the river.")

Anyone who has been in a network for some time is, obviously, accepted by others in the network and is, hence, "one of the guys." This is not to say that being in a network is what makes a person "one of the guys." Being recognized as "one of the guys" depends upon tending one's gear and relating to other fishermen according to the prevailing norms. As the activity itself and the norms surrounding it impose a great deal of independence, there are individuals who are "all right," who are "one of the guys," but who are not actively participating in any network. Any two or more guys in a network may become friends and see each other socially, but while that relates to the network, it is a different issue. Networks are essentially made up of "acquaintanceships." These relationships are informal and friendly, but not quite personal or close. They are very much like the open general friendship among a bunch of children in a neighborhood.

## SELF-RESPECT AND RESPECTING OTHERS

The critical basis for respect of self and others in lobstering is the tending of gear. The issue is not *how much* gear you have, but *how well* you take care of it. As already noted, a lobster trap with line and buoy costs some $40. True, used traps are purchased more cheaply, but used traps are not always available and it still costs $40 to replace one in a hurry. A guy fishing 65 traps, the average number for a part-timer, is dropping some $2,500 worth of gear into the ocean. So a critical rule is that you must tend your gear. This means keeping gear in decent repair and "servicing" it—that is, hauling it up and tossing it back down every two or three days. Only the careless do not keep gear repaired, because a damaged trap is a trap lobsters can escape from. But as we do not inspect each other's traps, we only know how much a guy keeps his traps in shape by how much he might mention such in informal conversation, by seeing him fixing traps if we visit him, and by judging what kind of guy he is. On the other hand, we do know when someone fishing in our area is not tending his traps regularly, because we seldom see him out there and because his gear is not moved around very much. Such slackness is noted because it intimates that the fellow in question is not a "right guy," and for two practical reasons besides. First, gear not serviced on a regular basis means trapped lobsters eating each other up. This in a situation where we are all trying to make a buck off a limited resource, and where a lobster getting consumed could be an egg-bearing female. Second, since the fellow in question seemingly is not "serious" about lobstering, we find the navigational hazards that his buoys constitute unacceptable.

But it is not only not servicing gear which breaks the rules, but also not valuing your gear. In an average storm, gear from up to 300 or 400 yards out may get tumbled through the waves and tossed on shore, slightly to severely damaged, or completely ruined (depending on whether the trap traveled over sand or rocks). During Hurricane Bob in August 1991, traps were washed in and smashed from as far out as a mile to my certain knowledge, and perhaps from up to twice that distance. As that hurricane made its way north from the Caribbean, we all kept our ears on the marine weather stations. The problem is, you have to guess. If the storm hits full force, as this one did, you want all your gear (or at least all up to two miles out) back on shore. But if your boat will carry 20 traps and you have 160 out, you need to make eight trips, and you cannot do that in one day. If you are fishing out of Narrow River, you will likely need (with the sea coming up) three days to do it. So you have to decide early. But if the storm veers off and does not hit, you lose four to six days of fishing all your traps, because it is going to take a lot of trips to get them back out. When Hurricane Bob was beginning to hit, I bumped into Peter at the boat-launching ramp where he and I pulled our boats, and helped pull out others' boats as well. Arthur had told Peter that he had seen me pulling my traps in. Peter said to me, "Hey, you hot shit, you got your gear in!" I only had 20 or so traps at the time, so a few days later, I queried him about this matter:

> T:   It seems like you guys with lots of traps would think a guy with 100 was more into it than someone with 20, who might just really be fooling around.
>
> P:   It ain't a matter of numbers. What matters is how a guy does it. You respect a guy who tends his gear. Some guys put gear out and hardly ever tend it, or leave it in over the winter. I mean you wouldn't do a job outside and leave your tools out to rust in the rain, would you?
>
> T:   Well, you said you expected me to haul my gear before the hurricane and you were pleased that I did. But what if I had said that I didn't haul it because I didn't give a shit about the 100 bucks it cost me; that it wasn't that much money to me so I didn't need to be bothered?
>
> P:   I would have found out something.
>
> T:   What?
>
> P:   I would have found out that while I had thought you were all right, you were actually an asshole.

And here's Peter and me, very late in the season, talking about a guy who didn't tend his gear:

> T:   I know that guy. I tried to get hold of him. I know his boat is out of the water but he still has at least three traps out there.
>
> P:   I don't understand what he did. He put all that gear out there and then a lot of his traps never got hauled for weeks. I still don't understand why.

Clearly, one has a responsibility to service one's traps regularly, and to haul them in before big storms and at the end of the season. This responsibility is to oneself (in terms of money and self-respect), to other fishermen (gear should not be in their way without sufficient reason), to the resource (lobsters deserve to be trapped and harvested, not trapped and left to cannibalism), and to the gear itself (one doesn't abuse one's tools).

In addition to tending gear properly, a "right guy" helps out other guys when he should. One fellow I know told another that he had spotted a trap that the other guy thought was lost. But when the trap owner, whose boat was out of the water, asked if the guy would pick the trap up for him, the answer was, "Hey, I'm not in business to spend time picking up your traps!" This story (of course) moved through the network. The response I got when I told the story to another lobsterman was, "Wow! The guy's a real fucking asshole!"

So you tend your gear properly, you help out when you should, and you pay attention to learn about what you are doing and about how to do it right. On this dimension of knowing what you're about, there is Arthur's comment when I mentioned that I wanted to talk to a guy we both know who runs only a few traps very casually:

A: What do you want to talk with him for? He doesn't know shit about lobsters.

T: But he's been doing it for a few years, hasn't he?

A: Yeah, three or four I guess. But he doesn't pay attention, and he doesn't give a shit. He's just screwing around. He's in the goddamn way out there.

Another thing one is individually responsible for is keeping abreast of the weather. One guy complained that no one had warned that the hurricane was coming. I asked a guy in my network about this. His answer:

Jesus, it's not my job or yours to tell him about the weather. We got our own stuff to take care of. I mean he's got to keep up on the weather himself. If he wants to lobster, he has to do the whole job.

Lobstering means you listen to the marine weather station after you say your prayers at night and before you brush your teeth in the morning. You become continuously totally conscious of the weather. You know what weather system you are in the middle of, and what is predicted to move in next. When you can and cannot go out is governed by that. Further, you have to learn to judge how the weather will affect the mouth of the river and the part of the sea you are fishing. Many times I have called Peter at 6 a.m., and more than once Peter has called me, to get our heads together on reading the situation. Where we fish, wind direction means a lot. A stiff offshore breeze means we can shoot right out the mouth of the river and haul traps in a cool breeze (though the breeze will blow our boats around when we are working a trap). But make that breeze onshore from the southeast, and we are looking at working in swells of three feet of more (which is unpleasant after 20 traps or so), and the mouth of the river might not be passable. So the weather is always in your head, and your head is always on the mouth of the river and the sea. On hot summer days, I have walked out of the building I work in at the University of Rhode Island and had colleagues say, "Great, this breeze blowing will cool things down." "Right," I say. But I think, "Shit, right out of the southeast. If it picks up much at all, I can forget hauling traps in the morning." Never before (not even in Minnesota winters) has the weather been such a factor in determining what I did and did not do. And while I have lived in Rhode Island since 1969, I had little awareness of the intricacies of the interactions between the weather and the sea before I went lobstering.

Still, all this is not to say that one is left totally on one's own to figure out how the whole job ought to be done. Peter took pains to tell me what kind of radio I should buy, and Arthur gave me a lot of information on how useful electronic depth finders are and then took time to come with me to help me pick one out. And Frank showed me how to get in and out of the mouth when I first went out with him. If you are working hard, are interested, and are listening, guys teach you things. But if you do not seem to be hard at it, interested, or listening, they simply do not say much.

So doing "the whole job" is the source of respect from others and self-respect as well. With 500 or 50 or only 5 traps in the water, you ought to service them every two to four days (depending on the catch and the weather). You have to keep up on the weather continuously. You have to keep your traps and boat in order, and tend to the many other details which keep your lobstering going. And you have to at least try to give as much help to others as you would ask of them. Lobstermen are, in an important way, very much like many independent professionals. They offer to their fellows, and ask from them, a certain level of help, and they assume a certain level of competence. Finally, you do not have to prove that you are "one of the guys," or a "right guy," to gain entrance to a network. When you start lobstering with any seriousness, entry is open. But if you transgress, prove yourself to be an "asshole," you get cut out.

## WORK: LENDING A HAND

Within a network on shore, or with anyone on the water, a right guy lends a hand. While there are times that one asks for help, such only occurs when one is really in need and up against it, or when one has a real friendship relationship within a network. Indeed, I did ask Peter if he would get me to the airport for a 6 a.m. flight. I was going to a professional meeting to present this paper, and he is the only person I know who rises at 5 a.m. daily. (Peter knits heads, the net funnels that lobsters crawl into traps through, for about an hour most mornings while he has coffee with his wife, Wendy. Making traps, like lobstering itself, can be a demanding business.)[6] Yet a ride to the airport, even to deliver a paper on the subject, still is not part of lobstering. But it is also true that Peter, with his truck, helped me get my boat in and out of the water at the beginning and end of a season, and also took the boat out and in with me in the middle of the season for motor repair, hauling it back and forth 10 miles. I never directly asked him for this favor. Instead, when my boat broke down, Pete heard the news from Arthur (who had towed me home from outside the mouth of the river), and when I saw him, Pete asked what I was going to do. I replied, "Haul it up to East Greenwich and get it fixed." Pete then said, "When do you want to do it?" obviously offering his truck since I had no truck of my own at the time. At another time, I had to move two very heavy floating dock sections I had built from my lawn down a five-foot stone wall to an eight-foot wide terrace, and then down another five-foot stone wall and into the water, where I had to attach them to the end of my existing dock. Friends (including two guys from my lobstering network) told me I would need three men to do it, but no one said "Give me a ring." So when the time came, I built a ramp and slid the things from the top wall over the bottom wall and into the water, and managed to do so with the excellent help of a 105-pound woman. There is, in short, an ethic of doing things oneself. This is based in part on an ideal of independence, but is also based in a reality of people being very busy and time being at a premium. Other people's time is seen as valuable, and one does not take it unless necessary. Further, when one is at a particular task, one wants to finish if one can, because with lobstering there is usually another job to do. Sometimes, as with so many conventions, the ethic gets in the way. Knowing Peter could use my help taking his boat out at the end of one season, I spent days nosing around to determine when he was going to do the job so I would not miss helping out.

Over the long run, we do help one another a lot. A number of times, when I have stopped by Pete's, there was a load of wire on his truck to be unloaded, or a roll of wire that needed two men to carry it into the workshop. Clearly, on days that I do not show up to lend a hand, someone else usually does. If no one does, Pete does the job himself—with a hand truck and with great effort, but he does it.

Arthur also drops by Peter's on and off and, surely, lends Pete a hand. Peter reciprocates as opportunities occur, as when Arthur took two crab traps over to Pete's workshop and they fixed them up together.[7] While one guy may help another out more than gets reciprocated during one season, it all seems to even out over the network and over the years, so no one feels put upon. Cases do, however, come up where an individual needs (or worse, asks for) too much help. Such is experienced and talked about (gently but openly) in various terminologies as "not right." Such individuals usually get gently but definitely "brushed off" within a network, and if they do not get the gentle message, they may be sent packing quite openly and directly. As it is, conditions and tradition make being on the sea fishing for lobsters an essentially solitary pursuit, and this emphasis holds sway on shore as well.

## TROUBLE: LENDING A HAND

As we all know, sooner or later everyone's automobile breaks down. It is the same with outboards, so sooner or later we all get in trouble on the ocean. The problem there, of course, is you cannot walk. It is an obvious principle that you always check on anyone out there who seems to be in trouble. We all carry tow lines to use on someone else, or to have someone else use on us. We watch out for one another and we watch out for other people. Most every year someone dies in our waters, usually through carelessness and/or lack of knowledge of how dangerous the sea can be. We are air-breathing, land-walking creatures. That is what we are built for. We may love the sea, but it is never "home." It is essentially "other," essentially alien. We go out often, and we go out in heavy weather, sometimes when it turns out we shouldn't have, so we are wary of the sea. And the more we go out, the more wary we become. Winds, swells, tides, curents, shoals, deeps, surf, shores— these combine in myriad ways so that it is almost always a little different than anything we have seen before. And now and then, it is amazingly and sometimes frighteningly different. Peter (who, along with Arthur, has years and years more experience than the rest of the twelve) says:

> You don't ever turn your back on it, you don't ever inside of you—you know, how sometimes inside of you you say, "If I speed up my pace I'll be across the street before that car gets here." You don't ever do that with the ocean. You don't ever do it. . . . You have to stay aware. On a beautiful day when you *know* there isn't a big swell within a thousand miles, it can lift you and throw you into the rocks.

So we keep an eye on other people when we can, knowing that things can happen. Boats have been swamped, or sometimes lifted six or more feet and deposited on huge rocks, right in the mouth of the river.

As to troubles that I and guys in my network have run into, Peter once slipped on a deck slick from bait, fell against his motor, and got tossed right out of his boat. He was treading water with boots on, with his 17-foot skiff doing circles around him at about half throttle. One of the guys, Bob Eddy, hauling traps nearby, noticing the boat going round and round with no one at the wheel, came over and pulled him out. Further, I broke down just outside the mouth of the river and Arthur came along and towed me through the mouth (a tricky maneuver) and then all the way home. A more minor example is the day I went out to service traps and saw Bill Poirier hanging over the back of his boat. He had drifted over a buoy while servicing a trap and had caught the rope up in his shaft when he threw his boat into gear to steam to his next trap. This meant his motor was stalled by the many turns of line wrapped amazingly tightly around his propeller shaft. So I backed up to the stern of his boat, and while he held the two boats two feet apart (it was a calm day), I unwrapped the line. Then there was the day I dropped my lobster gauge into the sea. I got Peter on the radio and he was only about a mile away. So I motored down to borrow one from him because, while the chances of being boarded by agents of the Rhode Island Department of Environmental Management were very slim, if they nailed me with an undersized lobster, they could have taken my boat. And finally, if I was happy that I had the radio which Peter had encouraged me to buy on the day I lost the gauge, I was even happier about it the night I radioed him at home from a half mile out on the ocean, with my motor not running and the sun about ready to set.

So, just as it is possible for an automobile to go out of control on an icy road despite the most careful driving, a boat can get tossed, or swamped, or whatever despite the greatest care. You might well need your boat saved. You might well need your life saved. That may be part of why a guy who is an "asshole" is so upsetting—he is not a person you would want your life to depend on. Again, we are reminiscent of neighborhood kids. But we are not up in a treehouse, we're on the Atlantic Ocean.

## EXCHANGING INFORMATION

Information means lobsters and lobsters mean money, so an important function within a network is the exchange of tips on how, where, and when to fish. Such exchanges prove that relationships are reciprocal and solidify the connections between fishermen and so solidify the network. Here is Peter talking about exchanging information with Arthur and about another guy who doesn't reciprocate:

> P: Arthur and I exchange information. If I find a good spot, I tell him I'm having good luck at Bonnet so he'll bring a few pots down there.
>
> T: Will he put them right with yours?
>
> P: Well, you know, within 75 or so yards—not right on top of me. And he'll pass information to me. See, that's the thing with that other guy. I told him that we can help each other out. So I tell him about a good spot, but it never comes back. He's always cryin' that he ain't catching any lobsters. And then he comes over and he needs this or that and I sell it to him same as I do with anyone else—for a little more than I pay, but cheaper than you can get it anywhere else. And he bitches that I didn't *give* it to him, or that I'm charging too much, and then he runs up a bill with me. It keeps goin' on like that. I don't like to say this, but a guy like him you like to see get out of it. He's a pain-in-the-ass.

And here is Arthur, responding to a question about him and Pete exchanging information about details of the grounds:

> T: You and Peter exchange information, Art. So, Pete tells you he's doing good at Bonnet. How close to him will you fish?
>
> A: It depends. If he's got his pots 150 yards apart, I might drop one in between. On the other hand, I was fishing at Bonnet before him, and I had some spots—like at the corner of this or that rock—that I told Pete about. So Pete will grab them in the springtime before I have gear in the water. One of my spots that he grabbed this spring, he moved a pot out of it. Mine was there the next day. Stayed there the rest of the summer. *He moved his pot out.* I mean, there's nothing reserved out there. There

was another spot I had inside a rock near that jetty. He gets there early because he starts early, so he puts two or three pots behind it. Maybe I used to put one on the corner. So I just put one around the corner. Maybe it's only 30 or 40 feet away from him. So it varies. I used to fish that area all along.

T:    Wait. Let me get this straight. Pete gets out early. So Pete's pot is in a spot you always fish. So you put yours there.

A:    Yeah, I'll put one there. If he moves his out, I'll put three or four more in. He's tryin' somewhere else. I like that stretch because it's easy to work. I mean, Peter and I do that a lot. I get behind this boulder in shallow water. I move my pot out; he moves one in. He's gotta try it, you know?

And here's Pete again, talking about Arthur pointing out a big underwater boulder to him (lobsters like to hang around rocks):

So one day at Bonnet there were so many tiny crabs in the water you'd have sworn a duck could walk on them, and fish were breaking all over the place feasting on them. I mean they were all over, and Arthur was off Bonnet Point and he was casting a lure. I saw him hook up and start fighting a fish, so I went over and pulled up near him to watch. There were bluefish leaping all over but Art had a striper on—only about 24 inches, but he got real excited because it was a striper.[8] After he released it, he pointed down in the water between our boats and said, "There's something down there, you know." But his boat is higher than mine and I couldn't see what he could, so I said, "I can't see anything." And Art smiled and said, "But someday you will, you will." And one day, when the light was right, I did. He's always telling me about things like that.

Peter and Arthur are obviously on the brink of friendship, but as of this writing they remain warm acquaintances. They are both very busy and their limited free time is overcrowded. Meanwhile, they have something going between them which they have with no one else—a conversation of traps.

As a novice, I was (as the reader would expect) the person who needed the most information, and who would not be able to reciprocate very much. What happened was that Peter and Arthur essentially mentored me. They did this with obviously needed information, like good spots to drop traps. They did it with many bits of information which are not obvious at all. Peter explained that when the sea is up it picks up your buoy with every big wave. The buoy, in its turn, then yanks your trap along a foot or two. This rubbing on the bottom of sand or rocks would wear out the bottom of the trap quickly if you did not have two strips of wood (called skids) attached lengthwise to the bottom of the trap. (Some people use one-and-a-half inch wide strips of automotive tires woven in and out of the wire mesh of the trap as skids.) But a trap takes a lot of abuse, so the center of the bottom sags and rubs after a while. So when Pete makes traps, he uses a third skid in the middle. He explained all of this to me because as I fix a lot of really beat-up traps, I might want to put a third skid on them. Further, he knows I listen—indeed, he told me he had noted that. And he also knows that I know enough about making things to appreciate what he does. (This is one way I am able to reciprocate his favors—I validate his intelligence and craftsmanship.) My second example here is that Arthur taught me how to arrange the tunnel of net that lobsters travel through to get into the second compartment of a trap so as to minimize their chances of escape.

So information that means more lobsters in the boat is reciprocally exchanged, as is information about operating more safely and about dealing with other guys—in short, about aspects of the activity of lobstering involving relating to the lobsters, to the sea, to your boat, and to other lobstermen.

## RESPECTING THE RESOURCE

There are, as noted earlier, counting commercial and noncommercial lobstermen, about 1,500 folks out there with some 268,000 traps. As it happens, the State of Rhode Island has only about 30 conservation officers to police them. Moreover, these 30 officers are responsible for checking not only the 1,500 lobster fishermen and their 268,000 traps but also all other saltwater fishing (that includes fish and shellfish, commercial and noncommercial), all freshwater fishing, all trapping, and all hunting. Clearly, there is room to get away with a lot out there. Most lobstermen, however, are very aware of the pressure on the lobster population and are careful about what they do. Legal size on a lobster is now 3¼ inches from its eye socket to the end of its carapace. While I have been told of guys who will knowingly take shorts, I have no firsthand knowledge of such, although I am sure it happens, especially when a fellow is planning to have a few lobsters for his own dinner. (I did hear a story about a guy getting caught with 30 or 40 shorts, but I think that is apocryphal.) Further, it seems that no one would take an egg-bearing female. Unlike taking a short, taking an "egger" is too obviously self-defeating in terms of the future catch.

On the other hand, lobstermen are often casual about regulations that do not effect the supply of lobsters, such as the regulation requiring each trap and buoy to bear the licensee's name and identification number. I have seen many unmarked traps and buoys. This does not hurt the lobsters, but if conservation officers haul traps to check on identification, which they sometimes do, they will confiscate unlabelled traps. Still, with only 30 officers and 268,000 traps, it does not happen often, causing lobstermen to become careless sometimes. Another recently enacted regulation made it mandatory for all traps to be rigged so that if lost or abandoned, lobsters in it could eventually escape. This is easily accomplished by attaching an escape hatch with clips made of steel which rusts rather than stainless steel which doesn't. Did all lostermen fix all their traps? Well, it's still 30 officers and 268,000 traps. One final example is the regulation which forbids anyone to be in possession of lobster "parts." This is meant, one supposes, to prevent unscrupulous lobstermen from tearing a claw off eggers and shorts and packing them home for a fine, if illegal, dinner. In practice, it is hard for me to imagine anyone doing that. On the other hand, lobsters often tear claws off their fellows when in traps, and now and again, a lobster loses a claw when one is removing it from a trap and banding it. The lobster loses a claw, and its value drops by up to a dollar a pound. The claw is fresh and there is ice in the cooler. There is no problem in getting it home to cook. To do so is to break the law. But with the account (Lyman and Scott 1971) of "waste not, want not" so evident, I am sure some get cooked.

## RESPECTING TERRITORY

Anyone just beginning to lobster, or anyone already doing it but moving or expanding to different grounds, is well advised to go easy. One wants to drop a few traps to begin with, and no closer than 75 yards or so to other buoys. Further, it is a good idea to pull up and say hello to guys established in the area. They will want a reading of you, and further, it is easier to mess with the traps of an anonymous someone than with those of someone you know. Once in a territory, you have to follow the informal rules of sharing grounds. You do not fish too close to anyone unless invited. (We will sometimes tell another guy that a particular spot is really hot and that he ought to drop a pot of two fairly near ours.) You also must tend your gear regularly. Being constantly on the lookout so you do not wrap a line around your propeller is wearing, so traps that are not tended regularly every few days are, as already noted, not appreciated. Further, guys think about the lobsters in those untended traps eating one another. So untended traps sometimes get taken (as discussed earlier) or get separated from their buoys. (The chance of getting caught with someone else's trap on your boat is very small if you do not dock in a public place, and the chance of getting caught cutting a line out on the ocean is, with any care at all, zero.)

This brings us to the issue of what we might call lobster wars. Further "down east" in Maine, for example, folklore has it that lobstermen are extremely protective of their established territories (which often have been in the family for generations) and will cut all lines of those they see as intruders. Perhaps because there have been so many new people settling near the Rhode Island coast over the past 25 years, and thus many new people trying their hand at lobstering, such occurrences have become less common. But losing a few traps when trying a new ground still happens. Indeed, there are areas in Narragansett Bay where you can almost count on it happening. Further, it does happen that two guys will get seriously "on the outs" and play fast and loose with each other's gear. Overall, though, tampering with other's gear is the exception, not the rule. Other people's gear is usually respected, though the rules governing this turn out to be the informal ones of the networks rather than the formal ones of the legal code.[9]

## RESPECTING OTHERS' GEAR

The law says that you simply do not handle anyone's gear but your own at any time under any circumstances. Thus, we are required to have one of our own distinctively painted buoys displayed at least two feet above the highest point of our boats whenever we are hauling traps, so that a conservation officer in another boat can notice at a distance if we are hauling a buoy of a different color. Technically, I could be arrested for attempting to return one of my friend's traps I found on the beach. What we all say openly is that you never touch anyone else's gear. But actually, the informal rules that we live by allow us to handle other guys' gear in a number of different situations.

First, if there have been some high seas, so that traps have been washed in close to shore, you might be in close to some rocks hauling your own trap and notice a friend's nearby. If no one is around to see you, you are likely to haul the friend's trap and drop it further out (leaving any lobsters in it), especially if the friend in question has a boat that is hard to maneuver close to shore.

Further, there is the issue of traps which are carried by storms into your area from farther out, or which are abandoned over the winter by guys whose boats are out of the water. This is something noncommercial five-trap guys are given to (or so I and the guys I know well say) especially if they got their traps cheap or picked them up off the beach for free,[10] and if they only summer in our area. When Labor Day comes, they close up the summer house and take the boat out of the water. If the sea is too heavy to go retrieve their five traps, or they have not the time, they often just leave them, creating a hazard to navigation and to lobsters, as noted earlier. When gear is left for the winter, it is abandoned as far as lobstermen are concerned. It is, according to the informal rules, fair game. Let me add that in the waters I fish, nine out of ten traps will not make it through a winter. The lines will weed up so heavily that the buoys will be dragged under. Further, storms will carry many in and smash them on rocks. The only stipulation attached to taking such gear is that one be fairly sure that it is abandoned. For example, talking with a guy one autumn, I was asked if I knew if another guy still had his boat in the water. When I said that the guy was definitely out of the water for the season, I got this exasperated rejoinder:

> He's still got gear out there! If he's gonna leave his pots out there, I'm gonna grab them. Cause the chance of finding them in the spring is nil. It's like he's abandoning his pots out there.

Abandoned gear is fair game because it is a hazard to lobsters and navigation and because, by definition, it cannot belong to a right guy. And the definition is not completely self-serving, because anyone who really values his gear and is willing to do the job right will retrieve his gear, even if his boat is down and he has to get help.

In addition to abandoned gear, there is always gear washing up on the beaches after storms. Sometimes a big storm will throw traps on the beach which have been lost for some time. After the hurricane, one turned up which belonged to a guy who had not fished for two years. Legally, one should not grab any pot with any identification on it, and most that wash up have identification. But informally, the rule is that if you are willing to cart the trap out (a wet wooden trap will weigh up to 80 pounds or more), it is yours to dispose of. If you are a right guy, however, you will do so according to certain rules. Before I detail these rules, I want to relate some of the information and experience which went into figuring them out.

First, a long exchange between myself and one of the guys provided a lot of information, including some about another incident with a third guy, which I detail in the conversation:

T:   On what basis do you keep another guy's trap?

G:   If he leaves it out there for the winter, fuck 'im. By springtime, there will be so much shit on the buoy he won't know if it's his or not, and that's if the buoy is up at all. And a good winter storm will smash it on the beach.

T:   So if you leave it out all year it's abandoned gear?

G:   Yeah. Now, I don't know what the law states, but . . .

T:   Well, we're not talking about the fucking law.

G:   That's the way I feel about it. That's how I got the scup pot. [Scup are a fine eating fish, which are called porgies in Connecticut and New York.] I had two pots right where the scup pot was. I got to go out of my way to miss it. I want to fish the area but there's this piece of abandoned shit there. So I pull it up—right?

T:   Right.

G:   So, it was a scup pot. Nobody is fishing it. I kept the rope. I kept the buoy. I kept the pot. If it was a piece of shit, I'd just keep the rope and let it go back on the bottom; it's not going to hurt anything. The rope can be a pain in the ass.

T:   What's your feeling about the hurricane. Say you go down the beach and there's pots washed up and there's one of mine?

G:   I'd bring it to you.

T:   Yeah, anyone you know you'd bring it to, or tell them it was there.

G:   Yeah, cause it's not abandoned, it's washed in.

T:   What if there's a couple of others—you don't know the guys, but there are names on them?

G:   If the pot was a piece of shit, pretty well banged up, I'd keep it. If it was a good pot, I'd try to get in touch with the guy. I'd talk to Peter. He'd likely know the guy. If there was no identification on it, well . . .

T:   So, if it was banged up, you'd keep it?

G: Yeah, if it was like—most guys don't fish junk. Most guys like to keep their gear in decent repair. I mean, if it's a piece of shit, I say, "Hey, it doesn't have a future." Like that little pot I showed you. I don't remember where I got that one. It had no marker on it, but it . . it's fishable. But it doesn't have much life in it. Basically, it's a piece of shit. But I'll fish it in shallow water, and if I lose it or it disappears or whatever, I won't sweat it because I didn't buy the pot for $35. Chances are the other guy wouldn't even bother fixing it. It would be something he'd throw in his backyard or something.

T: I would do the same thing. Are we talking ourselves into something?

G: Yeah, we are.

T: Sure, because if it was our pot that washed up beat up we'd take it home and fish it in shallow water. So the other guy might do the same thing because there are other guys like us. So what we're saying to ourselves is, "It's some guy into it big or something, who doesn't care. The guy with 600 pots isn't gonna fix this." Right?

G: Yeah. Or . . . one time I lost a bunch of pots, and I found them up in the rocks. There were a few other pots in the rocks. I took 'em and I said, "No man in his right mind. . . ." I mean, I'm not going to carry the pot home for the guy. I mean, if I'm going to have to bust my hump, he's gonna have to bust his. Chances are he's just gonna say, "Fuck it, to carry it out and fix it ain't worth it." That's if he walks his ass down onto the rocks to begin with.

T: Yeah. After the hurricane I was on the beach with one of the guys. So there's a pot that washed up, and he says, "Hey, here's a good one for you. All you need to do is straighten this one a little, and fix the funnel, and put on a door, and it will fish." And he keeps walking. So I come up to the pot and there's a nametag on it. What's going on? Obviously, it's a guy he knows is an asshole, cause he's telling me to pop the tag and cop the pot. But he, at the same time, ain't watching me do it. About one trap, he says, "Wow, amazing, this guy stopped fishing five years ago." Then he sees another name on it and he says, "Well, this second guy didn't even take the old tag off, so it's yours now."

This exchange is remarkable because the fisherman says he would take another's trap under certain conditions, and I agree that I would too. Then I ask him if we are just talking ourselves into justifying what we would do, and he admits that we are. So what we have is two people talking themselves into a view of things which allows them to get away from the letter of the law, when that letter is far from the realities of the situation. Within limits then, the rules get stretched or bent.[11]

Next, a note on how I inadvertently revealed a way I could come into traps which belonged to someone I considered a right guy. I hauled two traps belonging to a guy who had pulled his boat out of the water. As I knew him, I took the traps home and called him. He thanked me and said he would come get them. Six months later the traps were still in my yard, so I asked Arthur how long I should wait. He answered:

Hey, a month, or six weeks at the most. You've had those things six months, and my advice is do him a favor and keep the traps in good order by fishing them. Wooden traps get loose if you leave 'em out too long.

Finally, and again from my notes on myself, there is the issue of "Brodeur A-frames." Brodeur designed these traps (shaped like an A-frame house with the top cut off) and built them for a while. He wasn't making them anymore when I started lobstering and I wanted one badly. After the hurricane, I found one on Narragansett Beach. It had no name on it. So I carried it a quarter mile to my truck, which was not easy as Brodeur's traps always run heavy because he uses bigger bricks to hold them down and extra braces to make them strong. I got it home and called Brodeur. He told me the trap had to belong to Bill Poirier because only he and Bill were using A-frames off Narragansett Beach. I did not know Bill, or of Bill then, so I asked Pete what kind of guy he was. Pete said he was a real good guy. Now, I was really stuck. I was kind of hoping Pete would say the guy was a real "asshole" who hardly tended his traps and who would never carry a trap a quarter mile off the beach, so I could easily talk myself into keeping the trap. I really wanted the trap. But it was supposed to go back to its owner. It took me two days to talk myself out of the trap and call Bill Poirier. (This was one of the moments when I recognized that I was "crossing over.") Keeping the trap was the wrong thing to do, because the way things were defined I would be stealing the trap. And I knew Brodeur would be waiting for me or Bill or someone to tell him the trap got back. He would not have to ask. That kind of thing gets around. And if I had just fished the trap without telling anyone I had found it? Well, if (as happened a year later), my motor broke down, and Brodeur helped me out by taking me along to haul his traps and then to pick up a bunch of mine, he would have asked me where I got the A-frame. And even if no one ever found out, I still would have felt like a thief.

Through conversations, observations, and introspections (those just recounted and similar others), five informal rules[12] governing the grabbing of gear came into focus. As already noted, gear that one feels comfortable in calling abandoned is fair game. As to gear washed up on the beaches or rocks, things are a bit more involved. As I said above, if you carry it out, it is up to you what you do with it. But self-respect keeps you within the rules, which revolve around the fact that at base you might well want to keep the trap, but you have to justify that. These rules as I list them here are as they came into focus in my mind. They no doubt would come out different in order and number from the minds of others in my network. Let me add, also, that I made a special effort to formulate these rules for this essay, and that within the network they were spoken to directly only when I asked questions about when one takes someone else's trap. The rules as I formulate them are:

1.  If the trap belongs to a friend, or a guy you know to be a good guy, you bring it to him, or give him a call and say you have it, or if you have not carried it, tell him where it is.

2.  If the trap belongs to someone you know is an "asshole," you can grab it.
3.  If the trap belongs to a guy you do not know, you can just leave it. If you take it, you have to make some effort to find out who he is. If you find out he is a good guy, you have to return it. But if he is said to be an "asshole" by guys you trust, you can keep it if you like. You do not have to kill yourself trying to find out—asking a guy or two who know a lot of fishermen is enough. (Some plastic tags for traps have the owner's name and phone number on them, making it easy to call if you want to return the trap.)[13]
4.  If the trap belongs to someone you do not know, and it is really beat up, you can talk youself into feeling that no one but you would bother fixing it, and you can grab it.
5.  If the trap has no identification on it, and no other distinction to the untrained eye,[14] you can fix it and fish it if you like.

The way all this works out then, is that friends return gear to friends, and right guys return gear to right guys (or calls get made to tell friends or right guys where their gear is washed up).[15] This in only mitigated by the fact that damaged gear, especially when found in barely accessible places necessitating long hard carries, is likely to be copped with the "no-one-but-me-would-fix-this" account (Scott and Lyman 1971). But as many guys grab unknown guys' damaged gear now and then, things obviously pretty much even out over time. But no one returns gear to an "asshole" or even tells him if they saw his gear on the beach. So the price of being ostracized is that you supply a few other guys with gear. And even if an "asshole" is doing his own gear grabbing, he is but one against many.

## SUMMARY AND CONCLUSION

Lobstering on a part-time basis, then, turns out to be easy enough to plunge into if one has some time and a bit of money. To keep the money at a minimum, the would-be lobsterman has to hustle to find a cheap used boat and to find used traps he might also have to repair. He will also have to get a license, paint his buoys, and find a souce of bait. If he does not get connected with one network or another getting ready to go lobstering, he will once he starts in most instances.

There are some 1,500 people fishing some 268,000 lobster pots in Rhode Island waters. The novice, wherever he fishes, is going to find himself fishing in the company of at least a few of these fishermen and a lot of their gear. The information needed to break into fishing grounds smoothly, as well as the basic information about how to fish for lobsters, one needs to gather from a network. (One could, of course, garner this information through trial and error, but a network is decidedly faster.) Virtually all part-time lobstermen work alone most of the time. But in some situations, like hauling boats or breakdowns at sea, one needs help, and being wired into a network is invaluable. Helping each other with heavy work, breakdowns, and exchanges of information, guys in a network make things easier and better for all concerned.

Further, there are the rules governing territory and gear which can only be picked up through a network, as the State of Rhode Island obviously does not put out pamphlets on how to ignore its laws. In this area, part-time lobstermen make their own way, puzzling out what does and does not seem right in conversations with their peers, or alone on the water or on the beach, with their own sense and conscience as their guides.

Finally, there is the issue of self-respect and the respect of others. If you do the job of lobstering right, you gain the respect of others and you respect youself. If you are "active" in your network, you help other guys when they need help, and gain their thanks and their willingness to help you. So the networks do a lot to keep people fishing for lobsters on a part-time basis. They also operate to protect the resource these men share, and to sanction the unsocialized and the unsocializable.

In conclusion, there are some 700 part-time lobstermen in Rhode Island, and none of them make significant money at it. But they are going out on the sea, again and again and again. As Arthur Johannis said, "Any reason to go out there is a good reason." It is a very different world out there. Simply, it's not the land, it's the sea. And there are lobstermen working the sea. They aren't just gliding over it; they are in a more intimate connection, And then there's history. People have been catching and eating lobsters in Rhode Island for many centuries—indeed, since Native Americans first arrived on these shores. And people have pulled lobster pots here for well over a century.[16] Some of us take a small pleasure in feeling we are a part of that. And finally, part-timers (and most full-timers too, for that matter) who persist, do so because they like work, and lobstering is good, hard, satisfying work. Karl Marx was quite right. It is good to own your own tools, and it is good to not be alienated from the means of production.

## NOTES

1.  As said above, getting into lobstering came first and writing about it came later. This is not to say I had no idea of writing about it when I started, I did. But that idea was vague at best and I recognized it for what it was—a justification for buying a boat, traps, a license, and so forth.

2.  Essentially, then, while there is a size range, these are all small boats. Fishermen who go further out (up to 20 miles) use decidedly bigger boats.

3.  These figures are for medium-sized traps. We are fishing at the mouth of Narragansett Bay. (On a calm day it seems almost like the bay—on a day less than calm, there is no mistaking that you are on the ocean.) Smaller traps with lighter gauge wire are used inside Narragansett Bay, and larger traps with heavier wire are made for fishing farther offshore. The materials used, of course, make for a difference in price.

4.  Lobstering for a living takes a lot of investment. The cheapest suitable boat will cost about $15,000 used, and the cost of 700 traps new, with line and buoys, would be about $31,000. You might luck out and catch the boat and a bunch of traps from someone who is getting out of the business, in which case you could get away with, say, $25,000 Further, traps are subject to wear and tear or piracy. And lines sometimes break. And sometimes the sea just smashes them. So you have to figure on replac-

ing one-fourth to one-third of your traps each year. As to your boat and motor, you pay dearly for fall-ing for the sea. Consider, a new 100 hp outboard engine costs $9,000 to $10,000. One can buy a whole car for little more, and that includes air conditioning. Controls for your outboard come separately at another $200 to $400. It is the equivalent of buying a car and finding that the gear shift and gas pedal are extra. Also, outboards are not built for the hundreds of hours a lobsterman does in a season, and so usually last no more than three years without complete overhaul. If you have an inboard, which you could well end up with in a bigger boat, you usually (though not always) get better service.

I must note that the estimates herein (as well as estimates further on in this chapter, under the head-ing, "How Many Lobstermen? How Many Traps?") were made with the help of Peter Brodeur. Brodeur has made his living by lobstering, or by lobstering and building traps, for some 20 years. He has fished 120 traps hauling by hand; he has fished 600 and more traps using an electric hauler. He has worked a 17-foot boat that he docked in Narrow River; and he has worked a 38-foot boat that he docked in the major fishing port in Galilee, Rhode Island. He is a past-president of the Rhode Island Lobsterman's Association.

5.    Few people established in lobstering will consent to be leaders. And even when the few are willing to lead, insofar as being the president of the Association means leading, there are few followers. Recall, the Association has only 65 members. A key element in the tradition of lobstering is to do it on one's own. Most of us prefer to go out alone, even though it is a dangerous activity and having someone along cuts down the danger. (For example, it is good to have someone else in the boat if one gets thrown into the water.) In any case, anyone looking for too much help, or asking for too much information, gets neither. In fact, he gets less of both than he would otherwise. But I address this issue below in the sec-tion "Self-respect and Respect for Others."

6.    There are people around, both men and women, who knit heads for people like Peter who build traps. One would not think it after casual observation, but heads can fit traps poorly, reasonably well, or really well, and this makes a difference in how many lobsters get caught. Peter tried buying heads from a local woman who has knitted heads as a sideline for years. And he went through a lot of trouble explaining just what he wanted and why. But they ended up fitting only reasonably well. Peter is an exacting craftman, and his traps catch well. People will wait for Brodeur traps instead of getting traps more quickly elsewhere. That is why Peter went back to knitting his own heads.

7.    Arthur uses green crabs, which he catches in the river, as bait to catch a fine eating finfish called tautog. If you know what you are doing, know when to do it, and are strong enough to pump a fish up from the bottom in a hurry, you can catch 100 to 150 pounds of them in three hours. They bring about a dollar a pound—sometimes a bit more or less.

8.    The population of striped bass, a great eating and fighting fish which can weigh as much as 60 pounds, was at a severely low level in the late 1970s and early 1980s. A size limit was instituted, Rhode Island banned the selling of the fish, and an information campaign was begun. The fish recovered much quicker than any biologist predicted. That is why Arthur was excited, and why he returned the fish to the ocean.

9.    Other lobstermen are not the only ones who tamper with traps. Yachters have been known to haul up a trpa or two and help themselves to a lobster dinner, and divers are notorious among lobster-men. I and two others in my network fish off a small military installation. Every summer, a training program for divers is run there, and every summer we find traps near the said installation empty.

10.    Unlike other flotsam and jetsam which washes up on the beach, lobster traps remain the legal property of the person whose name is on them.

11.    These rules getting "stretched" or "bent" is one reason I am not using the term "subculture" in this essay. Subculture is usually understood to designate something of permanence. But these rules of lobstering, which we in the network on our own, or with others, interpret as far as we can to suit our own designs (but not so far as to threaten our self-respect—we would not be thieves) seem too open for the usual notion of subculture. And it should be no surprise that interpretation is so open, as the time any guy spends with all other guys in his network combined, is, at most, only about one-fifth of the time he spends working alone on the sea, and fixing traps, and fixing his boat. A different way to con-

ceptualize this business might be with Russell Chabot's (1992) notion of "projects" or some variation thereof. Investigating local Rhode Island reggae bands in the 1980s, Chabot found the bands to have such continuous turnover in musicians that the term "group" made no sense. But each band had a leader who patched the band together, and patched it together again, and again, and so forth. Further, each band had an aim and/or theme, and/or distinctive sound. So Chabot conceptualized the bands as ongoing projects—something with an aim and with continuity, but much less defined and with much more turnover than groups. In the present case, we might say that every guy in the network has his own project—to keep lobstering. And doing that means everyone has procured, and is maintaining and using, a boat and some traps. To learn what he needs to know to begin with, and to keep up with information, the fellow is most likely to associate with a network. Networks are loose and informal, and there are people with lines to two or even three networks. Learning the basics you need to know, such as where to fish, what bait to use, where to sell, and how to run the mouth of the river, you will get only by having some connection to a network. So networks could actually be conceptualized as part of the project, or as the context where a lot of project information gets picked up, and a lot of project work which is not singular gets done. One other point worth making here is that networks validate (Stone 1970) identity and experience. One feels more the lobsterman in the company of peers, and you can only talk in full detail about being out there lobstering to others who have lived the experience.

12.   I call these "rules" in an everyday sense. And they are rules, but rules which are in some ways loose enough that one might more informally call them "guidelines." I have, however, stuck with "rules" because my discomfort in wanting to keep the A-frame was the strong discomfort caused by toeing a line that must not be crossed. The term "guideline" seems a little weak given my emotional reaction.

13.   The State of Rhode Island Department of Environmental Management, of course, has a list of all people with lobster licenses in any given year. These lists, however, are not easy to obtain unless one is involved in some research out of one or another state agency.

14.   I say "to the untrained eye" because Peter Brodeur can identify traps made by any trapmaker in Rhode Island and a few from Massachusetts as well. I have hauled traps with no tags off the beach and bought traps with no tags at yard sales and such. Then I have taken them by Brodeur's so he can tell me who built them and show me the small details by which he knows.

15.   On four occasions, I called people whose names and phone numbers I learned from tags on their washed-up traps. They were audibly pleased that I was calling and told me that such calls were rare, or that mine was the first. It is obviously a good way to contact lobstermen for possible interviews.

16.   Up through the early nineteenth century, it was possible to wade out and get as many lobsters as you would like with a hand net. By the middle of the century, it was getting a little more difficult, and simple traps were being designed and used. Sometime around 1860, a lobster tap of superior design was developed. It was made entirely of wood, including the funnel-shaped part for the lobster to enter through. The traps used today are refined versions of that very same design.

# REFERENCES

Acheson, J.M. 1988. *The Lobster Gangs of Maine*. Hanover, NH: University Press of New England.

Chabot, R.R. 1992. Local Version(s): Rhode Island Reggae Bands as Projects. Unpublished Ph.D. Dissertation, Department of Sociology, State University of New York at Buffalo.

Ballenger, B. 1988. *The Lobster Almanac*. Chester, CT: Globe Pequot Press.

Denzin, N.K. 1990. "Presidential Address on The Sociological Imagination Revisited." *The Sociological Quarterly* 31: 1-22.

Scott, M.B., and S.M. Lyman. 1968. "Accounts." *American Sociological Review* 23(February): 46-62.

Travisano, R.V. 1981. "Alternations and Conversions as Qualitatively Different Transformations." Pp. 238-248 in *Social Psychology through Symbolic Interaction*, 2nd edition, edited by G.P. Stone and H.A. Farberman. New York: Wiley.

Travisano, R.V. 1994. "Lobstering Out of Narrow River: Minding One's Grasp." Pp. 161-190 in *Studies in Symbolic Interaction*, Vol. 16, edited by N.K. Denzin. Greenwich, CT: JAI Press.

Van Maanen, J. 1988. *Tales From the Field*. Chicago: University of Chicago Press.

Weigert, A.J. 1991. "Transverse Interaction: A Pragmatic Perspective on Environment and Other." *Symbolic Interaction* 14: 353-363.

# FROM CASUALS TO CAREERS:
## THE PROFESSIONALIZATION OF
## REAL ESTATE SALESWORK

Carol S. Wharton

## ABSTRACT

This chapter examines the field of residential real estate sales as a case study of the process of professionalization. The focus is on the nature of the work and the construction of a professional identity for the workers themselves. The data come from in-depth interviews with 30 women from 17 real estate firms. The interviews included detailed descriptions of what attracted the respondents to the job, how closely their experiences matched their expectations, their work structures, how they did their work, and what they found satisfying and frustrating about their work. The data provide evidence that realtors perceive themselves as professionals and want the general public to perceive them in the same way. They recognize, however, that in general the public is not in agreement with this perception; there is widespread public skepticism. Respondents believe that increasing the training required and enforcing ethics standards will help them overcome this skepticism. Perhaps, also, public perception would be enhanced if more people were aware of the real nature of the work and the commitment of its practitioners to professional standards.

Current Research on Occupations and Professions, Volume 10, pages 115-134.
Copyright © 1998 by JAI Press Inc.
All rights of reproduction in any form reserved.
ISBN: 0-7623-0034-5

# INTRODUCTION

*[According to] the marvelous propaganda which real estate men made for the "good guys" in real estate . . . they were truly experts on land values and brokers; in short, professionals, not businessmen. . . . Here was a wonderful case where the professional attitude was fostered.*

—Hughes (quoted in Ritzer 1994, p. 112).

Seventy years ago when Everett C. Hughes studied the Chicago Real Estate Board, he described real estate salespeople as "casuals," meaning that they were engaged in what was by its nature a temporary occupation, almost a hobby. Salespeople themselves, however, preferred to see their work as a profession. A profession may be defined as "a high-status, knowledge-based occupation that is characterized by (1)abstract, specialized knowledge, (2)autonomy, (3)authority over clients and subordinate occupational groups, and (4)a certain degree of altruism" (Hodson and Sullivan 1995, p. 288).

As in Hughes' time, contemporary real estate salespeople refer to their occupation as a profession. Today, as the field becomes increasingly complex and regulated, it may be closer to achieving this status. It is no longer possible to enter as a "casual" or amateur. The amount of abstract, specialized knowledge continues to expand. As independent contractors, realtors possess a great deal of autonomy. They have authority over subordinate groups such as clerical assistants and, in the sense of selecting or approving them, loan companies, house inspectors, repair or revision contractors, and surveyors. Realtors define their work as altruistic in helping people find houses that suit the customers' needs. They also evidence altruism in a wide array of charitable and community projects that they participate in as individual companies or as members of the local and national boards. Realtors, in short, are striving to improve the standing of their occupation through emulating the characteristics of a profession (Hodson and Sullivan 1995, p. 302).

This chapter examines the field of residential real estate sales as a case study of the process of professionalization. The focus is on the nature of the work and the construction of a professional identity for the workers themselves.

Selling residential real estate is a form of contingent work, performed by independent contractors. Contingent work, defined as jobs lacking long-term contracts or guaranteed minimum hours, represents an increasing number of workers and is predicted to grow to perhaps one-half of the U.S. workforce by the year 2000 (Polivka and Nardone 1989; Judd and Pope 1994; Reskin and Padavic 1994). Contingent workers include part-timers, temporaries, and independent contractors. Most of the recent attention paid to contingent work has focused on its disadvantages, which include a lack of health care and pension benefits, sick leave, vacation pay, workmen's compensation, unemployment insurance, and social security. Many workers are reluctant participants, whose jobs have been converted to contingent status as employers seek to cut costs and make their workforce more adaptable to fluctuations in demand. About two-thirds of contingent workers are women, and the majority of them would prefer a permanent, full-time job (Judd and Pope 1994). However, there is also a segment of the workforce which prefers or chooses contingent work. These are most likely to be independent contractors, who may earn more per hour than salaried workers but also lack employee benefits. Realtors belong to this latter group of independent contractors, although they categorize their work as permanent and full-time.

This chapter draws on a study of women in residential real estate saleswork. Again, Hughes provided the first sociological look at women in this occupation:

> It is as the sub-division salesman [sic] that the woman finds her place in the real estate business. The characteristic thing about the woman who looks for a job is that she hasn't one already. She is the casual par excellence. Like the hobo, she is likely not to want her job long, or to devote herself to it too completely. Some crisis may throw her into the labor market, to grasp at any passing driftwood (Hughes 1928, p. 150).

Although the situation for women in real estate has changed dramatically since 1928, Hughes' own example belied his stereotype of the woman realtor as temporarily committed to her work. He cited a woman, whom he described as typical, who was a widow and worked full-time as a salesperson, building her clientele and establishing a place for herself in the Chicago real estate market. Clearly, she was not a "casual par excellence," nor was she any less devoted to her job than her male counterpart.

The gender composition of real estate saleswork has shifted dramatically in recent years, causing this occupation to change from a predominately male field to one in which the majority of salespersons at the residential sales level are women. Men continue to constitute the bulk of managerial (broker-manager) positions and to dominate nonresidential specialities, but residential real estate is becoming defined increasingly as "women's work" in a pattern of gender resegregation that appears similar to that which occurred in the nineteenth century in the clerical and teaching professions (Strober 1984; Fine 1990; Reskin and Roos 1990). Rapid growth of a field, because of changes in population or technology, creates a labor demand greater than can be met by the existing pool of workers. Wages and other benefits do not increase with the increasing labor demands, while at the same time other fields are competing for the same labor pool and offering better incentives. Concurrently, the labor pool is expanding as other categories of workers enter the labor force—due to immigration or ideological changes in the definition of workers suitable for an occupation.

There are four major structural reasons that residential real estate sales became more accessible to women beginning in the 1970s. The first of these structural changes also contributed to the professionalization of the field: legislation enabled firms to shift salespersons from the status of employee to that of independent contractor, thus altering the pay structure from salaries to commissions and therefore reducing the risks of employing inexperienced salespersons who might be less productive. Second, economic fluctuations made commission-based earnings less attractive to primary wage earners and resulted in large declines in the number of men in residential real estate sales, at the same time that large numbers of married women with children were entering the labor force. Concurrently, national franchise firms such as Century 21 emerged and recruited large numbers of new salespeople, focusing their recruitment efforts on women. Finally, men in real estate shifted to nonresidential specialties and left the field of residential sales more open to women than at any time in history (Thomas and Reskin 1990.)

Given the opportunity, women choose to enter the real estate field because it is attractive to them. There are several explanations for this attraction. The literature suggests that women perceive residential real estate sales as an occupation that will allow them easy entry with low educational requirements, flexible working hours, autonomy, high income, and opportunities for self-employment (House 1977; Thomas and Reskin 1990.) One of the reasons most frequently cited by women for choosing real estate sales as an occupation is that it provides great flexibility, allowing them to combine their work and family obligations in a relatively comfortable way compared to most other jobs.

The data come from in-depth interviews with 30 women from 17 real estate firms. I used a snowball sampling method, starting with five women in five different firms, whose names were given to me by acquaintances, and asking each person whom I interviewed to give me the names of two or three other women.

The interviewees were women living and working in one metropolitan area. I selected respondents with differing numbers of years of experience and from different types of real estate agencies, based on size, location, and type of ownership—independent local or national franchise/affiliate. The 30 women in my sample represented 17 real estate firms, ranging in size from five to over 400 salespeople, with a median size of 20. The firms included 13 that were locally owned and four that were franchises or affiliates of nationally owned companies.

The interviews were open-ended and conducted privately with each woman, usually at her office. Tape-recorded and later transcribed verbatim, the interviews lasted from one to two hours and included detailed descriptions of what attracted the respondents to the job, how closely their experiences matched their expectations, their work structures, how they did their work, and what they found satisfying and frustrating about their work.

## CHOOSING TO SELL REAL ESTATE

I was interested in what reasons the women gave for choosing a career in real estate and whether these reasons were related to a perception of the work as professional. Subjective factors such as intrinsic job rewards, peer cohesion, social support, and role conflict and ambiguity correlate with objective working conditions, including salary level, job title, workload, and supervisory responsibilities, in workers' ratings of jobs (Phelan et al. 1993). Autonomy and control in the work setting are important determinants of job attractiveness and satisfaction for workers in a wide variety of occupations, from domestic service to high-level administration (e.g., Collins 1988; Romero 1992). Autonomy includes being able to determine the content of one's work, setting one's own standards, and controlling the pace and routine of the work (Statham et al. 1988, p. 34). Emotional labor can give workers satisfaction in doing "good work" (Hood 1988) by helping others.

Recent findings indicate that, in general, women and men workers rank occupations on the basis of the same criteria (Feldberg and Glenn 1982; Jencks et al. 1988). Although family-related issues are often cited in explaining women's work aspirations and attitudes, job-related issues such as working conditions, income, prestige, interesting work, job security, and policies of recruitment and promotion are equally important influences on women's responses to work opportunities. Women move into occupations that have been dominated by men when the opportunity arises, simply because in general those occupations are preferable to most traditionally female occupations. In particular, the earnings potential for these jobs is higher than for female workers as a group, and the desire for high wages is a primary selection factor for women as well as men (Padavic 1992; Reskin and Roos 1990). In addition, women with families may choose specific occupations which they perceive as compatible with their domestic obligations. This type of compatibility is based on features such as flexible hours, innovative scheduling, on-site childcare, and job-sharing. As another factor, Hochschild (1983) contends that emotion work is more important to women than to men, that women make a resource out of their capacity to manage feelings and relate to others.

Studies of direct selling organizations (DSOs), such as Mary Kay Cosmetics and Tupperware, are instructive in understanding a preference for real estate saleswork, since both occupations are contingent forms of interactive service work that attract women "of modest education and few credentials in families of middling income" (Biggart 1989, p. 50). Attractive features of DSOs include a sense of control over work, a belief in unlimited income potential, the ability to fit work around family obligations, and workers' limited alternatives—other jobs available to them are lower paying, dull, and routine, with little opportunity for advancement. In addition, the jobs are structured to give salespeople a sense that their work is meaningful, that they are providing an important service to their customers (Biggart 1989; Connelly and Rhoton 1988; Leidner 1991).

The features of residential real estate saleswork which respondents in this study cited repeatedly as being the most attractive can be grouped together as providing autonomy, flexibility, and feelings of self-worth. Real estate was more flexible and promised higher earnings than any other jobs they could imagine for themselves. Thus, the explanation of respondents' job satisfaction lies partially in the limited alternatives available. Prior to entering the real estate field, they had held a range of jobs in relatively low-paying, high stress fields, including clerical, teaching, social work, and other sales. Compared to these experiences, real estate was higher paying and offered greater autonomy and satisfaction.

This sense of relative improvement came up repeatedly in the interviews, and for women coming from these occupations, greater income potential was the most significant factor in their attraction to real estate. For others, who had been making fairly good wages previously, the ability to set their own hours and otherwise gain greater autonomy were more important. The flexibility of real estate saleswork allows its practitioners to interweave family-related tasks with their paid work more easily than do other types of work that occupy a fixed set of hours of every day (Wharton 1994). Recent findings indicate that workers in a wide variety of occupations desire greater flexibility in their work schedules (McHenry and Small 1997).

## QUALIFYING FOR THE JOB

Salespersons must be licensed by their state, according to the particular regulations of that state. In most cases, salespersons must work under the supervision of a broker or be brokers themselves. All of my respondents were Realtors, which is a registered trademark for real estate sales agents who have joined the National Association of Realtors and subscribe to its code of ethics.

In order to be come a Realtor, each of the women had gone through a training period in preparation for taking the state licensing examination. The training ranged from several weeks of classes once a week to two weeks of classes every day. A few took a course at a local community college or university; others took courses offered independently by for-profit operations. Regardless, after passing the class, everyone had to take the state licensing examination. Many also took what they called motivational classes and classes on specific issues, such as writing sales contracts and qualifying customers financially.

In addition, most of the subjects had gone on to acquire further training and technical designations, such as Certified Site Agent (CSA) or Certified Residential Specialist (CRS). A few had completed the requirements for a broker's license, which includes several of the above types of designation, a specific amount of experience as a sales agent, further classes, and an additional licensing examination. Recently, the state had passed a new requirement that all salespeople had to renew their licenses every two years by taking a refresher course and an exam.

In short, the training requirements for this occupation are becoming increasingly demanding, and many of the respondents believed that this was a step in the right direction. They felt that in the past it was too easy to enter the field, and that contributed to the occupation gaining a negative reputation. One woman explained why she supported the adoption of more stringent entry requirements:

It helps us realtors. The image of realtors is, well, maybe a step up from the used care salesman, and we work so hard to keep our image high, and to let people know that we subscribe to the national code of ethics. And as we increase the educational requirements, it's just going to keep increasing the image.

## JOINING A COMPANY

Brokers are also licensed by their states, and may either operate their own business or work under another broker. Brokers are responsible legally for the actions of the agents who work under them. As independent contractors, agents are responsible for most of their expenses, including advertising, withholding and social security taxes, health insurance and retirement plans. The broker may provide them with office space, telephones and other office equipment, and limited clerical support. In this type of arrangement, salespersons share office duty, which means being present during certain hours to take phone calls and walk-in customers. They often work cooperatively in showing and promoting each other's listings, although only the listing agent and whoever makes the sale will share the agent's commission, while another percentage of the commission goes to the broker. More rarely, usually in cases of site agents working exclusively in a new housing development, agents work as employees of a broker, meaning that they are paid a salary and benefits.

As independent contractors, real estate agents are not *hired* by brokers or companies. Instead, the process of affiliation consists of becoming acquainted with different firms and choosing one that best suits the individual agent's preferences. Respondents were vague on how this process works. Some joined a company because its broker/owner approached them and asked them to join. Others said they contacted a specific company because of its reputation. In some cases, they had been with another company previously and had become dissatisfied with it. Companies do not advertise openings; indeed, there seems to be no set number of agents in most companies. Size might be determined by available space—offices, or even desks in a common room. One respondent explained:

> It's not like there's a certain number of slots available and you have to wait for one to open. There are certain guidelines set by the Board of Realtors that companies, brokers, have to meet. One phone can only be used by two agents. So you couldn't have, say, 10 phones and 50 agents. They don't have to provide an office for each agent. I happen to have an office with two other agents. We each have a desk, and two phones between the three of us. But I've run into situations, agencies, where there was one big room and a lot of phones and desks.

In 1996, the National Association of Realtors (NAR) conducted a survey of real estate agents to find out how agents select a firm. The NAR found that the following 10 attributes, in order of importance, were cited as determining factors in choosing a firm: (1) the company's image in the marketplace, (2) the office support staff, (3) computer access to Multiple Listing Services (MLS) provided by the company, (4) the company's market share, (5) the advertising provided by the company, (6) the company's referral network affiliation, (7) sales manager support, (8) franchise affiliation, (9) company-provided education and training opportunities, and (10) company-prepared marketing materials (Richmond Association of Realtors 1996).

There is a high turnover rate in this business, in terms of both people entering and leaving the field altogether and people switching companies. Respondents insisted that switching companies was fairly easy to do, and so commonplace that there were rarely any hard feelings. The major requirement in switching seemed to be that an agent could not take her listings or, although this would be harder to enforce, her customers to another company:

> You have to close out all your listings with a company. Any ongoing sales you can keep until they close, and then you get paid. Any listings would stay with the company. So it's prudent to dissolve all your business before you leave.

Most of the switching occurs as a result of an agent's preference for a new setting. As independent contractors, agents are not "fired" by companies, although some companies require a minimum level of sales in order to provide the office amenities. Agents who do not meet that minimum might be asked to leave or, more often, would feel more subtle pressure in the form of loss of collegiality or otherwise being shut out of communication channels.

## NEGOTIATING COMPENSATION

Realtors must navigate through a complex system of rules in negotiating their commissions. While 6% is the standard commission on residential sales, that amount is divided in a variety of ways by different companies and even within the same company. It is also sometimes the case that a seller might negotiate a lower commission when listing her or his house, or accept a lower offer from a buyer in exchange for a lower sales commission for the realtor. In any event, the commission must be divided between the listing agent (the agent who "lists" the house, meaning obtains the initial agreement with the seller and advertises the house) and the agent who actually sells the house (the agent who handles negotiations with a prospective buyer). This division is usually 50% of the commission to each agent's company.

The most variation occurs in the way agents and brokers split the commission. The simplest split would be 50/50, meaning that the broker and the agent each receive 50% of their agency's share of the commission. Thus, the listing agent's broker, the listing agent, the selling agent's broker, and the selling agent would each receive 1.5% of the total 6% commission. For most of my respondents, how-ever, the broker/agent split was never 50/50. More common was the graduated split: the more sales an agent makes, the higher the split. In the following example, the split is reset every year:

> The more income you make during the year, the higher your splits get, and you [could] go all the way up to 90/10 [agent/broker's shares]. Once you have made $30,000 of income, you're at 70%. Then every year you start back at 70, so you don't have to go back [further than 70/30]. Generally, we start at 60/40, but if an agent would rather take a 50/50 split and have the company pay a certain part of their expenses, they have that option.

Although none of my respondents worked for a company that did not require a split, several described an arrangement offered by one company which is part of a national franchise. In this arrangement, the agent pays the company a monthly fee and then is entitled to keep 100% of his or her commissions:

> [The agent] pays [the company] $1000 a month or something like that. And then you still have to pay for your advertising and everything—your stationary and all the office expenses. That's just to give you a roof over your head. But that $1000 a month is tax deductible, it's a business expense. So if you're a big producer and make 100% of your commission, it would be good.

In this type of arrangement, the "roof over your head" refers to having office space and, more importantly, to using the company's logo in advertising.

In exchange for their share of the commission, brokers provide legal protection for agents (although agents also must carry their own "errors and omissions" insur-ance), as well as office space, telephones, varying levels of clerical assistance, and other material benefits. However, as indicated by the quotation above, some com-panies may require agents to pay at least partially for even these services.

In any case, agents must pay their own social security taxes and arrange for their own health insurance and retirement, as does anyone who is self-employed. Other expenses include advertising, gasoline and all car expenses, monthly fees for the multiple listing service (MLS), and dues to realtors' associations. Gasoline and car upkeep expenses are substantial, since agents spend a great deal of time on the road. They usually drive as large a car as they can afford, in part because they must often drive clients around to see houses, and haul their FOR SALE signs and other equipment. Several mentioned also that they believed it was important to drive a "nice" car—an expensive model—to convey an image of themselves as successful businesspeople. Their clothing budgets were high for the same reason. The fol-lowing account describes many of these expenses:

> I pay for advertising, all my car expenses. It's funny, we have the magnets on our car doors that say [company name] and people will think that we're provided a company

car. What a joke! When we go on vacation, first thing my husband does is take them off. But the biggest expense is the advertising. It can cost $40 or $50 for an ad in the Sunday paper. The company provides us with our FOR SALE signs and our lead-in directional signs. We provide the riders with our name and phone number and any other riders that you want to put on the sign, which I do a lot. Especially in these old houses, I might have a rider that says "Renovated/Air Conditioned" and I'll put it on top of the sign. Just another little marketing tool. Any marketing expenses, any kind of brochure that we might want to do or any promotional gimmick—refrigerator magnets, anything like that. Postage, legal pads, everything.

This is a fairly typical arrangement, although some brokers pay for stationary, or copying, or other office-related expenses. The general rule seems to be, the more amenities provided by the broker, the higher the broker's share of the commission, and vice versa. The following is one respondent's assessment of the broker's contribution:

> He provides the administrative staff, the computer system, microfilm for research on city and county properties. The signs, the name, reputation. If we run a full page ad in the front of the Real Estate section of the Sunday paper, we're all required to run a minimum of two ads. If we don't run them, we get charged for them anyway. But still, all of that does not pay for the page. The company pays part of it.
>
> For me it [the broker's fee] is worth it, because I don't want the hassle of being responsible for everything. Even though I am an independent contractor, I still feel like it's his business. And this company has a good reputation in the community, that it has had for years and years, and I feel like because of that I get a lot of business. I think reputation of the company is important.

Most of the respondents stated that they felt that their brokers were justified in retaining a portion of the commission. I expected to find some resentment of the policy, since as independent contractors, it seemed reasonable that salespeople should not have to share their commissions with anyone else. However, such resentment was never expressed to me, and I did probe for such reactions. Similarly, I found several subjects who had earned or were in the process of earning their broker's licenses, but only two planned to start their own companies. The rest saw the broker's license as a further step in their professionalization but did not want to leave the protection of a larger company or take on the responsibilities of managing a company.

## ATTRACTING CUSTOMERS

Residential sales agents use a stock of tried and true measures as well as developing new strategies constantly for attracting new clients and customers.[1] One of the standard strategies is called "farming." This consists of selecting a particular geographic area—several square blocks, or a distinct neighborhood, or a housing development—and concentrating one's energies in that area. The realtor comes to know what's for sale in that area, the price range, who the builders are, where the schools, shopping and other area services are. She or he focuses on meeting the residents and maintaining a visible presence there. For example, a realtor who is farming a neighborhood might visit all of the homes and leave her business card as well as some sort of token—a cookbook, a refrigerator magnet, a telephone address book—with her name and company logo. She might publish a neighborhood newsletter or send seasonal homeowners' tips to the residents. One respondent described the farming process as follows:

> I picked an area that was a price range at that time from about $55,000 to $90,000. Single family. it appealed to me because they were little colonials, everybody's yard was neat. It was just a well-maintained area, a good location. So what you do is, you get out and knock on doors. The initial contact is through a letter. There is a whole procedure that you follow. I had a farming book that tells you exactly what to do. The first step is just to write a letter explaining who you are and that you are going to be specializing in that area, and that you will know everything that's sold, and what it sells for. It just makes a seller feel comfortable with somebody that's really keeping up with everything that's going on in that particular area.

> Q:  How would you know someone else wasn't farming it?

> A:  Oh, you don't! But there's nobody to say that you can't do it anyway. It's a big area and there are about 300 houses in it, so I figured even if there were somebody else farming it, they couldn't do it all. So I didn't let that deter me.

> And then after the letter, you take something, like a little keychain—I took these rubberized things that help you get tops off—and there's always the company logo. And then after you do that you write a thank you note, if you were able to make contact with a person, if they were at home. You know, you just chit chat with them. People are usually very nice and pleasant if you are, and just say "I want to take a few minutes of your time and did you receive my letter?" You know most of them just trashed it, and kind of laugh about it and say "I'm here to introduce myself." I've had people invite me in and show me their house that day. That doesn't mean they're going to list it.

> Q:  They're trying to figure out what it would sell for?

> A:  Sure! So maybe they're taking a little bit of advantage of you, but you're aware of that, they're aware of that. Then you follow up once every six weeks to two months with a newsletter that kind of updates what's going on, like in the mortgage world. And I'd always include a few tips, like in the spring about planting. Or if it was fall, ways to winterize your home. There are so many things like that that realtors are sent and you can type them up and run them off, and either mail them out or put them in a mailbox or inside the door. I took cookies at Valentine's Day. It's a lot of work.

> Q:  Did you find it worthwhile?

> A:  I certainly got listings. I started it in October, and I got my first listing in February. It is a wonderful technique for a new agent.

Other listing procurement techniques include contacting people who are trying to sell their own homes ("FIZBOs"—For Sale By Owner) and trying to entice them to list with an agent, advertising one's selling record, and networking at all kinds of social events—churches, schools, sports facilities, and so on. The point is to be highly visible so that when people decide to sell their home, they will think of a particular agent with whom to list it.[2]

Techniques for attracting buyers for a listing involve advertising in newspapers, real estate magazines, and other trade publications, the multiple listing service (MLS), and with house signs. Another popular technique is hosting open houses for prospective buyers and for agents. This latter includes "caravanning," in which several agents from a firm visit sale houses together, often on a particular day each week.

Salespeople also work to attract customers who are looking for a house to buy and need an agent to help them find houses. Some realtors consider themselves primarily buyers' agents and concentrate most of their energies on working with buyers, while others consider themselves primarily listing agents and focus on obtaining listings. The following is an explanation of the preference for working with buyers:

> I spend most of my time finding houses [for prospective buyers]. That is what I prefer to do. There are people that are wonderful with listings; it's not my cup of tea, for one particular reason. Most people get very antsy when they're trying to sell their house, and you can imagine, it's a lot of stress. You don't know when it will be, how much it will be; you don't know what you're going to do. And as they get stressed out, they tend to pick on people, and who is the most likely person? I don't care who they are, they're going to be annoyed at somebody and the realtor is the most likely candidate. That's one of the reasons it's not my favorite thing to do. I do it, but you've gotta really learn how to hang tough.

Another drawback to handling listings is that the agent has to get the house ready for market—in other words, encourage the owners to clean up the house, inside and outside, and make sure that the house is in good shape whenever prospective buyers come to see it. As the following remark indicates, this is not always easy:

> If you're listing a house, then you have to do all kinds of stuff about getting the owners to make it look good. And sometimes it's hard to talk to people. It's difficult when you go in and the house is a pigpen, and believe me, I have seen them that are pigpens and you have to say, very tactfully, "You want top dollar for this, it's got to look, da da da da." And they just can't see it.

Agents who prefer to work with listings have equally strongly held reasons for their preference, as illustrated in this respondent's answer:

There are lots of reasons [why I prefer to do] listings. You can be more in control than you can if you're working with buyers. With buyers you might go all over town and then they have buyer's remorse, walk into an open house and buy without you, whatever. It's hard to have a warm fuzzy relationship with a buyer. Also, if you are doing listings, the work is different. People who don't like paperwork are probably not going to like listings. I don't have to work as hard, because I can choose open houses, I can do my paperwork. Whereas with buyers you get them in the car from 9 in the morning until 6 at night, all day Saturday, all day Sunday, and it's more of a commitment of time and I'm not willing to do that.

Most salespeople engage in both listing and buying activities, although recently it has become legal in some states for agents to contract specifically with a client as a buyers' representative. This means that the agent represents the interests of the buyer and is under no obligation to a seller. Otherwise, an agent's legal obligation is always to represent the seller's interests, a fact which has caused confusion for buyers who believed that an agent was looking out for their interests first. As one respondent explains:

> The real estate agent represents the seller, because the seller pays our commission.
>
> Q:  So even if you are looking for houses for someone, and you've never met the seller, you still represent the seller?
>
> A:  Right, and it's very hard, because the people that you're with are always saying, "Well, what do you think?" or "What would you offer?" And we cannot do that. All I can say is, "This is what it's listed for, these are the comps [comparables] for the area. It's up to you. You have to decide what's the right thing to offer." There is a new disclosure form that we're supposed to get them to sign when we're first meeting somebody. Sometimes I have to wait a little longer to do that. I mean, it's a little awkward when you're first meeting somebody to say, "Oh, hi, how are you, what are your real estate needs? Oh, by the way, I want you to know I only represent the seller." It doesn't particularly endear you to a person.

In the newer concept of the buyer's agent, the agent and client sign a contract binding the agent to the buyer's interests. I asked one respondent how she would represent a buyer if the buyer were interested in a house which she had shown previously and knew from the seller's perspective:

> It is a conflict and you cannot do that. If you came to me and said, "I want to buy a house, and I want you to be my representative," the first thing I would do is show you all the listings of this company, and if you didn't like any of them, then we would enter into the buyer's broker contract. Now, what would happen if we'd been looking for several months and a new [her company] listing came on the market. I couldn't represent you. And you'd have several choices there. You could get another agent, or go to an attorney, or just know that I don't represent you.

In any case, when a salesperson is assisting a client in finding a house, she or he spends time with the client, to determine what the client's preferences are and, hopefully, to insure the buyer's loyalty throughout the process (this is a serious concern for many salespeople: over and over my respondents told me stories of clients in whom they had invested countless hours, who had then found a house on their own and signed a contract without consulting the agent, thus precluding the agent from any commission for her work.) The buyer's agent also "previews" houses for clients, driving around and touring many houses until the agent compiles a list of those which meet the client's expressed preferences. Agents also acquire clients through referrals, both from other clients and from out-of-town referral services hired by individuals who are relocating to the area or by employers who are transferring employees into the area. In these cases, agents often spend a whole day or more driving the out-of-towners around, perhaps buying them lunch or dinner, and showing them a large number of houses in the amount of time they have in town to locate housing.

Another "type" is the site agent, who works at one housing development, either exclusively or primarily—some site agents also work off site part-time. Site agents are working with builders selling new houses. They are not the same as listing agents. In some cases, the builder is the listing agent; in others, there is a development company coordinating sales for several builders. These agents are paid a salary plus a fee for each house they sell. Some of their expenses, such as hosting open houses for other realtors, are paid by the builders. The developers pay for some advertising, although the site agent may find it necessary to buy more advertising. Most site agents are new to real estate sales; they work at a site for a time while they get established and develop contacts. Few choose to stay in site sales indefinitely, since the income-earning opportunities are limited.

I was surprised by the volume of houses that most of my respondents had sold in the year prior to our interviews. The volume ranged from 12 to 70 houses in the previous calendar year, with a median volume of 26 houses. Sixteen of the women had sold between 20 and 30 houses in the previous year. At the two extremes, the two women who sold 12 houses each said that they were selling at the upper price range, that is, houses costing $350,000 or more, while the three who sold over 50 houses each were selling at least some that were in the $50,000 to $60,000 range. One of the highest volume respondents was the site agent for two new home sites, but the other two were selling all types of listings. None felt that the previous year had been unusually productive, although three of those in the lower volume (under 20 homes) said that it had been an unusually bad year and that they usually sold more. For example, one woman had closed 28 sales[3] the previous year and saw that as a low number:

> Last year was a bad year. I think it was because I did so much new construction, and new homes were down some. On the average I usually do 35 to 40 [closings], and really you should be able to do more than that. This motivational class I just went to,

the man thinks you ought to be able to do three or four listings a week. But I would have to have my own private secretary. I mean, there's so much paperwork involved with every transaction. Like right now I'm working with an out-of-towner that bought a house, and I've done everything. I mean, because they're not physically here to do it, they took the loan application, but I've met the fence man—this is a brand new house, they want a fence up, I've had to do that. I've met the wallpaper people; I've met a carpenter to have a dog door put in.

This account sounded unusual to me, and perhaps it was because the clients were moving from out of town. However, several other respondents described similar processes of high involvement with the buyers, including arranging for house inspections, being physically present for these and for utilities workers, and doing some of the leg work on the loan applications. In short, being the agent on a closing involves more than just negotiating a contract of purchase and accepting a commission at the closing.

## FACTORING CLASS AND RACE

The residential real estate market is, of course, stratified socioeconomically. Neighborhoods tend to contain houses of similar prices, within a wide or narrow range. Predictably, realtors also tend to concentrate on particular price ranges in their inventories of listings and the purchasing power of their clients. Although any agent may occasionally sell a house far above or below that range, most habituate their own niche in the market. Thus, some agents predominately sell houses in the $50,000 to $100,000 range; others sell in the $350,000 and higher range; others tend to focus somewhere in between. Where an individual locates her or himself depends on the company, its location, the agent's "sphere of influence" (people with whom the agent comes into regular contact and thus draws referrals), and personal inclination. A respondent described how an agent tends toward a particular price range:

> Your listings tend to plug you into a price range. For example, I'm getting ready to list a house for probably about $250,000, and when I hold it open I will be having people, serious buyers, coming in who are going to know that house is $250,000 and they're not gonna be paying serious attention to it unless they have that kind of money. And if that house is not for them, part of my job is to say, "I would love to keep my eye open for something for you." And we have a reputation of dealing in better houses, I would say. That's not to say we don't sell small houses when we get the opportunity.

Some agents are very deliberate about targeting a specific price range. For example, one respondent explained that she had chosen from the beginning of her career in real estate to "work smart," meaning she decided to focus on establishing a niche for herself in the higher price range so that she would not have to sell as many houses to make an acceptable income:

> I probably sell an average of 15 to 20 houses a year. Because I sell in the higher price range, I don't have to sell as many to get the big numbers, whereas a lot of agents are working in the 50[000]s, 60[000]s, 100[000]s. They might sell 60 houses. But I don't do that. . . . I sort of got my little niche in selling the high-priced houses so I don't have to sell as many, and I can make good money and have good numbers.

Agents who select the higher price ranges are comfortable working with wealthier clients, and probably have access to social situations which place them in proximity to prospects who can afford their inventory. One of the earliest tasks in working with new clients is to "qualify" them—to determine what they can afford. [Other early tasks include determining what the clients' preferences are in neighborhood, amenities, and housing features] Some agents work primarily with white clients, some with black clients, although most respondents insisted that the real estate market is blind to race, ethnicity, or other ways that neighborhoods were "redlined" previously. This is a sensitive issue; part of the Realtors' code of ethics states that they will not make any distinctions based on race or ethnicity. Thus, the white respondents would not acknowledge any race awareness at all. When asked, they all responded in a similarly flat tone that race was irrelevant to their business. However, four of the five African-American respondents perceived at least some racism in their experiences as agents. Each of these women felt that they were discriminated against by their brokers and by white agents. They said that they did not get a chance at listing higher priced houses or working with customers looking for that price range. When they did succeed, their efforts were less likely to be recognized by the company. One of the women cited this example:

> I would be top agent in the company, but they're trying to come in and put this white woman in as the top agent. They tried to say she sold more real estate than I did last year, but I know she didn't. But they want to put her up front because she's white and they want to be identified as white.

One of the four women who perceived racism in real estate worked in a black-owned company and said that almost all of her company's customers were black. The other three women worked in white-owned companies, but said that most of their (the individual agent's) customers were black. For the most part, they felt that this arrangement was expedient for themselves and their customers as much as for the company, because they could establish good rapport with their customers and build a good network of referrals. At the same time, they recognized that the fact that they worked predominately with black customers meant that the industry is much more race conscious, if not racially segregated, than it acknowledges. Race was not a focus of my research, and I do not have enough data for further analysis, but this is an important issue for future research.[4]

## ARRANGING THE WORKDAY

The daily structure of residential saleswork is typical of many types of independent contracting. The women have to decide how to arrange their workdays since they are subject only to themselves for reporting to work. Part of the task of accomplishing professional status is achieved in the ways they structure their work. Generally, they try to keep a schedule of regular working hours even though their work demands vary from day to day.

To begin to get an understanding of the women's actual workdays, I asked them how many hours they worked per week, on the average. Most of them had some difficulty answering that, and all explained that the number of hours varied from week to week, depending on the pace of the market and their own particular client load. All of them had to estimate the number of hours they worked, and the summaries are therefore susceptible to varying self-perceptions.[5]

With this in mind, however, the most frequent response was over 40 hours a week, with some attempts to specify this ranging up to 80 hours. For example, Sandra, who had been selling for two-and-a-half years, calculated her worktime as follows:

> I don't know. O.K., a normal work week would be 40, plus I try and only work one day on the weekend. Sometimes I only have one or two showings on a weekend, sometimes I work all day. Sometimes I work both days . . . so say eight hours a weekend, and then another eight hours of nights. Let's say 60 hours a week.

Tess's week was similar to several others' descriptions. She tried to work Monday through Thursday, take off on Friday, and work Saturday and Sunday. On weekdays, she ran errands in the mornings—dropping off contracts, picking up materials, meeting housing inspectors. She did paperwork and phone work at her office in the afternoons. Also during the day, she spent time previewing homes for prospective buyers. In the evenings, she did a lot of work at home by telephone, talking to buyers and sellers. When she was working with buyers, she picked them up after they finished work and, she said, "It's not unusual for me to get home at 11 or 11:30." On Saturdays she usually worked about half a day, and on Sundays she often hosted an open house from 12 to 5:00. This schedule adds up to 50 or more working hours a week.

While I have included here only the activities that were related to her real estate work, Tess and most of the other interviewees actually interspersed their descriptions of work-related and family-related activities. The description above represents only a "skeleton" of Tess's week—the real-estate part. Her household labor provided the flesh that filled out her weeks. For a description of the complexity of the arrangements for both spheres, see Wharton (1994).

All of the women had developed regular routines for themselves to get their real estate work done efficiently. Many, like Tess, felt that they needed to be in their offices, to be in contact with other agents and to be available to drop-in clients, or because they were too easily distracted by housework at home. Some women preferred to work more at home, but they also disciplined themselves to focus on real estate and exclude household responsibilities for certain intervals (although the advantage of being able to do both at once was attractive also, see Wharton 1994). In summary, all of the women had developed regular routines for themselves to get their real estate work done efficiently. They had also made arrangements to manage their household and family responsibilities.

## CONCLUSION

This chapter has studied residential real estate sales as an occupation in the process of professionalization. This process includes the following characteristics: (1) full-time work, (2) formal training and credentials, and (3) a national association that will work to achieve legal recognition of the occupation as a profession and (4) will create a code of ethics (Ritzer 1994, p. 126; Wilensky 1964). For most salespeople, the work has become a full-time job. Although few engaged in the occupation have attended a college-accredited training school, many believe that this will become necessary in the near future, and they already must complete training approved by the state board. The National Association of Realtors represents the interests of the occupation and strives to achieve the rights and privileges of a profession. Finally, the National Association has created a code of ethics to ensure that its members behave in a professional manner.

The data provide evidence that realtors perceive themselves as professionals. They believe that the field provides autonomy, flexibility, and a sense of doing worthwhile work, as well as a high income potential. They are able to select a company with which to affiliate, and to negotiate the terms of that affiliation, within the broad limits of the organization. They maintain a professional appearance in their dress and in the ways they arrange their work.

Realtors want the general public to perceive them in the same way, as responsible professionals. They recognize, however, that in general the public is not in agreement with this perception; there is widespread public skepticism, which serves as an obstacle to professionalization (Hodson and Sullivan 1995, p. 305). Respondents believe that increasing the training required and enforcing ethics standards will help them overcome this skepticism. Perhaps, also, public perception would be enhanced if more people were aware of the real nature of the work and the commitment of its practitioners to professional standards.

# NOTES

1. Respondents used these two terms—customer and client—in contradictory ways. Some used them interchangeably; others specified that customer meant someone looking for a house while client referred to someone selling a house; others used the terms in the opposite way. Still others added the term "prospects" to refer to people who were looking for a house (see House 1977, p. 42, for a similar discussion of these terms). Because of the confusion, I have opted to used the terms interchangeably. Thus, clients, customers, and prospects all refer to people engaged in buying or selling a home through a real estate agent.

2. House (1977, pp. 21-22) describes several other techniques: a telephone canvas of all home-owners in a specific area, checking the local Real Estate Board's notices of expired MLS listings and contacting those sellers about listing their homes, and connections developed over years of living in a community.

3. Respondents explained that the sale was not completed until it had closed, and thus they might have had more contracts pending at the end of the year, but their sales volume was based only on completed closings.

4. A recent study by a private fair-housing group found that housing in this metropolitan area is racially segregated, with only one in eight residents living in integrated neighborhoods. The group attributed this pattern to discriminatory practices, including redlining and steering. Responses from the local realtors' board denied discrimination while acknowledging that housing patterns are racial and attributing these patterns to historical arrangements.

5. House (1977, p. 20) found the same problem with his subjects trying to specify how many hours they worked a week. He noted also that the hours fluctuated seasonally—with fewer hours worked in the winter and summer, more in the spring and fall—and with market conditions. My interviews took place in the winter and early spring, and several respondents mentioned that their hours would increase after daylight savings time started, and slacken again when the weather got extremely hot in midsummer.

# REFERENCES

Biggart, N.W. 1989. *Charismatic Capitalism: Direct Selling Organizations in America*. Chicago: University of Chicago Press.

Collins, S.K. 1988. Women at the Top of Women's Fields: Social Work, Nursing, and Education. Pp. 187-201 in *The Worth of Women's Work: A Qualitative Synthesis*, edited by A. Statham, E.M. Miller, and H.O. Mauksch. New York: State University of New York Press.

Connelly, M., and P. Rhoton. 1988. "Women in Direct Sales: A Comparison of Mary Kay and Amway Sales Workers." Pp. 245-264 in *The Worth of Women's Work: A Qualitative Synthesis*, edited by A. Statham, E.M. Miller, and H.O. Mauksch. New York: State University of New York Press.

Feldberg, R.L., and E.N. Glenn. 1982. "Male and Female: Job versus Gender Models in the Sociology of Work." Pp. 65-80 in *Women and Work: Problems and Perspectives*, edited by R. Kahn-Hut, A.K. Daniels, and R. Colvard. New York: Oxford University Press.

Fine, L.M. 1990. *The Souls of the Skyscraper: Female Clerical Workers in Chicago, 1870-1930*. Philadelphia: Temple University Press.

Hochschild, A. 1983. *The Managed Heart: Commercialization of Human Feeling*. Berkeley: University of California Press.

Hochschild, A. 1989. *The Second Shift: Working Parents and the Revolution at Home*. New York: Viking Press.

Hodson, R., and T.A. Sullivan 1995. *The Social Organization of Work*. New York: Wadsworth Publishing.

Hood, J.C. 1988. "The Caretakers: Keeping the Area Up and the Family Together." Pp. 93-107 in *The Worth of Women's Work: A Qualitative Synthesis*, edited by A. Statham, E.M. Miller, and H.O. Mauksch. New York: State University of New York Press.

House, J.D. 1977. *Contemporary Entrepreneurs: The Sociology of Residential Real Estate Agents.* Westport, CT: Greenwood Press.

Hughes, E.C. 1928. A Study of a Secular Institution: The Chicago Real Estate Board. Ph.D. dissertation, University of Chicago.

Hughes, E.C. [1951] 1971. "Work and Self." Pp. 338-347 in *The Sociological Eye: Selected Papers on Work, Self & the Study of Society.* Chicago: Aldine-Atherton.

Jencks, C., L. Perman, and L. Rainwater 1988. "What Is a Good Job? A New Measure of Labor-Market Success." *American Journal of Sociology* 93: 1322-1357.

Judd, K., and S.M. Pope. 1994. "The New Job Squeeze: Women Pushed into Part-Time Work." *Ms. Magazine* IV(6): 86-90.

Leidner, R. 1991. "Serving Hamburgers and Selling Insurance: Gender, Work, and Identity in Interactive Service Jobs." *Gender & Society* 5: 154-177.

Leidner, R. 1993. *Fast Food, Fast Talk: Service Work and the Routinization of Everyday Life.* Berkeley: University of California Press.

McHenry, S., and L.L. Small 1997. "Who Decides When Money Equals Time?" *Ms. Magazine* VII(6): 34-37.

Padavic, I. 1992. "White Collar Work Values and Women's Interest in Blue-Collar Jobs." *Gender & Society* 6: 215-230.

Phelan, J., E.J. Bromet, J.E. Schwartz, M.A. Dew, and E.C. Curtis. 1993. "The Work Environments of Male and Female Professionals." *Sociology of Work and Occupations* 20: 68-89.

Polivka, A.E., and T. Nardone 1989. "On the Definition of 'Contingent Work'." *Monthly Labor Review* 112: 9-16.

Reskin, B.F., and I. Padavic. 1994. *Women and Men at Work.* Thousand Oaks, CA: Pine Forge Press.

Reskin, B.F., and P.A. Roos. 1990. *Job Queues, Gender Queues: Explaining Women's Inroads into Male Occupations.* Philadelphia, PA: Temple University Press.

Richmond Association of Realtors. 1996. "Study Shows What Top Producers Look for When Selecting a Firm." *The Richmond Association of Realtors* (December): 9.

Romero, M. 1992. *Maid in the U.S.A.* New York: Routledge, Chapman, and Hall.

Statham, A., E.M. Miller, and H.O. Mauksch. 1988. "The Integration Work: A Second-Order Analysis of Qualitative Research." Pp. 11-35 in *The Worth of Women's Work: A Qualitative Synthesis*, edited by A. Statham, E.M. Miller, and H.O. Mauksch. New York: State University of New York Press.

Strober, M.H. 1984. "Toward a General Theory of Occupational Sex Segregation: The Case of Public School Teaching." Pp. 144-156 in *Sex Segregation in the Workplace: Trends, Explanations, Remedies*, edited by B.F. Reskin. Washington, DC: National Academy Press.

Thomas, B.J., and B.F. Reskin. 1990. "A Woman's Place Is Selling Homes: Occupational Change and the Feminization of Real Estate Sales." Pp. 205-223 in *Job Queues, Gender Queues: Explaining Women's Inroads into Male Occupations*, edited by B.F. Reskin and P.A. Roos. Philadelphia: Temple University Press.

Wharton, C.S. 1994. "Finding Time for the 'Second Shift': The Impact of Flexible Work Schedules on Women's Double Days." *Gender & Society* 8: 189-205.

Wharton, C.S. 1996. "Making People Feel Good: Workers' Constructions of Meaning in Interactive Service Jobs." *Qualitative Sociology* 19: 217-234.

Wilensky, H.L. 1964. "The Professionalization of Everyone?" *American Journal of Sociology* 70: 137-158.

# MODELS OF CULTURE IN PHYSICIAN GROUP PRACTICES

Elizabeth Goodrick, Ann Barry Flood, and
Allen M. Fremont

## ABSTRACT

This paper integrates insights from the professional socialization and organizational culture literatures into a conceptual model specifying the multiple paths through which homogeneity of values, beliefs, and practices can occur in professional organizations. Our model focuses on two dimensions along which professional cultural development can vary: the source of shared understandings—professional training or organizational inculturation—as well as the primary process through which these understandings develop—via interaction or in response to the nature of the work itself. We illustrate our arguments by examining current patterns of beliefs and practices in a large U.S. multi-site, multi-specialty practice of physicians.

Current Research on Occupations and Professions, Volume 10, pages 135-156.
Copyright © 1998 by JAI Press Inc.
All rights of reproduction in any form reserved.
ISBN: 0-7623-0034-5

# INTRODUCTION

Within a local community or small area of the country, considerable evidence suggests that physicians tend to employ similar practices, particularly with respect to the likelihood of hospitalizing patients or operating on them (c.f., Wennberg, Freeman, and Culp 1987). Yet, these same studies also document wide variation in treatment styles from region to region for most care. There has been little research to explain how this apparent paradox—local homogeneity in practices with wide variety across localities—can arise. Are there "schools of thought" which permeate a region? Do such orientations develop because of basic training in skills and values received during medical school or residency? Is it the nature of the work associated with the types of problems generally treated by each type of specialty which brings about similarities? Is it the current organizational setting and its culture which encourages homogeneity in approaches to care? Or do such variations arise simply as a function of differing availability of facilities, technology, and support personnel—that is, differing capacity to provide services to patients?

In the past, efforts to explain why physicians might differ in their attitudes about treating patients have generally focused on the long professionalization process and its capability to inculcate values and practices. This approach tends to ignore the importance of organizational context and ongoing adaptation to norms and practices of colleagues in the same setting. In contrast, students of organizational culture, confronted with the same basic question about how values and practices arise among workers, tend to assume that similar practice orientations arise from shared understandings developed through organizationally based interactions (c.f., Trice and Beyer 1993; Van Maanen and Barley 1985). This approach to the question tends to deemphasize the importance of people having shared values *before* they join the organization (such as those values and practices which originate during training periods).

In this paper, we integrate the professional socialization and organizational culture literatures to develop a conceptual model specifying the multiple paths through which homogeneity of values, beliefs, and practices can occur in a large medical practice. In our model, we focus on two dimensions along which professional cultural development can vary: the source of shared understandings—ranging from originating exclusively during professional training to being dominated by the organizational setting—as well as the primary process through which these understandings may develop—via interaction or in response to the nature of the work itself. While the primary contribution of this paper is theoretical, we also use data we collected as part of a larger project to illustrate our arguments. Our strategy permits us to illustrate how scholars might empirically examine culture in the unique professional context instead of relying on ethnographic description alone. We are able to obtain a snapshot of these cultural processes by examining current patterns of beliefs and practices in one organization: a large U.S. multi-site, multi-specialty practice of physicians.

## ORGANIZATIONAL CULTURE IN PROFESSIONAL ORGANIZATIONS

Most recent work on organizational culture has not focused on professional organizations. One of the few exceptions, Tolbert's (1988) study of American law firms, built on one of the hallmarks of the professions-the long, intense period of training which many have argued socializes neophytes into particular values, beliefs, and practices as well as providing them the necessary knowledge and skills. Tolbert argued that homogeneity of values in professional organizations could arise via a process she called "transplantation." She observed that law firms achieved homogeneity by recruiting lawyers trained at the current partners' alma maters, thereby increasing the likelihood that the recruits' values resembled the firm's culture because they had undergone the same basic socialization as had the partners during their own training.

The special importance of socialization and initial training for explaining American physician culture dates from a much older literature (e.g., Carr-Saunders and Wilson 1933; Goode 1957; Parsons 1951) and a subsequent debate over the impact of professional socialization. Merton and his colleagues (Merton, Reader, and Kendall 1957), for example, examined how physician-students learned to cope with uncertainties in medicine and thereby internalized values and practices which would last throughout their careers. On the other hand, when Becker and his colleagues (1961) sought to explain the typical change from idealism to cynicism as student-physicians progressed towards graduation, they argued that such attitudes represented a situation-specific response which would "persist only when the immediate situation makes their use appropriate" (p. 433), implying that professional attitudes and practices could be further adjusted to new situations (Light 1980). Because Merton and his colleagues focused on why physicians develop similarly over time, socialization was depicted as exerting a stable influence on professionals. Becker and his colleagues and Tolbert, in contrast, focused on changes in values and attitudes and concluded (explicitly in the first case and implicitly in the second) that factors in the *specific training situation* were critical to explaining the development of professional values and practices.

More recent work has continued this debate about the extent to which medical training is a transformative and thus lasting experience. For example, anthropologists such as B. Good (1994) and M.D. Good (1995) focus on the acquisition of lifelong skills and perspectives. B. Good (1994) argues that the process of learning medicine consummates in physicians gaining knowledge and experiences that go far beyond simple understanding; it permits them to acquire a distinctive "lifeworld" through which to understand and interpret clinical events. M.D. Good (1995) argues that the process whereby medical students learn to recognize clinical competence is nonprogressive, unsettling, and erratic. It is unsettling because clinical training teaches them that the habits and competencies that made them successful in basic medical science are largely irrelevant to success in clinical encounters. It is erratic because their formal clinical training rotates them to work with different specialities every few weeks, repeatedly returning them to the bottom level of medical competency. Together, these factors underscore the nonprogressive nature of gaining competency and make physicians sensitive to the informal, emotionally laden settings by which they learn to recognize (in)competence.

In contrast to this emphasis on the transformative nature of medical training, scholars such as Freidson (1970, 1986, 1989), Miller (1970), and Mechanic and Schlesinger (1996) have embraced Becker's emphasis on situational adjustment. Freidson, for example, has consistently argued for the importance of the specific practice setting in understanding professional values and practice. His work has emphasized the implications of the social organization of practice and the situational nature of medical judgments. Similarly, Mechanic and Schlesinger (1996) argue that situational factors such as financial incentives, utilization review, and the referral process influence how treatment decisions are made.

Our arguments seek to meld several insights from the literature regarding the stability of professional values and the role of socialization and development of skills during training with those regarding the on-going influence exerted by an organization's culture and the practices of peers. Indeed, the juxtaposition of the contrasting sides of this debate leads to several interesting observations regarding the sources for and stability of professional culture. First, some professional values and practices may indeed be very stable for an individual while others may be more easily influenced by an organization's culture. For example, stability may reflect the fundamental quality of the values, the nature of the work, or the certainty associated with choosing a course of action, all of which transcend the organizational setting. On the other hand, when the suitability or effectiveness of practices are uncertain due to a lack of medical consensus, they may be more subject to influence by peer review, more easily swayed by financial or other incentives, or more willingly adapted to fit practices common in the professional "community" or based on the availability of facilities or support staff. Second, similarities may occur within an organization for several reasons—some intraorganizational, such as in the above examples, and others extra-organizational, including the process of recruiting concordant individuals at any of several critical stages—all of which are controlled by the profession: entry into a medical school, a particular specialty, a hospital staff, or a specific practice.

## ORGANIZATIONAL CULTURE AND MEDICAL PRACTICES

A basic premise of the organizational culture literature is that organizations, like nations, have different cultures. Organizational culture thus implies that members of an organization have values and beliefs which are distinct from the larger society. And, similar to nations, organizations can have subcultures which revolve around shared background and tasks.

There is no single agreed-upon definition of organizational culture. Instead, culture has been referred to as a family of concepts capturing many of the soft or informal aspects of organizations such as values, beliefs, symbols, rites, rituals, and assumptions (Pettigrew 1979). Culture is also commonly thought to be: (1) holistic, that is, it encompasses broadly defined aspects of interactions in the organization; and (2) both historically and socially determined, that is, it evolves over time from social interactions (Hofstede et al. 1990).

Disagreements persist primarily regarding what constitutes the theoretical core of organizational culture, rather than how it is manifested. For example, Schein and others argue that basic beliefs and assumptions about how the organization and its environment operate constitute the core of organizational culture (Ott 1989; Schein 1992). Values, symbols, and other artifacts then are the expression of those assumptions which in turn influence actual practices. Others, more cognitively oriented, argue that socially derived classification schemes or taxonomies form the basic core of culture so that shared values, practices, and beliefs are all derived from individuals assigning similar meaning to situations (Barley 1983; Van Maanen and Barley 1985).

The interpretation that taxonomy forms the core of organizational culture is particularly well suited to physician groups because it permits medical decision making as well as more conventional values and beliefs to be understood in cultural terms. Many authors have argued that the basis of much of medical decision making is driven by classifications; that is, a primary goal of clinical practice is to classify the underlying disease based on patient signs and symptoms in order to subsequently identify an appropriate therapy. However, the many uncertainties inherent in these processes, including the changing context of medical knowledge, give rise to variations in interpretations of both the underlying causes of the patient's problems and the most appropriate course of action (Gerrity et al. 1992). These sources for legitimate variations in clinical interpretations render physicians likely to adopt the traditions and orientations of physicians with whom they interact (Eddy 1984) and suggest that physicians are likely to be influenced by the organizational culture where they practice (Kralewski, Wingert, and Barbouche 1996). Indeed, a number of studies suggest that physicians, when provided with feedback that their practice pattern is not typical, will alter their practices to resemble the average practice (Eisenberg 1986; Wennberg 1988).

# HOW DOES ORGANIZATIONAL CULTURE DEVELOP?

The question of how culture arises in organizations has not been the focus of organizational scholars' attention; most research on organizational culture takes the existence of culture as a given. Most models of organizational culture, however, either assume or explicitly state that shared understandings arise and are maintained through organization-based interactions. When explicitly discussing cultural development, these theorists tend to focus on a particular form of interaction in organizations: that between people engaged in shared problem solving (c.f., Schein 1992).

Expanding on this logic, theorists such as Van Maanen and Barley (1985) argue that differential interaction, such as that which could arise from proximity (e.g., being in the same branch for a multi-site organization) or from departmentalization, leads naturally to processes whereby subcultures or "multicultures" develop in an organization: an individual will identify as a member of a subgroup within the organization and take action on the basis of collective understandings unique to that specific subgroup. For these theorists, subgroups are more likely to arise in certain structural contexts, such as in larger or more complex organizations. While rarely applied to professional organizations, this approach can easily be extended to encompass features of professional work, such as specialization, and structural considerations in a professional organization such as organizing work at multiple sites. This approach to explaining culture tends to deemphasize the importance of people having shared values before they join the organization.

A few organizational scholars, however, have explicitly argued that shared understandings occur in an organization because similar individuals were recruited or "transplanted" into an organization. One group of these scholars has focused on the role of selection processes in cultural development (e.g., O'Reilly, Chatman, and Caldwell 1991; Schneider 1987) and has generally argued that selection and intraorganizational socialization are complementary means to ensure the development of a shared culture. Another group, concentrating on particular professional and/or occupational groups in organizations (e.g., Tolbert 1988; Van Maanen and Barley 1984), have (like the professional literature) assumed that socialization during training influences members to adopt the standards of the occupational group, making organizationally-based interaction a less important source of shared understandings.

These ways that organizational theorists explain how culture arises in organizations differ on the source of "key" interaction (one group focusing on current organizational interactions and the other on prior training and socialization). They are alike, however, in their assumption that interaction with others is necessary for shared values, beliefs, and practices to develop. Neither approach explicitly considers that shared understandings may result from the organization of the work itself rather than from interaction. In this case, individuals independently invent shared values as a result of their similar interpretation of their work environment.

Recall that this latter explanation is a part of the professional literature. In this literature (e.g., Merton et al. 1957), some similarities among physicians were explained in terms of their responses to task uncertainties. Perhaps the reason these scholars tended to assume that shared beliefs and values arise from the demands of the work per se rather than interactions with colleagues and mentors is that solo practice settings dominated most medical care in the United States during this period. However, even in a group setting, work is organized around patients and most of a physician's time is spent with them instead of with colleagues. Consequently, the organization of medical work provides fewer opportunities (and need) for physicians to interact than in many other types of organizations, making interaction a less important mechanism for developing shared understandings.

The idea that the organization of the work itself may be responsible for shared values parallels institutional arguments about the source of similar structural arrangements across organizations (e.g., Tolbert and Zucker 1983; Orru, Biggart, and Hamilton 1991). In this perspective, the regulatory and normative environment organizations face determine their structural arrangements. Consequently, similarity in structure across organizations is not due to interaction among them but rather to similarities in the institutional environment. In much the same way, similarities in the work environment individuals face may lead to shared values, beliefs, and practices without direct interaction among them.

Shared understandings may result from similar interpretation of the task environment both during training and after individuals join an organizations. For example, compared to surgeons, physicians trained in family practice are more likely to take the patient's social situation into account when planning therapy—in part due to the types of therapies being considered and in part due to differences in the basic philosophical approaches of these specialties. In addition, differences in the context of work at different organizational sites also contribute to shared values and practices (Flynn 1992). An example here would be differences in work which arise in a branch location due to having fewer resources readily available, thereby changing branch physicians' thresholds of when it is necessary to order sophisticated diagnostic technology. Finally, the institutional context of both training and practice may produce broad similarities among physicians embedded in the same market and regulatory environment.

## SOURCES OF BELIEFS AND PRACTICES IN PROFESSIONAL ORGANIZATIONS

These two lines of reasoning to explain culture in professional organizations—the theories and concepts underlying the professional model and the organizational culture literature—together suggest two separate dimensions, illustrated in Figure 1. The first dimension, *Modes of Development*, portrays the process

|  | Institutional Source: | |
|  | Professional | Organizational |
| Modes of Development: | | |
| Interaction | (A)<br>TRANSPLANTED;<br>SUBCULTURES<br>DEPEND ON<br>INTERACTION | (C)<br>ARISE IN ORGANIZATION;<br>SUBCULTURES DEPEND ON<br>INTERACTION |
| Independent Invention | (B)<br>TRANSPLANTED;<br>SUBCULTURES<br>DEPEND ON<br>SIMILIAR WORK | (D)<br>ARISE IN ORGANIZATION;<br>SUBCULTURES DEPEND<br>ON SIMILIAR WORK |

***Figure 1.***   Origins of Culture in Professional Organizations

through which shared values arise. While this dimension is most appropriately considered as a continuum, we emphasize the polar extremes. At one extreme, similarities develop solely as a result of the interaction of professionals with other professionals. At the other extreme, similarities arise solely because individuals independently invent shared values due to the organization of the work.

The second dimension, *Institutional Source*, portrays the source of shared institutional values. At one pole on this dimension, the source is primarily extra-organizational—for example, the long training period physicians undergo or the professional meetings they attend. At the other pole are intraorganizational influences—for example, the attitudes of other professionals in the same clinic or branch and the incentives within the specific professional organization.

Next, we use our cross-sectional data to illustrate a quantitative approach to assessing culture and explore the usefulness of our taxonomy for explaining differences by addressing the following general question: which source is a better predictor of physicians' values, beliefs, and practices—the professional group into which they have been socialized or their organizational membership in different departments or branches? In the analyses we present, we assume that there is a time-related hierarchy—that is, the effect of socialization during training necessarily must precede the effect of current organizational setting.

While both sources of culture can operate independently, we expect the effect of organizational setting to be limited to several specific conditions. First, we expect to observe organizational effects when the current organizational setting does not duplicate the conditions of training. This would occur both when the issues involved are not relevant to residents as well as when the conditions contradict those of training, for example, a physician trains in a fee-for-service setting but then works in a wholly owned HMO. Second, the current organizational setting is more likely to have an effect when there is no medical consensus on the suitability or effectiveness of practices. Under such conditions, physicians are more likely to be influenced both by the norms of their local professional "community" and the availability of facilities and support staff.

# METHODS

## The Research Setting and Sources of Data

The research setting was Carle Clinic Association Clinic of Urbana, Illinois: a large, for-profit, multi-specialty group practice. At the time of our data collection, Carle Clinic was among the 15 largest practices in the United States and had a main site and seven branch clinics spread over a radius of 50 miles. It employed over 150 physicians and treated approximately 2,000 outpatients daily. It was affiliated with a 350 bed nonprofit hospital and offered its own Health Maintenance Organization (HMO) insurance plan to employers. About one third of its patients had HMO insurance and the remainder were reimbursed for their care on a fee-for-service basis.

The sources of data we used include interviews, archival data, and physician questionnaires. Semi-structured interviews were conducted with administrative and medical staff at Carle Clinic to gain both a historical perspective about the organization and insight into its culture. The archival data came from clinic records and included demographic information about the physicians such as specialty, gender, and years in practice and work-related information such as years at Carle Clinic, and physician-specific performance measures related to treatment of HMO patients—for example, use of ancillary services such as x-ray and receptivity to using less expensive treatment alternatives.

After pretesting and some subsequent modifications, the questionnaire was administered in 1988 to 160 physicians at Carle Clinic. In this paper, we focus on the 132 physicians providing direct patient care in the Clinic; omitted groups include radiologists, pathologists, anesthesiologists, and emergency room physicians. Among the direct care physicians, 91% returned a questionnaire.

## Measures of Culture

Because part of the original intent of the research was to assess whether physicians' attitudes influenced their receptivity to administrative change, our questionnaire was designed to measure a broad range of beliefs, values, and practice perceptions that we thought would reflect salient aspects of physician culture in a large medical practice. Our reading of several literatures including the organizational culture (i.e., Van Maanen and Barley 1985; Schein 1992), HMOs (i.e., Mechanic 1976, 1989; Barr and Steinberg 1983, 1985), strategic health care planning (i.e., Shortell, Morrison, and Friedman 1990), and sociological influences on physician decision making (i.e., Eisenberg 1979, 1986), as well as literature addressing changes in the autonomy of physicians (i.e., Freidson 1975; 1989; Haug 1988; Flood and Scott 1987; Shortell 1983; Eisenberg 1979), led us to focus on portraying four major dimensions of culture in a large medical group practice: *Professional Autonomy, Peer Relationships, Practice Orientation,* and *Strategic Orientation.*

*Professional Autonomy* refers to beliefs about the extent to which physicians should practice medicine without external interference while *Peer Relationships* refers to the type of relationship colleagues in the same practice setting have with each other. *Practice Orientation* is a physician's approach to treatment, such as whether aggressive intervention is considered as the initial treatment whenever appropriate. Finally, *Strategic Orientation* refers to beliefs about the proper orientation of a medical groups and encompasses both perceptions of the health care environment and beliefs about what strategic stance the organization should take in response to this environment.

We thought that both *Practice Orientation* and *Professional Autonomy* would be heavily influenced by professional training because they are central to a resident's experience, while *Peer Relations* would reflect both professional and organizational influences. In contrast, our expectation was that *Strategic Orientation* would be influenced little by training since it represents a set of issues not relevant to residents.

Our measures of culture included administration assessment of physician behaviors, actual physician behaviors, and attitudes, beliefs, and self-reported behaviors measured by the questionnaire. We used principal component analysis to reduce the number of questionnaire variables (Kim and Mueller 1978). Initially, we selected 18 items from the questionnaire that were consistent with the four dimensions of culture we sought to portray. These items included values and practices which could potentially develop because of the nature of the work performed (i.e., aggressive intervention as the initial strategy) as well as those which could reflect values learned through interaction (e.g., discussing patient cases with colleagues). Based on the convention of retaining factors with eigenvalues greater than one, (Kim and Mueller 1978), we initially retained seven factors, including two which only had one variable load heavily on them. The two factors explaining the least amount of variance (each less than 5%) were subsequently eliminated because of difficulties in discerning the meaning of their factor loading patterns. The overall measures from each of the remaining factors were scored by adding up the individual items and dividing by the number of answered items. Individual items were included in a scale if they had a factor loading greater than .50. Scored

***Table 1.*** Measures of Culture

---

**Autonomy**

Freedom from Administration
- I believe that a well-run clinic should not have any voice in how I practice medicine.

Cooperation with Managed Care
- Scale of the following five items ($\alpha = .90$):
  - Responds to phone calls and pages to discuss patient management issues.
  - Cooperates with HMO in pre-certification of patients for hospital services.
  - Is receptive to suggestions which will reduce health care costs through alternative uses of health care services.
  - Assists in the transfer of patients into a care facility when approached.
  - Works with HMO to assure member stays within the provider system.

**Peer Relations**

Collegiality
- Scale of the following four items ($\alpha = .74$):
  - Doctors in my division try to get along with each other.
  - Doctors in my division strongly influence policies and operations.
  - My divisional head decides divisional problems without consulting others (reverse coded).
  - Doctors in my division usually resolve divisional problems by consensus of the majority.

Interdependence
- Scale of the following four items ($\alpha = .71$):
  - Doctors in my division help each other (e.g., trade on-call hours).
  - Doctors in my division informally discuss care related to a specific patient.
  - Doctors in my division discuss organizational issues related to the clinic or division.
  - Doctors in my division discuss economic, social or ethical issues related to health care delivery in general.

**Practice Organization**

Interventionist Practice Style
- Doctors in my division aggresively treat illnesses at an early  stage.

Ancillary Resource Use
- Average of the age-sex adjusted per HMO enrolle rates for anesthesia, laboratory and x-ray services ($\alpha = .98$). Not available for surgeons.

**Strategic Orientation**
- Scale of the following four items ($\alpha = .72$):
  - Control any costs which are not directly reimbursable.
  - Monitor doctor utilization patterns.
  - Attract and retain effective administrators.
  - Increase quality assurance activities for the clinic.

---

in this way, the measures were found to have adequate internal consistency ($\alpha > .70$). The measures are described below and are summarized in Table 1.

The first two measures assess *Professional Autonomy*. The first measure, **Freedom from Administration**, is a single questionnaire item asking the extent to which a physician believes that a well-run clinic should have no voice in how the physician practices medicine. This item had a five-point scale anchored by "strongly disagree" and "strongly agree." **Cooperation with Managed Care** addresses autonomy by focusing on a physician's performance, as assessed by the HMO administration, on five items evaluating physicians' cooperation with HMO managed care requests. These variables were measured on a four-point scale anchored by "never" and "always" and were averaged to create an overall scale of HMO cooperation ($\alpha = .90$). The specific items used in this scale are detailed in Table 1.

*Peer Relationships* were assessed using two measures derived from the principal components analysis of questionnaire items. The first measure, **Collegiality**, consists of four five-point "strongly disagree" to "strongly agree" items assessing division-level administrative decision making ($\alpha = .74$). The second measure, **Interdependence**, is comprised of four items assessing interdependence related to professional responsibilities ($\alpha = .71$). These items were measured on a five-point scale anchored by "never" and "several times a day." The specific items used in both of these scales are detailed in Table 1.

Two measures of *Practice Orientation* were used. The first measure, **Interventionist-Oriented Practice**, is a single item asking a physician to assess the proportion of physicians in his or her division or branch who generally prefer to use all appropriate medical tests or services as an initial strategy. This item had a five-point scale anchored by "few (0-20%)" and "most (80-100%)." For this question, we asked physicians about others in their division rather than about their own behavior to avoid getting "politically correct" assessments. **Ancillary Resource Use** is a behavioral measure of the physician's practice orientation. It consists of the physician's per enrollee rate (age-sex adjusted) for ordering three types of ancillary services for HMO patients: anesthesia, laboratory, and x-ray services. This measure was available only for physicians who were "gatekeepers," that is, responsible for care, and who treated at least 100 HMO enrolles a month. Consequently, this measure was not available for surgeons since they could not serve as HMO "gatekeepers." The three adjusted rates were averaged to produce a measure of total ancillary resource use ($\alpha = .98$).

*Strategic Orientation* was assessed using a measure derived from the principal component analysis of questionnaire items. The four items ($\alpha = .72$) all address the importance of strategies related to administrative issues, for example, retaining effective administrators and controlling nonreimbursable costs, and were measured on a five-point scale anchored by "not important at all" and "essential." The specific items used in this scale are detailed in Table 1.

## Analytic Approach

Our interest was in investigating whether there is evidence, in addition to the influence of professional training, that the division or branch in which a physician currently practices affects his or her attitudes and practices. Our argument assumes that culture is manifested in the shared attitudes, values, and practices individuals hold and thus implies that there are differences between groups of physicians that are, on average, greater than those among physicians in the same group. To explore this issue empirically, we used analysis of variance (ANOVA) (Winer 1971).

We used three different analyses to address the question of the source of physicians' beliefs and practices: the professional group into which they have been socialized or their organizational membership in different divisions or branches.

- We used one-way ANOVA to confirm that the professional group physicians are socialized into has some lasting effects on their attitudes and practices.

*Rationale and Methods*:    Significant differences between physicians specializing in medicine and surgeons on the cultural variables would be confirmatory evidence that the professional group a physician is a member of influences his or her current beliefs and practices. We expected professional training to be most strongly related to differences in the value placed on autonomy and style of practice. More specifically, we expected that surgeons would value professional autonomy more and would be more oriented toward an interventionist-oriented practice style than those trained in medicine because surgical training promotes the lifeview that the surgeon must be the unquestioned leader of the surgical team and glorifies invasive solutions for medical problems.

For these analyses, we included all physicians providing direct care to patients and classify physicians as to whether they are in medicine or surgery. Medicine included all primary care physicians and medical specialists—for example, adult medicine, family practice, oncology, and cardiology—while surgery included all general and specialty surgeons—for example, obstetrics and gynecology, and ear, nose, and throat surgeons. We used this classification scheme because it is well accepted that medical specialists and surgeons differ in their beliefs and practices both because of differences in the types of problems faced and because of self-selection.

- We used a nested ANOVA to analyze physicians by practice locations, nested by medicine or surgery.

*Rationale and Methods:*    A nested approach allowed us to control for any overall differences between medical and surgical specialties when examining physicians in different practice locations and in this way permitted us to compare the "true" effects of being in different division or branches. Thus, significant differences between physicians in different practice locations on the cultural variables would be evidence that subcultures do develop in a large practice. In the organization we studied, physicians are organized into divisions in the main clinic and into branches in the surrounding communities. For the organizational analyses reported here, we required that at least five physicians returned our questionnaire, leaving us with eight divisions in the main clinic (the medical specialties were pediatrics, adult medicine, family practice, oncology, and cardiology; the surgical specialties were orthopedics, ob/gyn, and ear/nose/throat surgery) and three multi-specialty branch sites.

- We used one-way ANOVA to examine the effect of family practice physicians being in different practice locations.

*Rationale and Methods:* By restricting our analysis to family practice physicians, we performed the most stringent "control" for specialty training while examining the effect of organizational membership in different branches. Thus, significant differences between family practice physicians in different locations would be further evidence that subcultures do develop in a large group practice. We selected four family practice locations on the basis of the site having four or more respondents to our questionnaire (main clinic and three branch locations).

Because physicians' values and practices can be influenced by factors other than professional training and organizational interactions, we also examined the effect on our results when several other control variables were introduced, such as age of physician, gender, and years at Carle Clinic Association. In our preliminary analyses, we found that gender had no significant impact (possibly because of the small percentage of women—13.3%) and that physician age and tenure were highly correlated. For the results reported here, we retained organizational tenure, measured as a trichotomy: less than five years, five to 10 years, and more than 10 years at Carle Clinic.

## RESULTS

Table 2 presents the results of the ANOVAs related to the question of intraorganizational versus professional sources of cultural development, while the corresponding means are in Tables 3 and 4. The first group of ANOVAs in Table 2 provide some confirmatory evidence of the lasting influence of professional training on physicians' attitudes and practices. Controlling for organizational tenure, there were significant differences between medically trained physicians and surgeons for two of the three variables we expected to most strongly reflect professional training: **Freedom from Administration** and **Interventionist-Oriented Practice**. In interpreting these results, it is important to note that the $\eta^2$ (Winer

**Table 2.**  ANOVA of Professional and Organizational Effects on Cultural Dimensions, Controlling for Organizational Tenure

| | Medicine versus Surgery (n = 119) | | Locations Nested within Medicine and Surgery[a] (n = 54) | | Family Practice Sites Only (n=19) | |
|---|---|---|---|---|---|---|
| | F | $\eta^2$ | F | $\eta^2$ | F | $\eta^2$ |
| Autonomy | | | | | | |
|   Freedom from administration | 4.816*** | 0.04 | 1.550 | — | 4.501** | 0.60 |
|   Cooperation with managed care | .437 | — | 1.920* | 0.23 | 4.136** | 0.55 |
| Peer Relations | | | | | | |
|   Interdependence | .261 | — | 1.591 | — | 3.733** | 0.53 |
|   Collegiality | 2.634 | — | 1.682 | — | 3.926** | 0.54 |
| Practice Orientation | | | | | | |
|   Interventionist-oriented practice | 5.666** | 0.05 | 1.133 | — | 1.064 | — |
|   Ancillary resource use[b] | | — | — | 7.524** | 3.135* | .48 |
| Strategic Orientation | .752 | — | 1.281 | — | 3.677** | 0.52 |

Notes:  ** $p < .05$; * $p < .10$.
    [a] Nested ANOVA; all others are one-way.
    [b] Medicine only.

1971) indicates that only 4% to 5% of the variance in these variables is accounted for by the medicine versus surgery distinction.

Turning to Table 3, the variation in mean values observed for medical specialists and surgeons lends some face validity to the argument that professional training imparts a basic, enduring philosophy of practice and values. As would be expected, we observe that surgeons, in contrast to medical practitioners, were more likely to report an interventionist orientation and to be less sanguine about the appropriateness of the administration having any voice in how they practice medicine. Interestingly, however, both groups were low on this latter global measure of autonomy (1.98 and 2.44 on a five-point scale), suggesting an overall organizational culture accepting of administrative influence in their practice.

Having found some confirmatory support for the broad influence of professional training on physicians' attitudes and practices, we turned to the second set of ANOVAs in Table 2 to determine whether there was evidence of any subcultures within these broader groups, controlling for professional training. In these analyses, practice location is nested within the medicine versus surgery distinction, allowing us to examine the effect of practice location while controlling for professional group membership. Note from Table 3, however, that for five out of eight practice locations within medicine and for all three surgical

**Table 3.** Means of Attitudes, Values, and Practices by Practice Location and Medicine versus Surgery

| | Medicine | | | | | | | | | Surgery | | | |
|---|---|---|---|---|---|---|---|---|---|---|---|---|---|
| | All Med | Adult | Fam | Peds | Cardio | Hem/Onc | Mixed Branch1 | Mixed Branch2 | Mixed Branch3[a] | All Surg | OB/Gyn | Ent | Ortho |
| **Autonomy** | | | | | | | | | | | | | |
| Freedom from administration | 1.98 | 1.87 | 1.60 | 1.57 | 2.00 | 2.00 | 1.83 | 2.57 | 1.33 | 2.44 | 2.80 | 3.00 | 1.88 |
| Cooperation with managed care | 3.41 | 3.39 | 3.52 | 3.31 | 3.91 | 3.18 | 3.29 | 3.77 | 3.00 | 3.34 | 3.16 | 3.31 | 3.29 |
| **Peer Relations** | | | | | | | | | | | | | |
| Interdependence | 3.15 | 2.93 | 3.50 | 3.50 | 3.50 | 2.88 | 3.04 | 2.90 | 2.50 | 3.23 | 3.54 | 3.50 | 2.89 |
| Collegiality | 3.51 | 3.37 | 4.13 | 3.55 | 3.75 | 3.25 | 3.25 | 3.43 | 3.33 | 3.69 | 3.75 | 3.62 | 3.66 |
| **Practice Orientation** | | | | | | | | | | | | | |
| Interventionist-oriented practice | 2.84 | 3.71 | 3.25 | 2.69 | 3.20 | 3.00 | 2.80 | 2.50 | 2.33 | 3.38 | 2.66 | 2.75 | 3.44 |
| Ancillary resource use | 5.13 | 7.62 | 5.66 | 4.01 | 6.92 | 6.47 | 3.70 | 5.37 | 3.87 | — | — | — | — |
| Strategic Orientation | 3.48 | 3.43 | 4.20 | 3.34 | 3.67 | 3.63 | 3.04 | 3.50 | 4.16 | 3.62 | 4.04 | 3.38 | 3.25 |

*Note:* [a] Physicians in mixed branches represented different specialities, usually all primary care and general surgery or ob/gyn. Only those in medicine are included here.

150

practice locations, "practice location" also implies that the physicians shared the same type of subspecialty training—thereby limiting our ability to tell whether any differences were due to subspecialty training rather than organizational subculture.

In these ANOVAs in Table 2, we found very little evidence that practice location (or perhaps subspecialty training), controlling for the medicine-surgery distinction and tenure, predicted any patterns of beliefs or practice. The one notable exception was **Ancillary Resource Use**, one of our measures of *Practice Orientation*, which differed significantly among practice locations (and an $\eta^2$ of 52%), although there were also marginally significant differences among practice locations in physicians' **Cooperation With Managed Care**. The means displayed in Table 3, however, provide some face validity to our argument that differences can be due to organizational influences, not just specialty training. In the case of **Ancillary Resource Use**, for example, the three mixed specialty branches (with a mixture of primary care specialists) differed not only from each other but also from the other primary care divisions (adult medicine, family practice, and pediatrics). For **Cooperation With Managed Care**, as well, the three branch locations represented wide variation, from having the lowest mean score—considerably lower than the other primary care divisions—to having one of the highest scores of any practice location.

The results become much stronger and more consistent when we turn to the third set of ANOVAs which include only family practice physicians. By examining only one specialty in four practice settings, we exercise our most stringent control to remove variation which might be due to professional training. Taken together, the results of the third set of analyses on Table 2 provide support for the argument that physician interaction and/or similar interpretation of the task environment can result in the organizational unit being the site of cultural development. These analyses indicate that family practice physicians in different practice locations, controlling for tenure, differed significantly on five of the seven cultural variables and marginally significantly on a sixth, including three variables which we expected professional training to heavily influence. In addition, the $\eta^2$ indicates that between 48% and 60% of the variance in these variables can be explained by practice location. The only dimension that these family practice physicians did not differ on by practice location was in their tendency to report an **Interventionist-Oriented Practice**—which was low in every location.

The means for family practice physicians in each practice location are displayed in Table 4. Although we controlled for specialty effect in these analyses, a different question arises in trying to compare the main clinic and three branches—namely, are the differences we observed due to physicians self-selecting into a particular type of practice represented by being at the main clinic (with over 100 doctors) versus at a branch (with 10 or fewer doctors)? To examine this question, we used the means in Table 4 and applied Tukey's

**Table 4.**   Means of Attitudes, Values, and Practices by Practice
Location for Family Practice Physicians Only

|  | Main | Branch 1 | Branch 2 | Branch 3 |
|---|---|---|---|---|
| Autonomy |  |  |  |  |
| Freedom from administration | 1.60 | 2.00 | 1.67 | 3.00 |
| Cooperation with managed care | 3.52 | 3.37 | 3.06 | 3.67 |
| Peer Relations |  |  |  |  |
| Interdependence | 3.50 | 3.25 | 3.17 | 2.25 |
| Collegiality | 4.13 | 3.81 | 3.25 | 3.33 |
| Practice Orientation |  |  |  |  |
| Interventionist-oriented practice | 3.25 | 2.00 | 2.00 | 2.33 |
| Ancillary resource use | 5.65 | 5.18 | 4.14 | 5.86 |
| Strategic Orientation | 4.20 | 3.81 | 2.67 | 3.25 |

studentized range test (HSD) (Winer 1971) to evaluate the contrast between the four practice locations. We found little evidence that the variation really is a main-branch effect. For example, using **Freedom From Administration**, Table 4 shows that Branch 3 physicians are the outliers (3.00 on a five-point scale compared to 2.00 and 1.67 for the other two branches and 1.60 for the Main Clinic), although Tukey's test supports as significant only the contrast between Branch 3 and the Main Clinic. Similarly, the significant differences in *Strategic Orientation* were due to the contrast between the Main Clinic and one of the branches (4.20 for the Main Clinic and 2.67 for Branch 2 on a five-point scale), with the Main Clinic family practice physicians placing more importance on strategies related to administrative issues. Again, however, there was considerable variation in mean values for the three branches, with means ranging from 2.67 to 3.81 on a five-point scale.

Other patterns occurred as well. In the case of **Cooperation with Managed Care**, only two branches are significantly different (3.06 and 3.67 on a four-point scale), while the Main Clinic family practice physicians have an intermediate value. Finally, the evidence indicates that the family practice physicians in the Main Clinic had closer relationships with their peers than did those in the branches, perhaps because the branch locations were mixed specialty, in contrast to the Main Clinic, which was single specialty. For both *Peer Relations* measures (**Collegiality** and **Interdependence**), the Main Clinic physicians had the highest mean value and Tukey's test indicated that the only significant contrast among family practice locations was between the Main Clinic and a branch location (Branch 3 for **Interdependence** and Branch 2 for **Collegiality**).

# IMPLICATIONS AND CONTRIBUTIONS

There are two main contributions of this paper. First, we present a taxonomic scheme, melding the professional socialization and organizational culture literatures, which portrays the major sources and processes by which shared culture can occur in professional organizations. We argue that, in order to blend the insights from these two literatures, it is necessary to recognize two dimensions. One focuses on the source of shared values, beliefs, and practices: professional socialization or organizational inculturation. The other portrays the mode of developing these shared understandings: through interaction or by independent invention due to the nature of the work. Second, we illustrate a quantitative approach to measuring organizational culture. In particular, using a single but complex professional organization, we examine the usefulness of our taxonomy for explaining differences in values and beliefs of physicians in an organization with many divisions and sites.

Several limitations of our data make our analyses only illustrative. With only one medical practice, we were limited to examining intraorganizational differences (i.e., subcultures) and were unable to compare across organizations to examine differences in organizational culture. In addition, our data were cross-sectional, which made it difficult to eliminate self selection as an alternative explanation for our findings. For example, it may be that physicians attracted to a large practice share certain characteristics, compared to small groups or solo practice. A longitudinal study of multiple group practices would be ideal to ferret out the relative importance of both the source of and mode through which culture arises in explaining why physicians differ in their attitudes about and styles in treating patients.

Within these important limitations, our analyses suggest that basic professional socialization may play a lasting role in the development of professional practice orientation and views of professional autonomy. While the differences we found were relatively small, they are consistent with the professional literature and suggest that the organizational culture literature alone is not adequate to understand the complexities of cultural development in physician organizations. We also found, however, that organizational membership in different divisions and branches did explain some of the values, beliefs, and practices of physicians, supporting the existence of "true" organizational culture in a large medical group. This evidence was most clear when we were able to refine our definition of professional group membership to include only physicians trained in the same specialty—family practice. Since insufficient cases prevented testing whether these patterns held for other specialty groups, an important next step will be to examine whether these patterns do generalize to other specialty groups embedded in complex organizations.

The ideas embodied in our taxonomic framework have implications for both the professional and organizational culture literatures. The organizational culture literature can be enriched by incorporating the insights from models of professional training and modes of organizing, which highlight the importance and durability of professional socialization in organizational settings. In emphasizing the cultural differences between organizations, the organizational culture literature—particularly the "how to" literature on cultural change—has been too quick to assume that organizations have tremendous power to transform the behavior of their workforces.

On the other hand, the professional literature can be faulted for neglecting to revamp its models of physicians' behavior to include organizational influences. During the 1950s and 1960s, researchers tended to focus on the influences of the profession in shaping individual behavior—a focus well suited for understanding the solo practitioner who dominated medicine in the United States during this period. Since then, notwithstanding the examples and admonitions of scholars such as Elliot Freidson and David Mechanic and despite the increasing importance of physician groups in medicine, the fateful impact of professionalization on the values and practices of physicians has continued to be emphasized. We argue that, instead, these two approaches need to be blended into a new model which makes obvious the need to investigate both the impact of professional training and the circumstances under which professional values can be modified by subsequent organizational arrangements and interaction with colleagues.

## ACKNOWLEDGMENTS

This work was supported by a grant from the Agency for Health Care Policy and Research, HS06159 and has benefitted from support by the Department of Organization and Human Resources at SUNY—Buffalo and the Center for the Evaluative Clinical Sciences at Dartmouth Medical School. We gratefully acknowledge the contributions and advice of Robert Henrickson, Robert C. Parker, Jr., John Pollard, and Gerald Tresslar. An earlier version of this paper was presented at the Inaugural Meeting of Australian Industrial and Organizational Psychology Conference, Sydney, Australia, July 1995.

## REFERENCES

Barley, S.R. 1983. "Semiotics and the Study of Occupational and Organizational Cultures." *Administrative Science Quarterly* 28: 393-413.
Barr, J.K., and M.K. Steinberg. 1985. "A Physician Role Typology: Colleague and Client Dependence in an HMO." *Social Science and Medicine* 20: 253-261.
Barr, J.K., and M.K. Steinberg. 1983. "Professional Participation in Organizational Decision Making: Physicians in HMOs." *Journal of Community Health* 8: 160-173.
Becker, H.S., B. Geer, E. Hughes, and A.L. Strauss. 1961. *Boys in White: Student Culture in Medical School*. Chicago, IL: University of Chicago Press.
Carr-Saunders, A.M., and P.A. Wilson. 1933. *The Professions*. London: Oxford University Press.
Eddy, D.M. 1984. "Variations in Physician Practice: The Role of Uncertainty." *Health Affairs* 3: 74-89.

Eisenberg, J.M. 1986. *Doctors' Decisions and the Cost of Medical Care: The Reasons for Doctors' Practice and Ways to Change Them.* Ann Arbor, MI: Health Administration Press Perspectives.

Eisenberg, J.M. 1979. "Sociologic Influences on Decision Making by Clinicians." *Annals of Internal Medicine* 90: 957-964.

Flood, A.B., and W.R. Scott. 1987. *Hospital Structure and Performance.* Baltimore, MD: Johns Hopkins University Press.

Flynn, R. 1992. *Structures of Control in Health Management.* New York: Routledge.

Freidson, E. 1989. *Medical Work in America: Essays on Health Care.* New Haven, CT: Yale University Press.

Freidson, E. 1986. *Professional Powers: A Study of the Institutionalization of Formal Knowledge.* Chicago, IL: University of Chicago Press.

Freidson, E. 1976. *Doctoring Together: A Study of Professional Social Control.* New York: Elsevier.

Freidson, E. 1975. *Profession of Medicine: A Study of the Sociology of Applied Knowledge.* New York: Dodd, Mead and Company.

Gerrity, M.S., J.A.L. Earp, R.F. DeVellis, and D.L. Light. 1992. "Uncertainty and Professional Work: Perceptions of Physicians in Clinical Practice." *American Journal of Sociology* 97: 1022-1051.

Good, B.J. 1994. *Medicine, Rationality, and Experience.* Cambridge, UK: Cambridge University Press.

Good, M.D. 1995. *American Medicine.* Berkeley: University of California Press.

Goode, W.J. 1957. "Community within a Community: The Professions." *American Sociological Review* 22: 194-200.

Haug, M.R. 1988. "A Re-examination of the Hypothesis of Physician Deprofessionalization." *Milbank Quarterly* 66: 48-56.

Hofstede, G., B. Neuijen, D. Ohayv, and G. Sanders. 1990. "Measuring Organizational Cultures: A Qualitative and Quantitative Study across Twenty Cases." *Administrative Science Quarterly* 35: 286-316.

Kaye, C. 1986. "Managing Clinicians." In *Management Budgeting in the NHS,* edited by R. Brooks. Keele, UK: Health Services Manpower Review.

Kim, J., and C.W. Mueller. 1978. *Introduction to Factor Analysis.* London: Sage.

Kralewski, J.E., T.D. Wingert, and M.H. Barouche. 1996. "Assessing the Culture of Medical Group Practices." *Medical Care* 34: 377-388.

Light, D. 1980. *Becoming Psychiatrists.* New York: W.W. Norton & Company.

Mechanic, D. 1976. *The Growth of Bureaucratic Medicine.* New York: John Wiley and Sons.

Mechanic, D. 1989. *Painful Choices: Research and Essays on Health Care.* New Brunswick, NJ: Transaction.

Mechanic, D., and M. Schlesinger. 1996. "The Impact of Managed Care on Patients' Trust in Medical Care and their Physicians." *Journal of the American Medical Association* 275: 1693-1697.

Merton, R.K., G.G. Reader, and P.K. Kendall. 1957. *The Student-Physician; Introductory Studies in the Sociology of Medical Education.* Cambridge, MA: Harvard University Press.

Miller, S.J. 1970. *Prescription for Leadership: Training for the Medical Elite.* Chicago, IL: Aldine.

O'Reilly, C.A., III, J. Chatman, and D.F. Caldwell. 1991. "People and Organizational Culture: A Profile Comparison Approach to Assessing Person-Organization Fit." *Academy of Management Journal* 34: 487-516.

Orru, M., N.W. Biggart, and G.G. Hamilton. 1991. "Organizational Isomorphism in East Asia." Pp. 337-360 in *The New Institutionalism in Organizational Analysis,* edited by W.W. Powell and P.J. DiMaggio. Chicago, IL: University of Chicago Press.

Ott, S.J. 1989. *The Organizational Culture Perspective.* Chicago, IL: The Dorsey Press.

Parsons, T. 1951. *The Social System.* New York: Free Press of Glencoe.

Pettigrew, A.M. 1979. "On Studying Organizational Cultures." *Administrative Science Quarterly* 24: 570-581.

Schein, E.H. 1992. *Organizational Culture and Leadership.* San Francisco, CA: Jossey-Bass.

Schneider, B. 1987. "The People Make the Place." *Personnel Psychology* 40: 437-453.

Shortell, S.M., E.M. Morrison, and B. Friedman. 1990. *Strategic Choices for America's Hospitals: Managing Change in Turbulent Times.* San Francisco, CA: Jossey Bass.

Shortell, S.M. 1983. "Physician Involvement in Hospital Decision Making." Pp. 73-101 in *The New Health Care for Profit: Doctors and Hospitals in a Competitive Environment,* edited by B.H. Gray. Washington, DC: National Academy Press.

Tolbert, P.S. 1988. "Institutional Sources of Organizational Culture." Pp. 101-113 in *Institutional Patterns and Organizations,* edited by L. Zucker. Cambridge, UK: Ballinger.

Tolbert, P.S., and L.G. Zucker. 1983. "Insitutional Sources of Change in the Formal Structure of Organizations: The Diffusion of Civil Service Reform, 1880-1935." *Administrative Science Quarterly* 28: 22-39.

Trice, H.M., and J.M. Beyer. 1993. *The Cultures of Work Organizations.* Englewood Cliffs, NJ: Prentice-Hall.

Van Maanen, J., and S.R. Barley. 1984. "Occupational Communities: Culture and Control in Organizations." Pp. 287-365 in *Research in Organizational Behavior,* Volume 6, edited by B.M. Staw & L.L. Cummings. Greenwich, CT: JAI Press.

Van Maanen, J., and S.R. Barley. 1985. "Cultural Organization: Fragments of a Theory." Pp. 31-53 in *Organizational Culture,* edited by P.J. Frost, L.F. Moore, M.R. Louis, C.C. Lundberg, and J. Martin. Beverly Hills, CA: Sage.

Wennberg, J.E. 1988. "Improving the Medical Decision-making Process." *Health Affairs* 7: 99-106.

Wennberg, J.E., J.L. Freeman, and W.J. Culp. 1987. "Are Hospitalized Services Rationed in New Haven or Over-utilized in Boston?" *Lancet* (May): 1185-1189.

Winer, B.J. 1971. *Statistical Principals in Experimental Design.* New York: McGraw-Hill.

# CROSSING LEGAL PRACTICE BOUNDARIES:
## PARALEGALS, UNAUTHORIZED PRACTICE OF LAW, AND ABBOTT'S SYSTEM OF PROFESSIONS

Jona Goldschmidt

## ABSTRACT

Theory development in the sociology of professions was greatly advanced by the publication of Andrew Abbott's book, *The System of Professions* (1988). His approach to professions focuses on the work actually engaged in by members of the professions, and the jurisdictional claims regarding the work of the dominant professions. This paper describes the invasion of the boundaries of legal practice by independent paralegals, one form of an occupational group created by lawyers themselves which is now in open competition with them. Using Abbott's system of professions as a theoretical framework, the paper analyzes the work of paralegals and their conflicts with lawyers through an examination of unauthorized practice of law (UPL) complaints filed with the State Bar of Arizona for the period 1988-1994. This case study gives us a deeper understanding of the nature of the interprofessional conflicts between lawyers and independent paralegals, the specific boundaries at which these conflicts arise, and the possible future settlements of the jurisdictional claims represented by these complaints.

Current Research on Occupations and Professions, Volume 10, pages 157-191.
Copyright © 1998 by JAI Press Inc.
ISBN: 0-7623-0034-5

# INTRODUCTION

In this paper, I report the results of an examination of 550 unauthorized practice of law (UPL) complaints received by the State Bar of Arizona during the period 1988-1994. When beginning this study, my immediate objective was to provide additional data on a little-studied aspect of the legal profession. I was only tangentially interested in the sociology of professions.

It was only after reading Abbott's *The System of Professions* (1988) that I realized that the data I had collected had greater value. Abbott explicates a novel and profoundly complex, yet in some ways simple, theory of professions which he tests by a study of three professions: law, medicine, and information science. He, too, uses UPL complaint data to illustrate the jurisdictional conflicts between lawyers and nonlawyers over the work and services traditionally provided by the legal profession, such conflicts and their settlement being at the heart of his theory. The data I report, which includes interviews of Arizona State Bar officials and representatives of paralegal organizations, gives us a better understanding of the interactions both within the legal profession and between the legal and paralegal professions.

This case study, beyond testing its fit with Abbott's theory, portends the landscape of the jurisdictional skirmishes we are likely to see in the future between lawyers and nonlawyer practitioners. As a study conducted in a natural laboratory, it has policy implications for the questions of whether to abolish or modify unauthorized practice of law regulation, and whether to regulate nonlawyer practice.

# TRENDS IN THE SOCIOLOGY OF PROFESSIONS

## Deprofessionalization

Before reviewing Abbott's theoretical contribution, two central concepts in the study of professions should be noted. One is deprofessionalization. As the attributional model of professions came to be supplanted by the process model, the concept of deprofessionalization emerged. It was a recognition that social forces had begun to threaten the base of professionalism (Haug 1973; Torren 1975). Deprofessionalization is said to be a consequence of social, economic, demographic, and political trends that undermine professions' claim to autonomy, monopoly, and social privilege. Factors considered by some to be sources of deprofessionalization are: (1) narrowing of the competence gap (public education); (2) routinization of expert knowledge; (3) the consumer movement; (4) specialization; (5) encroachments from allied professions, called "boundary maintenance" by Kronus (1976); and (6) the demographics of a profession. Rothman (1984, p. 202) notes:

> Autonomous, monopolistic professions may indeed become an anachronism—a form of social organization rendered obsolete by changing conditions, as were the medieval guilds.

Professional dominance may be replaced by a narrower, more clearly circumscribed client-expert relationship that permits the exercise of skill and judgment within a context of accountability to client and public.[1]

The latter prediction may be borne out by the increasing number of paralegals, as well as other paraprofessionals in medicine, information sciences, and other professions. The data reported in this study suggest that certain paralegals are now providing this new, predicted "client-expert"—rather than "client-professional"—relationship.

The factors cited above as contributing to deprofessionalization are still at work, as are other social and cultural influences. For example, the "customer service" or "user-friendly" and related "paradigms" (e.g., TQM, or Total Quality management) have swept the business and governmental management worlds and can be expected to affect the professions. (I distinguish these trends from the consumer movement, which in my view refers more to fairness and public safety in such matters as the regulation of health, food, product, and lending practices).

Another new and important influence—upon the legal profession in particular—is the movement toward greater "access to the court." In the past 10 years, bar associations and courts have initiated many statewide and local blue-ribbon commissions or task forces (often called "futures" commissions ) to evaluate and suggest improvements for the administration of justice. A major theme of these commissions and their investigations, conferences conducted on court reform, and the written products of this movement has been enhancing "access to justice" (Goldschmidt and Pilchin 1996; Pilchen and Ratcliff 1993; American Bar Association 1994a).

Self-representation in court, once thought to be exclusively undertaken by prisoners or indigent persons, is also increasingly popular among the middle class. The growing number of persons appearing "pro se" due to the high cost of legal services or anti-lawyer sentiment, or both, is now well-documented and of great concern to lawyers and judges (American Bar Association 1994b; Sales, Beck, and Haan 1993). These individuals still need assistance in order to navigate the complexities of litigation, even in "simple" cases. That assistance is now often provided by nonlawyer practitioners at a lower cost (Rhode 1990). This, too, is another social trend influencing lawyer-nonlawyer relations not previously considered by deprofessionalization theorists.

## Segmentation

A second relevant concept in the sociology of professions is segmentation. Segmentation is a profession's internal division of labor (horizontal or vertical). It is especially well known in the medical profession as a consequence of a high degree of specialization (Halpern 1992). Segmentation can also be found in other professions such as law, nursing, the clergy, architecture, information sciences, and others. According to Everett Hughes:

The division of labor among lawyers is as much one of respectability (hence of self concept and role) as of specialized knowledge and skills. One might even call it a moral division of labor if one keeps in mind that the term means not simply that some lawyers, or people in the various branches of law work, are more moral than others; but that the very demand for highly scrupulous and respectable lawyers depends in various ways upon the availability of less scrupulous legal problems of even the best people. I do not mean that the good lawyers all consciously delegate their dirty work to others (although many do). It is rather a game of live and let live; a game, mind you, hence interaction, even though it be a game of keeping greater than chance distances (Hughes 1958, p. 71).

Segmentation within the legal and paralegal profession continues to exist—not on the basis of moral conduct but on the work performed and clients served—and was found to be a factor in this case study.

## Abbott's System of Professions

Abbott (1988) proposes a new theoretical approach to professions using a "social history, comparative historical sociology, or whatever else we may wish to call this kind of analysis" (1988, p. 115). He justifies his new theory by noting that the "work professions do . . . is unmentioned in theoretical studies of professions, although it makes obligatory appearances in case studies" (1988, p. 17). The central problem, he suggests, is the prior exclusive focus on structure rather than work. His method, therefore, is one that goes beyond merely studying a common process of "professionalization" to a means of studying the content of professional life. The current process approach, he states, ignores "who was doing what to whom and how, concentrating instead on association, licensure, and ethics codes" (1988, pp. 1-2).[2]

Abbott uses a loose definition of "professions": "exclusive occupational groups applying somewhat abstract knowledge to particular cases" (1988, p. 8).[3] His "systematic" view of professions takes into account "jurisdictional boundaries [that] are perpetually in dispute" (1988, p. 2). By "jurisdiction" Abbott refers to "the link between a profession and its work" (1988, p. 20). His theory analyzes professional development by determining how this link is created in work, how it is anchored by formal and informal social structure, and how the interplay of jurisdictional links between professions determines the history of the individual professions themselves. By studying interprofessional conflict, "we can set the successful professions in their real context and correct our theories of development" (1988, p. 247).

The evolution of professions "in fact results from their interrelations" which are in turn "determined by the way these groups control their knowledge and skill" through control of technique (or "crafts") and abstract knowledge (1988, p. 8).

Any occupation can obtain licensure (e.g., beauticians) or develop an ethics code (e.g., real estate). But only a knowledge system governed by abstractions can redefine its problems and tasks to defend them from interlopers and seize new problems—as medicine has recently seized alcoholism, mental illness, hyperactivity in children, obesity, and numerous other things. Abstraction enables survival in the competitive system of professions . . . what matters is abstraction effective enough to compete in a particular historical and social context, not abstraction relative to some supposed absolute standard (1988, pp. 8-9).

Abbott posits three ideal types of conflict that are anticipated in interprofessional jurisdictional disputes: (1) "excess jurisdiction" occurs when potential jurisdiction is expanding relative to potential professional output; (2) "insufficient jurisdiction" occurs when current jurisdictions are insufficient to support the profession (i.e., the profession is looking for work); and where there is (3) an invasion of a "settled jurisdiction" by groups providing equivalent services at lower prices.

He refers to the "common cases" that make up the majority of the work of professionals, and that make particular jurisdictions vulnerable to attack:

> Most of these problems have trivial diagnoses, and for each the profession involved has a purely conventional treatment. Often the treatment is delegated to subordinates . . . [R]esults that are too easily measured lead to easy evaluation from outside the profession and consequent loss of control. They may also make it easier for competitors to demonstrate treatment superiority if they have it (1988, pp. 45-46).

Jurisdictional claims, on Abbott's account, are made in three settings: (1) public opinion; (2) legal; and (3) the state (i.e., the legislature, where the public and legal settings "run together"). He also refers to these as "audiences" of jurisdictional claims—the battlegrounds, so to speak, of boundary conflicts.

Jurisdictional conflicts are resolved by "settlement" of such disputes. According to Abbott, these settlements take a variety of forms: (1) "full and final" settlement; (2) subordination of one profession by the other; (3) division of labor, where a jurisdiction is divided into two independent parts (or share an area without a division of labor); (4) "advisory control" is retained by the dominant profession over certain aspects of the work; and (5) by client differentiation. The emergence of paralegals, according to Abbott, is an example of the "limited settlement" of subordination (1988, p. 71).

## Client Differentiation

Client differentiation has an important role in professional boundary disputes. It exists not only between professions but within them. As Abbott accurately points out, "In American law today, two virtually separate legal professions handle on the one hand the problems of large corporations and a few wealthy individuals and on the other hand the problems of small corporations and the mass of individuals" (1988, p. 77; Heinz and Laumann 1982; Carlin 1962; Hughes 1958). Such divisions offer:

> a potentially disastrous critique of the public and legally legitimate systems of interprofessional relations. They are jurisdictional claims implicit in an actual division of labor. It is obviously essential to the incumbents of full jurisdiction that the public and legal arenas

remain completely unaware of such client differentiation . . . [which appears] when demand
suddenly outstrips available professional numbers (1988, p. 77).

The highest status professionals are those "who work in the most purely professional environment . . . who do not sully their work with nonprofessional matters.
. . The frontline service that is both their fundamental task and their basis for legitimacy becomes the province of low-status colleagues and paraprofessionals."
This, in turn, weakens jurisdiction, "opening a potential vacancy in the system"
which is irreversible (1988, pp. 118-119).

One means by which a new group threatens a dominant profession's jurisdiction
is a form of increased demand, but one caused by the dominant profession itself.
This occurs when "A powerful profession ignores a potential clientele and paraprofessionals appear to provide the same service to this forgotten group" (1988,
p. 91).[4] New technologies and organizations arise to provide these professional
tasks; "Correlatively, the two are the central destroyers of professional work"
(1988, p. 92).

Abbott's theory contains many more theoretical statements, a fuller exposition
of which is beyond the scope of this paper. The numerous categories, dimensions,
and levels of analysis of his system of professions will keep researchers busy for
many years to come. Abbott, to be sure, has been criticized by some (Macdonald
1995) and praised by others (Kellermann 1990), and certain gaps in his theory
have been filled (Dingwall 1995). But, with some exceptions (Monahan 1994;
Anleu 1992), little empirical study applying his theory has been undertaken. The
system of professions, despite its flaws, appears to have direct applicability to the
case of professional conflict between lawyers and nonlawyer practitioners.

# JURISDICTION OF THE LEGAL PROFESSION

## The Practice of Law

One of the first questions to address when studying the relationship between the
legal profession and nonlawyers is, just what is the "practice of law"? Although I
have found eight definitions of this phrase in court opinions involving the prose-
cution of persons for engaging in the unauthorized practice of law, they are so
vague that it is fair to say that no one really knows what the practice of law really
is (Currier 1991).[5] The situation is similar to the Supreme Court's attempts to
define obscenity, where Justice Stewart declared that "I know it when I see it"
(*Jacobellis v. Ohio*, 1964, p. 197).

The definitions of the practice of law in current court opinions are: (1) the
"requires the knowledge and application of legal principles" test (*R.J. Edwards,
Inc. v. Hert* [Okla. 1972]).; (2) the "activities lawyers have traditionally performed" test (*State Bar of Arizona v. Arizona Land Title and Trust Co.* [(Ariz.
1961]).; (3) the "service incidental to principal business" test (*Ingham County Bar*

*Assoc. v. Walter Neller Co.* [Mich. 1955]); (4) the "knowledge beyond the average citizen" test (*State Bar of New Mexico v. Guardian Abstract and Title Co.* [NM 1978]); (5) the "balancing of interests" test (*Conway-Bogue Realty Investment Co. v. Denver Bar Assoc.* [Colo. 1957]); (6) the "activities which are incidental to appearance in court" test (*Akron Bar Ass'n v. Greene* [Ohio 1997]); (7) the "professional judgment of a lawyer" test (*In re* Burson [Tenn. 1995]); and (8) the "fair intendment of the term" (i.e., "practice of law") test (*Attorney Grievance Comm'n v. Hallmon*, 681 A.2d 510 (Md. 1996]).

According to the American Bar Association and some courts:

> It is neither necessary nor desirable to attempt the formulation of a single, specific definition of what constitutes the practice of law. Functionally, the practice of law relates to the rendition of services for others that call for the professional judgment of the lawyer. The essence of the professional judgment of the lawyer is his educated ability to relate the general body and philosophy of law to a specific legal problem of a client; and thus, the public interest will be better served if only lawyers are permitted to act in matters involving professional judgment (American Bar Association 1983).

Despite disclaiming the need for a specific definition of the practice of law, the ABA appears to have adopted the "professional judgment of a lawyer" test of what constitutes the practice of law. In any event, to the extent that the definition of the practice of law is kept as broad and vague as possible, the situation fits Abbott's theory insofar as a profession's need to establish an abstract body of knowledge that can only be understood through professional training.

## Fields of Law Practice

The next question we must ask is: what are the legal profession's jurisdictions? One answer is found in the enumeration of the various fields of practice that constitute the boundaries of the legal profession. The best source I have found for such a list is in Heinz and Laumann's (1982) study of the social structure of the Chicago bar.[6] They present data on variables such as type of client, proportion of time spent in various substantive areas of law practice, size of firm, and so forth. The data tables in their book include between 26 and 30 legal practice areas, depending on how they are grouped (1982, pp. 24, 29, 32, 46, 48, 50). These and other data evidence the segmentation of the bar noted earlier.

Their conclusion is: "Any profession will surely include disparate parts, but we doubt that any other is so sharply bifurcated as the bar. . . The difference between serving corporations and serving individuals is, for a lawyer's work and career, a difference that has important, highly predictable implications . . . In sum, the Chicago bar consists, to an extent that is quite striking, of two professions, quite different in type and content and both of substantial size" (1982, p. 170).[7] The same division of labor and client differentiation was earlier found by Carlin (1962, p. 18): "The practice of most metropolitan individual practitioners is consequently confined to those residual matters (and clients) that the large firms have not pre-empted: . . . the undesirable cases, the dirty work, those areas of practice that have been associated with them an aura of influencing and fixing and that involve arrangements with clients and others that are felt by the large firms to be professionally damaging. The latter includes local tax, municipal, personal injury, divorce, and criminal matters."[8]

Another perspective on the dimensions of the jurisdictions of the legal profession arises from this study. The data described here are not limited to unspecified activity within certain areas of law as in Heinz and Lauman (1982) and Abbott (1988); rather, they include specific activities and conduct within the various fields of law which illuminate the actual work about which the jurisdictional battle lines between lawyers and nonlawyers have been drawn.

## UNAUTHORIZED PRACTICE REGULATION AS A JURISDICTIONAL ENFORCEMENT MECHANISM

The use of exclusionary legislation to keep certain classes of people from practicing law arose in colonial times. English colonial governors used these measures to bar from court those nonlawyers who were believed to be stirring up litigation for profit, as well as the emerging American lawyer class which was in competition with the predominantly English members of the profession.[9] Ironically, after the birth of an organized bar, UPL laws were used by American lawyers to preserve their jurisdictional "heartland" of litigation (Goldschmidt 1994).[10]

The period 1960 to 1980, however, is noteworthy as demarcating a time when the development of UPL rules was "disrupted by events that have been largely beyond the control of the legal profession" (Christensen 1980, p. 197). These include: (1) a public movement in Arizona in reaction to the decision in *State Bar of Arizona v. Arizona Land Title abd Trust Co.* (1962)—which held that a real estate agent's preparation of certain real estate forms constituted the unauthorized practice of law—leading to a constitutional amendment overturning that decision; (2) U.S. Supreme Court cases that held intermediaries (such as the NAACP or labor unions) were constitutionally permitted to provide legal services for their members (*NAACP v. Button*, 1963; *Brotherhood of R.R. Trainmen v. Virginia* ex rel. Virginia State Bar, 1964; *UMW Dist. 12 v. Illinois State Bar Ass'n*, 1967); (3) the new possibility that UPL rules could be challenged as violating antitrust law due to the *Goldfarb v. Virginia State Bar* (1975) decision, which held that attorneys were not exempt from antitrust laws; and (4) preliminary investigations of the antitrust question by the U.S. Justice Department that led to an ABA repeal of interprofessional agreements that had been entered into with the real estate, accounting, and other professions. Court cases on UPL had, during this period, begun to develop into an incoherent and inconsistent body of state case law attempting to define what constituted the practice of law. Nevertheless, the organized bar's stated rationale for UPL enforcement during this time remained two-fold: competency and professional independence.

More recently, there has been a tremendous growth in the last decade of products and services that are, or border on, what some might view as UPL. Nolo Press in Berkeley and numerous other publishers sell "how-to" books or kits on legal subjects. Computer technology also has UPL implications by way of artificial intelligence (i.e., software that can prepare legal forms for consumers or answer legal questions for attorneys), legal data bases that can be accessed by nonlawyers, and computer kiosks used in some states to assist litigants in preparing court forms. In addition, franchise paralegal or document preparation services are opening in many states, and television commercials aired by nonlawyer collection agencies are offering to collect child support arrearage (a recent one featuring a woman in a business suit standing in front of a wall of law books). These events fit Abbott's assertion that technology and organization constitute primary threats to the jurisdictional claims of the legal profession. And these threats are primarily to the work of the lower-echelon or "personal plight" lawyers, rather than the lawyers serving corporate clients.

The literature reveals a dearth of empirical studies of UPL, particularly with respect to the question of public harm from nonlawyer practice. One approach to determining the existence of pubic harm from nonlawyer is by examining the reported court opinions involving UPL. Christensen (1980) did this for the period 1908-1969 in California, Colorado, Florida, Illinois, and New York. He found that only 8% of the cases involved an allegation of specific injury. Rhode (1981), in a survey of state bar associations, found that only 39% of the jurisdictions responding reported receiving UPL complaints from clients themselves, and only eight (21%) of these jurisdictions indicated that any of these cases had involved specific injury. Of the 1,188 complaints, inquiries, and investigations examined, only 2% arose from injured clients.[11]

Ten years ago the State Bar of California formed a Public Protection Committee that examined UPL and paralegal situation (State Bar of California 1988). It conducted a study of UPL in other jurisdictions and reported that, of the three-fifths of the jurisdictions that responded, about half (17 of 32) indicated they did not keep statistics on the number of complaints received. Six of the 32 states reporting, however, stated that a number (unstated) of complaints alleging monetary loss were received.

Another report by the State Bar of California Commission on Legal Technicians (1990) states that "anecdotal evidence" was received describing actual or perceived harm in specific cases from the services of "legal technicians" or paralegals. The commission concluded that "training, experience, examinations, and a client security fund will greatly relieve the harm currently experienced from some legal technicians" and recommended that the profession be licensed.

Beyond the Christensen (1980), Rhode (1981), State Bar of California Public Protection Committee (1988), and State Bar of California Commission on Legal Technicians (1990) reports, no other studies report the nature, frequency, or severity of public citizen UPL complaints. This study thus contributes to the prior, limited research on the subject of UPL and its implications for the lawyer-nonlawyer relationship.

# THE PARALEGAL PROFESSION

## Segmentation

I prefer to call the paralegal occupation a profession because, whether or not it really is, that seems to be the conventional practice (Green et al. 1990; Schrader 1991). Paralegal professional organizations differ on several fundamental political and ethical issues (Starsky 1993), but they all consider their members part of a new profession (Farinacci 1991; Nemeth 1995; Miller and Urisko 1995; Cannon 1996)). Their activities and body of knowledge are referred to as "paralegalism" (Farinacci 1991; Nemeth 1995).

There are two roles that can be assumed by those considered to be paralegals, broadly defined, reflecting segmentation within this profession. First, those who work under supervision of an attorney are usually members of the National Association of Legal Assistants (NALA). It requires its members to comply with a code of ethics that mandates such supervision. They are usually called "legal assistants" or simply "paralegals." "Freelance paralegal" refers to paralegals who also work under attorney supervision, but as an independent contractor and not as an employee of a law firm (Nemeth 1995). Those, however, who elect to provide paralegal services directly to consumers (or "clients," depending on one's view of their professionalism status), are referred to as "independent paralegals," "legal technicians," or "self-help legal service providers."

This study focuses on independent paralegals and other nonlawyer practitioners who deal directly with the public and work without attorney supervision. Surprisingly, the literature on paralegals has ignored this distinction (Johnstone and Hopson 1967; Johnstone and Wenglinsky 1985; Miller and Urisko 1995; Nemeth 1995). Even the most recent *Occupational Outlook Handbook* states that "paralegals work directly under the supervision of attorneys" and omits any reference to independent paralegals (U.S. Department of Labor 1994). The few empirical studies about paralegals show, interestingly, that they not only are able to provide document preparation services in many types of equivalent to those of attorneys, they also provide comparable advocacy services in administrative law matters (Kritzer 1992; Kritzer 1998; Humphreys et al. 1992).

## Subordination

Subordination is common in both the legal and medical profession. Paralegals, physicians' assistants, laboratory technicians, and other professional occupational groups have arisen to assist the "pure professions," but they remain subordinate to them. The advantages of subordination to the dominant profession are: (1) it "enables extension of the dominant effort without division of dominant perquisites;" (2) it permits delegation of "dangerously routine work;" and (3) it "settles the public and legal relations between incumbent and subordinate from the start" (Abbott 1988, p. 72).

Subordination of paralegals began in 1968, when the ABA first recognized them by the formation of the Standing Committee on Legal Assistants that took upon itself the role of "approving" paralegal education programs. After the decision in *Goldfarb v. Virginia State Bar Assoc.* (1975), holding that attorneys were no longer exempt from antitrust laws, the ABA prevailed in a suit challenging its paralegal program approval process (*Paralegal Institute, Inc. v. American Bar Association*, 1979).[12]

After being introduced into the legal profession's workplace, paralegals learned that many of the problems presented to their supervising attorneys were fairly routine. In fact, that is why they were brought into the workplace.[13] Abbott (1988, pp. 68, 72) introduces the term "workplace assimilation" to refer to the situation where (1) there is too much professional work, (2) some of it is delegated to non-professionals, (3) boundaries between professional jurisdictions blur, and (4) a knowledge transfer occurs.[14] This accurately describes the situation when paralegals were introduced into the legal workplace. Abbot pointed out that the introduction of subordinate personnel is an "inherently uneasy settlement partly because it is undercut by workplace assimilation and partly because subordinates become absolutely necessary to successful practice by superordinates" (1988, p. 72).

The paralegal profession is now one of the "fastest growing occupations" in the United States (Miller and Urisko 1995, p. 4). In 1984, it was estimated that there were about 81,000 paralegals under attorney supervision (Johnstone and Wenglinsky 1985, p. 4). One source recently estimated that there are 150,000 legal assistants employed by private law firms, government, and the corporate sector, and 10,000 nonlawyers providing legal services to the public (Cannon 1996, pp. xix, 35). There are two primary national paralegal associations (National Association of Legal Assistants and the National Federation of Paralegal Associations) and local associations in almost every state, the District of Columbia, and the Virgin Islands (Statsky 1993; Schrader 1991).

Nationally, the bar is extremely concerned about the growth of independent paralegals (Samborn 1993; Podgers 1994). The American Bar Association's Commission on Nonlawyer Practice recently concluded its investigation of the phenomenon and recommended, to the dismay of many ABA members, that: (1) increasing access to affordable legal assistance in law-related situations is an urgent goal; (2) protecting the public from harm from persons providing assistance in law-related situations is also an urgent goal; and (3) when adequate protections for the public are in place, nonlawyers have important roles to perform in providing affordable access to justice (American Bar Association 1995, p. 4).

## Paralegals and UPL Enforcement in Arizona

Let us now turn to a closer examination of the Arizona bar's efforts to challenge the growth of independent paralegals. In comparison to other states, the Arizona situation is unique. The Arizona Supreme Court decision of *State Bar v. Arizona Land Title and Trust Co.* (1961, pp. 8-9), which imposed severe restrictions upon real estate agents, defined the practice of law as "those acts, whether performed in court or in the law office, which lawyers have customarily carried on from day to day through the centuries."[15] The court elaborated on this definition by giving the following examples of law practice:

> one person assisting or advising another in the preparation of documents or writings which affect, alter or define legal rights; the direct or indirect giving of advice relative to legal rights or liabilities; the preparation for another of matters for courts, administrative agencies and other judicial or quasi-judicial bodies and officials, as well as the acts of representation of another before such body or officer . . . rendering to another any other advice or services which are and have been customarily given and performed from day to day in the ordinary practice of members of the legal profession either with or without compensation (1961, p. 14).

After this case was decided, a campaign was launched by the real estate industry to enact a state constitutional amendment overturning it. This led to the passage through a voter referendum of a constitutional amendment permitting real estate agents "to draft or fill out and complete, without charge, and all instruments incident thereto including, but not limited to, preliminary purchase agreements and earnest money receipts, deeds, mortgages, leases, assignments, releases, contracts for sale of realty, and bills of sale" (*Arizona Constitution* 1962, art. 26, ¶1). This movement exemplifies the well-known populist sentiment among the Arizona populace.

In 1985, the likewise populist and generally anti-lawyer state legislature sunsetted the statute establishing the criminal penalty against practicing law without a license. This left the state without any enforcement mechanism against the unauthorized practice of law. Therefore, the state supreme court promulgated a court rule that prohibited unlicensed persons from practicing law: "No person shall practice law in this state . . . unless he is an active member of the state bar" (Arizona Supreme Court Rule 31(a)(3)). This rule, however, is part of a set of rules governing the legal profession that does not extend to nonlawyers.

This regulatory void resulted in an unprecedented growth of nonlawyer practitioners in the state. Two paralegal associations came into existence during the time of this study (the Arizona Paralegal Association [APA] and the Tucson Association of Legal Assistants [TALA]), both of which were comprised of attorney-supervised paralegals. The former president of the APA, however, informed me that some independent paralegals "may have slipped in" to their organization's membership.

Another association, the National Association of Independent Paralegals (NAIP), based in California, also has Arizona members. It, unlike the APA, does not require its members to abide by an ethics code that requires that they work under attorney supervision. In fact, the NAIP is really a private company that sells courses on, as their brochures states, "how to provide legal services without becoming a lawyer."

Soon after repeal of the UPL law, the Arizona State Bar began received numerous complaints against nonlawyer practitioners. I now turn to an examination of these complaints, and the action taken by the state bar to fend off the widespread encroachment—or "poaching," in Abbott's terms—of its members' legal practice jurisdictions.

## METHOD AND DATA ANALYSIS

The State Bar of Arizona generously provided me with the complete file in each of the 550 UPL complaints it received from 1988 through 1994.[16] These data have significance because they are derived from a period of Arizona history in which there was in fact no legal sanction for engaging in unauthorized practice of law.

The files consist primarily of correspondence and internal memoranda of state bar attorneys. Each file was coded so that one would be able to answer the following research questions: Who files UPL complaints? Against whom are they filed? In what areas of law are these nonlawyer engaged? What specific nonlawyer activities are complained of? And what, if any, actual harm is alleged in these complaints?

I also interviewed two state bar counsel, a member of the bar's UPL committee, several paralegals who were officials with the Arizona Paralegal Association and the Tucson Paralegal Association, as well as an official of the National Association of Independent Paralegals, about the Arizona UPL/paralegal situation.

While collecting data from the UPL files, I kept abreast of the bar's efforts to re-regulate UPL. This information was drawn from the state bar journal and local legal newspapers, as well as reports of the UPL committee and task force of the state bar.

## RESULTS

### Who Files UPL Complaints?

As shown in Table 1, attorneys filed almost 70% of UPL complaints with the Arizona State Bar for the period 1988 to 1994. None of these stated that they were

***Table 1.***   Who Files UPL Complaints?

|  | Number (%) |
| --- | --- |
| Attorney/law firm | 381 (70) |
| Individual | 68 (13) |
| Judge | 24 (4) |
| County attorney | 16 (3) |
| Paralegal (supervised) | 15 (3) |
| Government employee | 11 (2) |
| Private corporation | 7 (1) |
| Anonymous or unknown | 6 (1) |
| County bar association | 5 (1) |
| Insurance companies | 3 (1) |
| Miscellaneous* | 10 (2) |
| Total | 546 (100) |

***Notes:***   Percentages have been rounded.
*See text for description of miscellaneous complainants.

made "on behalf of" a client, although some were initiated as a consequence of contacts an attorney's client had with a nonlawyer practitioner.

The next largest proportion of complaints, almost 13%, were filed by individuals. This is perhaps the most surprising aspect of these data, inasmuch as it has been assumed that attorneys themselves are the only ones complaining of UPL (Rhode 1981).

Table 1 also shows a wide variety of other UPL complainants, including judges (4%), county attorneys (3%), paralegals working under attorney supervision (3%), and employees of government agencies (2%). Additional categories of complainants in the "miscellaneous" category of Table 1 who constituted less than 1% of all complaints include: county bar association, insurance company, private investigator, physician, court clerk, the state bar, and stockbroker.[17]

These data show that county bar associations and county attorneys appear to have played an active role in the filing of UPL complaints with the state bar. The state bar itself took action on only one case (against an attorney) during the five-year period under study. Interestingly, two independent paralegals, usually respondents themselves, also filed UPL complaints.

## Against Whom Are UPL Complaints Filed?

The lengthy history of UPL reflects a profession that has been, in both perception and fact, continually under attack by its competitors. These competitors, however, appear to vary by time and place. As noted earlier, the eighteenth-century UPL respondents appear to have been predominantly corrupt public officials. Modern-day Arizona UPL respondents, however, include a broad spectrum of persons, in different settings, and in different roles.

***Table 2.*** Against Whom Are UPL Complaints Filed?

|  | Number (%) |
|---|---|
| Independent paralegal | 207 (38) |
| Document preparation company (general) | 103 (19) |
| Individual | 64 (12) |
| Unlicensed attorney | 31 (6) |
| Accident consultant | 30 (6) |
| CPA/accountant | 18 (3) |
| Homestead assistance company | 12 (2) |
| Collection agency | 9 (2) |
| Tax service | 8 (2) |
| Miscellaneous* | 68 (12) |
| Total | 550 (100) |

**Notes:**   Percentages have been rounded.
            *See text for description of miscellaneous respondents.

Table 2 shows independent paralegals (the term they use to describe themselves in brochures, advertising, or their own letterheads) are respondents in about 38% of all UPL complaints filed. The next largest category of respondents, document preparation companies (19%), were counted separately where the complaint was against a company providing a document service rather than against a particular individual. Employees of these companies could have been categorized as independent paralegals because they do not act under attorney supervision, but they do not hold themselves out to the public as "paralegals." If document preparers were so categorized, the proportion of complaints against independent paralegals as such would increase from 38% to about 56%.

Somewhat surprisingly, individuals (12%) make up the third largest category of UPL respondents. These are not persons affiliated with a document preparation company, nor do they provide paralegal services commercially. Rather, they are alleged to have committed individual instances of lawyer-like acts to assist themselves, or another, with respect to some legal problem.

It should be noted that one respondent, an independent paralegal—whom I will call "Mr. Jones"—is the respondent in approximately 27% of the 550 UPL complaints examined. His modus operandi is to obtain from court records the names of respondents and defendants in divorce and contract cases. If court records show they are not yet represented by an attorney, he directs a letter to them offering his services.[18] The high frequency of complaints against him is due in large measure to his systematic method of marketing, which often comes to the attention of attorneys who ultimately represent these parties.

Unlicensed attorneys are the respondents in 12% of the UPL complaints. They are typically accused of filing court papers in Arizona courts from outside the state without being licensed in Arizona. Several complaints were also brought against unlicensed attorneys who reside in Arizona but were trained and licensed in another state. These sometimes arose from the assistance they provided to someone in a federal proceeding, such as bankruptcy. The bar, however, takes the position that it has no jurisdiction to regulate the practice of law in the federal courts.

Numerous other categories of respondents constituting less than 1% of the total were the subject of UPL complaints. These included persons in the private sector (e.g., employees of a private corporation, an insurance company, a private investigation company, a law firm, a title company, and a nonprofit organization) and in the public sector (e.g., a state government employee, a former Justice of the Peace, and a court clerk). Complaints also include the following quasi-legal respondents: a Native American attorney (i.e., licensed on an Indian reservation by not by the state bar), a law school graduate, a jailhouse lawyer, and a mediator.

Some respondents have carved out areas of specialization. For example, "accident consultants" are those who offer their services to negotiate personal injury claims on behalf of an individual. These are distinguished from "damage appraisers," who specialize in the assessment of the value of damages to property. Others provide "homestead assistance" by way of preparation and filing of homestead (property tax) exemptions or tax services. There are also "immigration specialists," "bankruptcy representatives," "special education advocates," and "farm advisors" who offer specialized legal services. Complaints against these specialists are, however, rare, which may be a function of their low numbers in the community (or the satisfaction of their customers).

Some traditional categories of respondents were expected, such as accountants, collection agencies, insurance company employees, property management agents, and private investigators. The remainder of the respondents appear to come from a variety of walks of life, without any indication that the acts they are alleged to have committed are part of a commercial enterprise. These include government employees, a law school graduate, a court clerk, a massage therapist, and a nonprofit organization's employee.

## What Areas of Law Were Involved?

The areas of law in which the alleged UPL activities occurred are set forth in Table 3. The highest frequency of cases involve contract law (18%); but if the complaints against Mr. Jones are omitted, contract cases fall to seventh place (4.5%), thus showing the impact he had on the overall UPL caseload.

The broad range of legal subjects covered in the UPL allegations is striking. The predominant areas of law in which the UPL respondents are engaged as reflected in Table 3 are those of the typical sole or small firm practitioner—that is, divorce, wills and trusts, personal injury, bankruptcy, general consumer law, debt collection, real estate, and landlord-tenant. Likewise, the areas of homestead exemptions, income tax, powers of attorney, and insurance are all personal plight areas of law.

The miscellaneous category reflects an even broader spectrum of areas of law involved in the alleged UPL activity. In this category (which includes categories

***Table 3.***   Area of Law in which Alleged UPL Occurred

|  | Number (%) |
|---|---|
| Contract | 95 (18) |
| Divorce | 88 (17) |
| Wills and Trusts | 58 (11) |
| Personal injury | 56 (11) |
| Bankruptcy | 51 (10) |
| General consumer law | 40 (8) |
| Debt collection | 20 (4) |
| Real estate | 16 (3) |
| Landlord-tenant | 13 (3) |
| Homestead exemption | 11 (2) |
| Income tax | 8 (2) |
| Power of attorney | 8 (2) |
| Insurance | 7 (1) |
| Miscellaneous[*] | 56 (11) |
| Total | 527 (100) |

***Notes:***   Percentages have been rounded.
[*]See text for descriptions of miscellaneous areas of law.

with complaints numbering five or less (less than 1% of the total), we find some cases involving areas of law not typical within the domain of personal plight lawyers. These include corporations, patents, franchise, and securities. The majority of the other cases in the miscellaneous category, however, do reflect personal plight areas of law. These include civil rights, criminal law, child support, repossession, property tax, special education, unemployment compensation, immigration, conservatorship and probate, juvenile law, and injunctions. In addition, two other types of complaints included in the miscellaneous category that are not associated with an area of law as such are false claims to be an attorney and promoting money-making schemes.

## What Was the Activity Complained of?

The wide variation in types of UPL complainants, respondents, and areas of law affected is only exceeded by the variation in the kinds of activities alleged to constitute the unauthorized practice of law. These are listed in Table 4.

The most frequently complained-of activity is offering to negotiate a settlement of legal claims. Not surprisingly, this category is the one most affected by the activities of the aforementioned Mr. Jones, the independent paralegal who directs letters offering his services to persons whose names he obtains from court records. If we exclude cases against him, offering to negotiate falls to ninth place in the frequency ranking, or from 21% to 3% of all complaints. There are, however, other "accident consultants" and "damage appraisers" who also offer to settle claims and who are the subject of UPL complaints, but their complained-of activities involved actual negotiations or attempted negotiations rather than offers to negotiate.

*Table 4.*   Nature of Alleged UPL Activity

|                                                              | Number (%) |
|--------------------------------------------------------------|------------|
| Offer to negotiate settlements                               | 117 (21)   |
| Using attorneys' powers, while not licensed                  | 92 (17)    |
| General document preparation                                 | 87 (16)    |
| Giving legal advice (no misrepresentation)                   | 66 (12)    |
| Preparing pleadings                                          | 55 (10)    |
| Preparing living trusts                                      | 33 (6)     |
| Impersonating attorney                                       | 18 (3)     |
| Failure to return funds or documents (not UPL)               | 17 (3)     |
| Failure to render services due (not UPL)                     | 15 (3)     |
| Giving legal advice (with misrepresentation)                 | 10 (2)     |
| Document preparation (homestead exemptions)                  | 9 (2)      |
| Court appearance offered or made for another                 | 6 (1)      |
| Miscellaneous[*]                                             | 25 (4)     |
| Total                                                        | 550 (100)  |

**Notes:**   Pecentages have been rounded.
[*]See text for descriptions of miscellaneous activities complained of.

The second highest category is using the powers of an attorney while not licensed to do so (17%), which has many subcategories. These include: (1) appearing for another in a court or administrative agency proceeding; (2) representation out of court through a power of attorney form, or a form used by an administrative agency to designate a representative; (3) actual claims negotiations (as distinguished from offers to negotiate); (4) requesting medical records; and (5) using "attorney" below a signature in a published letter-to-the-editor.

As expected, general document preparation is the third most frequent activity underlying UPL complaints (16%). This category refers to activities such as preparation, or offers to prepare, a variety of common consumer documents, including bankruptcy schedules, contracts, powers of attorney, tax returns, and so forth. The category is distinguished from specialty document services included in the miscellaneous categhory, which include preparation of living trusts, homestead exemptions, corporate papers, and immigration and lien documents.

Interestingly, the fourth most frequent UPL activity is giving legal advice to another when not a licensed attorney without misrepresentation as to one's status (12%); that is, the complainant does not allege that the respondent misrepresented his or her status as an attorney. This category is in contrast to the act of giving unauthorized legal advice *with* misrepresentation of one's status, a subject of only 10 complaints during the period studied. In the category of giving another legal advice, one also finds the rare instance of a complaint based upon a claim that the respondent's acts were committed negligently. These include cases of independent paralegals giving erroneous legal advice, or failing to provide legal advice when the complainant expected it would be provided.

The preparing pleadings (i.e., court papers) category includes not only preparation of court forms, but also writing of letters on behalf of another to a judge, and submitting letters on behalf of another to an administrative agency. The preparation of living trusts category includes: (1) sale of living trusts by mail order, (2) preparation of codicils to living trusts, (3) preparation of probate forms, and (4) preparation of tax forms.

The category of impersonating an attorney includes impersonation by way of a false letterhead, telephone book entry, business, or verbal statement. The category of failure to return funds or documents in not technically considered unauthorized practice of law by the state bar. It includes, however, such acts as failure to pay a debt due to the complainant, retaining settlement proceeds, and withholding of personal property or documents. Likewise, failure to render services due is also not considered UPL, since it is more in the nature of a breach of contract.

In the miscellaneous category we find numerous other alleged acts of UPL, including: (1) filing of pleadings by an out-of-state or foreign attorney, (2) offers to split fees with non-attorneys, (3) signing pleadings as attorney when not licensed as one, (4) preparation of liens and incorporation and immigration documents, (5) being a jailhouse lawyer, and (5) "scouting" clients for an attorney. Additional non-UPL complaints received in this category include misrepresenting of services one is able to perform, proposing a money-making scheme (e.g., chain letters), falsely representing oneself as being with a government agency, failing to return funds (e.g., refusal give a refund to a dissatisfied customer/client), and failure to render services promised or paid for.

## Was Specific Harm Alleged?

As shown in Table 5, 71% of the UPL complaints received by the state bar contain no allegation of specific harm. Where money damages were alleged, these complaints are reported in separate categories at $500 intervals.

The second category, facing unnecessary litigation with unspecified money damages, consists of cases alleging the necessity of future corrective litigation without allegations of specific monetary harm. This is a large category that includes: (1) having to defend future, anticipated litigation, (2) having to pay for a debt that was supposed to have been discharged in bankruptcy, (3) being required to reinstate a corporation that had been dissolved due to the respondent's neglect in filing corporate reports, (4) receipt of a document that did not achieve a legal objective, (5) being required to refile a suit or administrative proceeding that was dismissed for failure to prosecute, and (6) being involved in further legal proceedings to undo a default judgment that was entered as a consequence of the respondent's failure to act. While no specific monetary losses are alleged in these complaints, there will presumably be some costs associated with the future litigation alleged to be necessary as a consequence of the respondent's neglect or omission.

The category of unspecified monetary loss with no allegation of future damages includes losses in unstated amounts as a result of such things as being required to

***Table 5.*** Harm Alleged in UPL Complaints

|  | Number (%) |
| --- | --- |
| No harm alleged | 389 (71) |
| Facing unnecessary litigation (unspecified money damages) | 42 (8) |
| Unspecified money damages (with no future damages) | 30 (6) |
| $101-$500 | 17 (3) |
| Erroneous legal advice (no harm alleged) | 14 (3) |
| $501-$1,000 | 9 (2) |
| Delay of litigation | 9 (2) |
| Denial of access to court | 9 (2) |
| $1,001-$5,000 | 7 (1) |
| Unspecified money damages (with future damages expected) | 7 (1) |
| Intimidation | 7 (1) |
| Miscellaneous* | 10 (2) |
| Total | 550 (100) |

***Notes:*** Percentages have been rounded.
* See text for description of miscellaneous harms alleged.

hire an attorney to complete or correct a matter started by the respondent, and having to pay for legal services to secure a refund of money paid the respondent after a request for a refund was denied.

Other types of harm noted in Table 5 may also involve some economic cost notwithstanding the fact that damages were not specifically claimed. For example, the category of delay of litigation includes delay of: (1) execution of judgments (i.e., collection of a debt), (2) personal injury settlements, (3) divorce, (4) injunctive relief, (5) a "stay order" from the bankruptcy court (ordering creditors to cease and desist from collection activities), and (6) obtaining a patent. The category of denial of access to the court may also involve indirect economic costs and includes such matters as: (1) complaints about the conduct of a jail official regarding notarization requests from prisoners, (2) nonfeasance in a legal matter, (3) dismissal of a bankruptcy petition, (4) loss of a trial transcript, and (5) nonappearance at a court hearing after receiving payment for such an appearance.

There are other damages that do not necessarily involve monetary loss; these include: (1) giving erroneous (but harmless) legal advice, (2) intimidation (which includes becoming "scared" or "frightened" due to a respondent's conduct), (3) breach of confidentiality, and (4) refusal to return legal papers.

# THE JURISDICTIONAL BATTLEGROUND IN ARIZONA

The Public, Legal, and State Settings

A professional association such as the state bar is known to have multiple functions:

It mediates between practitioner and profession, on the one hand, and, on the other, between practitioner and social environment, of which the most important elements, apart from clientele, are allied occupations and professions, the universities, the local community, and the government. . . The association also works to justify the scope, not infrequently the expanding scope, of the jurisdiction of the profession and to safeguard it against the rival claims of neighboring professions and technical occupations (Merton 1982, p. 207) .

Abbott posits that the jurisdictional conflicts between professions (or emerging professions, and other organized groups) take place in three settings: the public opinion, legal, and state settings. He does not, however, articulate the manner in which these conflicts arise and develop over time, except to suggest that different types of settlements are reached between those in conflict. This case study allows us to examine the manner in which these jurisdictional conflicts evolve, the politics of this process, and the internal and external interactions involved in these conflicts. In particular, this study gives us a deeper understanding of the role played by a professional association, in this case the state bar, both in the jurisdictional conflict between professions and their "license and mandate" claims (Hughes 1958; Akers 1968).

A genuine hue and cry by members of the state bar about nonlawyer practitioners arose during the period 1988 to 1994, as reflected by the steady stream of UPL complaints that poured into the state bar offices, and the numerous articles and letters to the editor in the state bar journal and the state's principal legal newspaper. Bar officials, lacking any UPL enforcement mechanisms, became extremely frustrated in their inability to stop encroachment by nonlawyer practitioners.

In one interview, the chief bar counsel described the history of UPL enforcement in Arizona, including the "little known fact," as she put it, of the disbanding of the state bar's UPL committee in 1988. Another bar committee was formed in 1990 to review the situation again, she advised, but it had done "virtually nothing." The committee was charged with ascertaining whether the public was being harmed by UPL, but it had only collected anecdotal data. A task force chaired by a nonlawyer was then created in 1991; it was concerned with the legal issue of whether the bar had jurisdiction over nonlawyers. The bar counsel insisted that the state bar did have jurisdiction over them under the *State Bar of Arizona v. Arizona Land Title Co.* decision cited earlier that found real estate agents guilty of UPL.

The chief bar counsel told me that "lots of lawyers get fired up" about UPL, but the bar keeps her from taking legal action "if its only an economic issue" that is involved. The bar wants the problem to "play itself out" and will pursue a legal action only if it has a case involving "demonstrable harm" to the public. In essence, bar officials believed that a legal challenge to suspected UPL activity, if unsuccessful, would "make things worse." At the same time, one bar letter to a UPL complainant acknowledged the existence of unmet legal needs in the state: "There is a general recognition that lawyers are falling short of providing needed legal services to members of the public at the lower end of the economic specter" [sic].

An assistant bar counsel told me that "all the UPL cases are still pending. We are still in an information gathering stage and are handling UPL complaints quite informally." She added, "We don't take action on ads that are sent to us by attorneys . . . because we don't look at ads as threatening." No legal action had yet been taken because "we are still assessing our jurisdiction on the matter."

The practitioners' frustration is evidenced by one attorney's letter to the chair of the state bar UPL Committee in which he described the activities of a local non-lawyer financial planner who was soliciting limited partnership and trust business from doctors. He wrote:

> The amount of UPL is getting out of control, with the pension plans, corporate work, and estate plans being done by nonlawyers all over town. . . . In addition, you are aware of the Arizona Banking Association proposals to allow trust companies to prepare wills and trusts. . . .When I was on the [UPL] Committee I contacted a number of the pension actuaries to advise them that they could not prepare pension plans with the threat of civil and criminal suit. Hopefully, this is still possible.

In fact, the state bar underwent a form of "identity crisis" after the loss of the UPL criminal statute; it began to doubt its own (legal) jurisdiction to regulate UPL. Responses to many of the complaints reflect the state bar attorneys' frustration with their inability to bring litigation against the UPL respondents. Jurisdiction to enforce the state supreme court's rules of professional responsibility, with which licensed attorneys are required to comply—and which included a prohibition upon an attorney "aiding" the unauthorized practice of law—was clear. But these rules were unenforceable against nonattorneys. Unable to invoke the power of the court to regulate UPL, bar staff felt increasingly emasculated and were reduced to making feeble written threats to stop the alleged UPL activity. For example, a January, 1988, memorandum from a bar legal assistant to investigator regarding a certain document preparation company stated:

> It is becoming increasingly apparent to all of us, I think, that dealing with UPL files concerning advertisements for document preparation is an exercise in frustration. Phone calls simply elicit the same canned response. Perhaps the next step should be a personal call without an appointment and the above named business might be a place to start.

A bar response to another UPL complainant articulated a new strategy being undertaken: "Our research has led us to conclude that it is the Supreme Court, and not the Legislature, which has the authority to regulate the practice of law, and this is why we are proceeding to draft new Supreme Court rules instead of lobbying the Legislature for a new statute . . . The unauthorized practice of law problem is complex and requires the Task Force to balance the need to provide legal services to persons with low incomes and adequate consumer protection." State bar officials urged attorney UPL complainants to collect data on harm to the public from nonlawyer practitioners: "We intend to use the accumulated complaints against the nonlawyers to demonstrate the necessity of the regulations that the Task Force ultimately will propose to the Board [of Governors of the State Bar]. Your letter certainly will help us document the problem."

Attorneys continued to bombard the state bar and its journal with letters complaining of nonlawyer practitioners:

> Each week I hear a lawyer comment about the fact that it is becoming more and more difficult to practice law because paralegal firms are springing up around the state usurping legal business away from licensed practitioners. . . . [I]t is becoming more and more difficult to make money at the practice of law . . . We pay dues to the Bar Association to help preserve, protect, regulate the practice of law, but the State Bar Association appears to sit idly by as more and more of the legal practice is being whittled away by paralegals. . . . It seems to me that many qualified and motivated individuals will pass up law school and thus avoiding [sic] high tuition and three years of no income (*Arizona Attorney* 1993, p. 7).

The state bar also made efforts to prepare for a jurisdictional battle in the state setting, that is, the legislature. Bar responses to attorney complaints urged lobbying: "We urge voters to advise the legislature of the need for criminal penalties for such conduct and the harm that occurs from nonlawyers practicing law. Without some form of penalty, nonlawyers will continue to provide legal assistance to the public, without the training and accountability required of lawyers." In the meantime, informal threats of sanctions continued to be made by the state bar to UPL respondents: "Further action of this nature may result in action being taken by the State Bar. To avoid action by the State Bar, you are hereby required to cease and desist from representing yourself or acting as an attorney or lawyer in this state."

## Paralegal Segmentation

In speaking with an official of the Tucson Association of Legal Assistants, I first learned of the segmentation of the paralegal profession between those who belonged to her organization or the Arizona Paralegal Association—whose members are required to work under attorney supervision—and the independent paralegals who refused to be so bound. The UPL problem is one that "puts our profession in a bad light," she said. While she did not consider document preparation services "in the same category" as paralegals, she said "they really scare me. . . . Legal questions must come up. How do they answer them?"

In checking with officials at the National Association of Independent Paralegals (NAIP), which at the time had over 2,700 members nationwide, I learned that they were well aware of the lack of UPL enforcement in Arizona. This organization markets courses that train people to operate their own paralegal services. Included in their training is instruction on what constitutes unauthorized practice of law. An NAIP official told me that the lack of enforcement against alleged UPL "does not show me that Arizona is liberal, they're just lazy!"

In 1993, a bill was introduced in the State Legislature reinstating the criminal UPL statute. It was overwhelmingly defeated. According to the special bar counsel on UPL matters, "The defeat of the bill was due in substantial part to the well-organized and effective lobbying efforts of nonlawyer document preparers. It is believed that those same interests are planning to propose legislation this year that would effectively open the courthouse doors to many nonlawyer document preparers and legal assistants—without any competency standards" (Shely 1994, p. 12).

After holding public hearings, the UPL task force issued a report in 1993 that recommended that the question of regulation of paralegals, and the form of that regulation, should be studied. This was somewhat of a surprise, in that it indicated a less entrenched perception of the UPL "problem" than what might have been expected.[19] The task force found that: (1) there was a growing UPL problem in Arizona, (2) nonlawyers were filling a gap in legal services to the public, and (3) document preparers were not going to go away.

The state bar's board of governors approved the task force's report and the "concept" of regulating nonlawyer practice, over the objection of many sole practitioners (Sallen 1993). The report of the meeting at which this decision was made indicates that the objectors, primarily sole and rural practitioners, were persuaded by a supreme court justice that spoke in favor of the proposal. He told the governors, "To do nothing is no solution. . . . If [nonlawyers are] going to be out there, isn't it better to bring them in, license them, and hold them to a code of ethics? . . . I'm afraid if we don't do anything, things will happen in the Legislature very detrimental to the profession" (Sallen 1993, p. 12). Several of the bar governors representing sole and rural practitioners ultimately stated that they would join other bar leaders from the corporate practice sector and vote for the proposition because of the analogous situation that exists in the medical community; there, "licensing started with doctors and expanded to include other practitioners." As one governor put it, "Everybody's licensed; there's a clear definition of what they can and can't do" (1993, p. 22).

After the vote, the controversy regarding the state bar's adoption of the "concept" of nonlawyer practice continued to rage (Simmons 1994; Kreutzer 1994; Greacen 1994). Nevertheless, bar counsel proceeded to draft a set of rules to regulate nonlawyers. The proposed rule would have: (1) established a new UPL enforcement mechanism, (2) licensed "Nonlawyer Legal Technicians," or NLLTs, and (3) expanded practices available to paralegals supervised by attorneys (the latter probably reflecting the quid pro quo to appease the attorney-supervised paralegals).[20] Reflecting the tension in bar-legislative relations, the cover page of the document submitted by the state bar establishing this licensing scheme states: "Proposed Amendments to the Rules of the Supreme Court To Be Implemented Only if the Legislature Passes an Unauthorized Practice of Law Misdemeanor Statute" (State Bar of Arizona 1994).

Some sole practitioner attorneys continued to object to the new proposed licensing rule for independent paralegals. As to the purported justification for paralegal licensing based on unmet legal needs, one attorney wrote the following letter to the editor of the legal newspaper, which evidences the acrimony between the sole-practitioner and corporate (defense attorneys) segments of the bar:

> Why are we being asked to bear the brunt of the proposals to provide legal services to the poor? If the goal is to make legal services more affordable, why shouldn't the large firms also shoulder some of this burden? I have an idea! Instead of permitting legal assistants to represent clients in court, why not permit claims adjusters to make court appearances for insurers in defending personal injury cases. With a little training and experience most would probably do very well in the courtroom on the run-of-the-mill rear-ender. Just think how much the insurance companies could save in defense costs if their own nonlawyer employees could represent them in personal injury cases. . . . What about removing the onerous, but time-honored restrictions on banks, utilities and other corporations from representing themselves in legal proceedings? (Levine 1994, p. 2)

Despite the bar's efforts, the State Legislature refused to enact a new criminal law prohibiting UPL. As reported by the bar journal, the bill again never made it out of committee, once again due to "the overwhelming opposition to it coming from independent paralegals" (*Arizona Attorney* 1995, p. 35). The same story, however, reported that a bill authorizing agents of business owners who are nonlawyers to represent them at unemployment compensation hearings did pass: "The Bar opposed this legislation on grounds that it was a legislative attempt to amend Supreme Court Rule 31 [the UPL rule]. . . . In spite of this, however, the bill has now been signed by the Governor."

# CONCLUSIONS

Abbott (1988) posits that the history of the development of professions cannot be understood in terms of identifying particular professional attributes or studying the process of professionalization. Instead, we must identify the actual work of professions, examine the manner in which they make jurisdictional claims to it, and how they fend off encroachments to the boundaries of these claims. Professions are to be treated at three general levels: (1) the "system of professions," being a "structure linking professions with tasks (movements of any one profession with respect to these tasks affecting others)"; (2) differentiation within the professions themselves, "below" the systems level; and (3) the effects of social and cultural forces upon individual professions under certain conditions. This study has enabled us to examine interactions at these different levels as Abbott theorized them.

In this case, the first level is reflected by the emergence, growth, and competition of paralegals who are able to provide the same services to the public in personal plight or consumer areas of law. The second is reflected first by the division of labor within the legal profession, with sole practitioners and small firms handling the same types of cases as are offered by independent paralegals. Differentiation is also observed in the paralegal profession itself; while some paralegals work under attorney supervision and follow a code of ethics, independent paralegals do neither. As to the third level of treatment of professions, social and cultural factors operate to encourage the growth of alternative legal services and self-representation.

Professions, according to Abbott, make three types of claims: (1) to classify something as a "problem" (to "diagnose"); (2) to reason about it (to "infer"); and (3) to take action on it (to "treat"). These "modalities of action" embody the "essential cultural logic" of professional practice, and "it is within this logic that tasks receive the subjective qualities that are the cognitive structure of a jurisdictional claim" (1988, p. 40). In this case, the essence of the practice of law is the diagnosis of problems. Attorneys are paid to reason about such problems as well as to "treat" them.

In this case study of lawyer-nonlawyer relationships in Arizona, we have observed evidence of two types of jurisdictional claims. Excess jurisdiction, in which the legal profession's work jurisdiction is expanding relative to its ability to satisfy a demand, exists in the form of unmet legal needs. Population growth always brings about a greater demand for personal legal services, but the growth of corporate legal business is greater than that of personal legal business (the primary domain of independent paralegals). The cost of legal services and the populist and anti-lawyer sentiment of Arizona residents contributes to the bar's apparent inability to satisfy unmet legal needs and to the growth of nonlawyer practice.

The second form of jurisdictional invasion articulated by Abbott, where settled jurisdictions are invaded by groups providing equivalent legal services, was also observed. Compared to other states, the case of Arizona is unique; the absence of any criminal sanction for UPL, the state bar's self-doubt about its jurisdiction over nonlawyers, and the respect these paralegals have gained in the state legislature, resulted in what Abbott would characterize as a "jurisdictional standoff" (1988, p. 73). The latter is a situation in which there is a "division of the jurisdiction into functionally interdependent but structurally equal parts" and where two groups "hold full jurisdiction in a particular task area."

In Arizona, this concurrent jurisdiction appears predominantly in the areas of family law and other personal plight practice areas. However, we have also observed 32 categories of respondents who are the subject of UPL complaints. Complaints against independent paralegals and document preparation companies are the most frequent, but other respondents include accountants, collection agents, insurance agents, and property managers. The wide range of legal practice areas in which the alleged UPL activities arose, and the numerous groups and individuals named in these complaints, establish the extent to which paralegals have encroached, and will continue to encroach, upon the jurisdictional link between the legal profession and its work. In addition, the data include complaints not only in expected practice areas such as divorce, wills and trusts, and bankruptcy, but also in such unexpected areas as patents, franchise, school law, and securities. Therefore, absent some form of settlement of these jurisdictional conflicts through a regulatory mechanism or an evolving division of labor, the lawyer-nonlawyer jurisdictional battles will continue.

Halpern (1992, pp. 1015-1016), in a study of the medical profession that recounts its jurisdictional conflicts with ancillary professions, refines Abbott's thesis by suggesting that "the dynamics among a profession's internal segments are critical in shaping its cross-professional boundaries. An emerging specialty's success at subordinating ancillary workers is contingent upon support from established segments and the broader professional community. . . . [W]hat happened between intraprofessional segments is also vital to a system of professions." In this study, we have seen such intraprofessional collaboration between the corporate and sole practitioner segments of the legal profession in its effort, albeit unsuccessful, to regulate and subordinate independent paralegals. In light of the broad spectrum of practice jurisdictions found in this study, the corporate segment of the legal profession may become increasingly interested in what has heretofore been a battle fought by the sole practitioner and small firm segment of the profession. The possible increased unification of the hemispheres of the legal profession portends increased UPL enforcement.

The Arizona conflict has taken place in all three settings of the system of professions: public opinion, the legal system, and in the state (legislative) settings. The public there, as elsewhere, is subject to the increasing costs of litigation. And many of those who can afford counsel elect not to do so for what they know are "simple," routinized personal legal matters (Abbott's "common cases"). Statements in bar correspondence and memoranda examined in this study concede as much.

The process of inference in a profession's work also affects its jurisdictional vulnerability. Inference is undertaken "when the connection between diagnosis and treatment is obscure." Abbott asserts that "the degree to which inference predominates, rather than the routine connection of diagnosis and treatment without inference," is the "most important aspect of professional inference in determining jurisdictional vulnerability." Thus, where a profession's work consists of tasks in which little or no inference is required by the professional, these tasks are an "obvious target both for poaching by other professions and for compulsory deprofessionalization by the state." The only defense to this is the argument "that such routine cases may be indistinguishable from the less routine ones" (1988, p. 51).

Inference in the context of the legal profession consists of the interpretation and application of law to a particular problem. Most would agree that inference is essential in the treatment of some legal problems. State supreme courts contribute to the intellectual jurisdiction of the legal profession by establishing and perpetuating vague and almost unenforceable definitions of the "practice of law." These contribute to the belief that inference is essential in order to connect the diagnosis of any legal problem and its treatment. In other words, these vague definitions of the practice of law make it appear that nothing is routine in the diagnosis and treatment of a legal problem, and that inference by a trained attorney is essential to treat all legal problems. It is clear, however, from data collected in Arizona and the recent growth of the independent paralegal profession, that there are many routine legal problems that require no "inference" for their treatment.

Ironically, many courts are now using paralegals to act as court "facilitators" to guide pro se litigants through the legal process in the absence of affordable counsel, adequate pro bono programs, or other civil legal services for the indigent and middle-income people (American Bar Association 1995). And, in this case, the populist, anti-lawyer Arizona Legislature was greatly influenced by the lobbying of independent paralegals, who succeeded in opposing reinstatement of a criminal sanction against UPL on grounds that they can provide equivalent legal services at a much lower cost.

The vague definitions of the practice of law are relevant to Abbott's conceptualization of abstract knowledge, which he describes as "the foundation of an effective definition of professions . . . knowledge is the currency of completion" (1988, p. 102). As jurisdiction expands and as the ideas unifying it necessarily become more abstract, jurisdiction attenuates. Paradoxically, however:

> Too little abstraction can make a jurisdiction weak as well. . . . Expert action, without any formalization, is perceived by clients as craft knowledge, lacking the special legitimacy that is supplied by the connection of abstractions with general values. While clients appreciate the simple case that shows them how professional knowledge works, they will not treat as professional a skill whose knowledge is *all* obvious to them. (Such a skill is also likely to be easily routinizable). . . . The higher the level of dominance, the more abstract the ideas holding the dominant jurisdictions together. This abstraction ultimately weakens the ideas' hold on the jurisdictions generally, and gives them less ability to define specific, peculiar jurisdictions. Thus, paradoxically, the bigger the dominant professions, both in structural and cognitive terms, the more numerous the jurisdictions open for attack. . . . [E]xcess expansion will lead to increased poaching and consequent division (1988, pp. 103-10).

The later aptly describes the situation of the legal profession and its maintenance of vague definitions of the practice of law. A certain degree of vagueness, as Abbott points out, is necessary to maintain the status of a profession, but expert knowledge that is too abstract, in the face of the public's knowledge that many of a profession's tasks are in fact routine, attenuates the profession's jurisdiction. This is, in fact, happening in the case of the legal profession; it now finds itself in somewhat of a conundrum.

The state bar in this case was on the front lines of the battle for the public and the state's acceptance of the legal profession jurisdiction as against nonlawyers. The data regarding public harm show that 71% of the UPL complaints failed to allege any harm from the acts complained of. Only 7% included a specific money damage figure, but other forms of damages did involve unspecific economic losses, including delay of litigation, denial of access to the court, the anticipation of unnecessary litigation, or other legal action necessary to correct a problem or accomplish a legal object. While attorneys comprised about 70% of all complainants, individuals filed 13% of the complaints.

These data can be interpreted in different ways, as the proverbial half-glass of water can be viewed as half empty or half full. On the one hand, they could lend support to the argument for (re)regulation of the practice of law in Arizona so as to exclude all of these UPL respondents from the legal profession's jurisdictions. Alternatively, they may be used to argue for the necessity of a means of certification, licensing, or regulation of nonlawyers on grounds that the "train has left the station," there is "no turning back," and attorneys can no longer maintain their traditional monopoly on the practice of law. Finally, this study lends support to the deprofessionalization theorists, who had predicted the waning of professions due to increased levels of public education, routinization of expert knowledge, the consumer movement, and other social forces (Haug 1973; Klegon 1978; Rothman 1984).

## ACKNOWLEDGMENT

The author wishes to publicly thank the Arizona State Bar for granting access to its unauthorized practice of law complaint files. I also thank my colleagues, Ray Michalowski at Northern Arizona University and Helena Lopata at Loyola University Chicago, for their interest and encouragement of this project.

## NOTES

1. Friedson (1994, p. 133) argues that the deprofessionalization thesis no longer provides a satisfactory explanation for the development of professions because "many of the [consumer] movements have simply collapsed; others have become so conventional that they are barely recognizable; and still others, if they have not disappeared entirely, have at least become considerably attenuated and altered by countermovements." The nature and extent of deprofessionalization in the legal profession found in this case, it will be shown, undercuts Friedson's criticism.

2. Abbott explains that his theory has "tried to handle what I regard as the classic problem of interactionism—its inability to explain the evident stability of many interactions over time . . . [by demonstrating] several layers of interaction, each operating at a different speed, such that the slower ones afford stability to the elements that are negotiated in the faster ones" (1988, pp. 112-113).

3. The most comprehensive historical account of the origin of the term can be found in Kimball (1992).

4. An "external source" of "system disturbances" is "enclosure," in which "a group claims jurisdiction over a task previously common to a number of professions" (1988, p. 91).

5. Currier (1991) administered a survey to attorneys and paralegals and asked them to state whether certain hypothetical actions by a paralegal constituted UPL. She reached the following conclusion:

There is much confusion and little agreement as to what is the practice of law. The definitions are generally unworkable, the cases contain only the most egregious fact patterns, and the survey results show a general lack of consensus. Nonetheless, as paralegals continue to increase both in numbers and in what they are trained and able to do, the question of what constitutes the unauthorized practice of law will continue to present itself as an ongoing problem. Until this problem is resolved, it will serve to hinder the growth of the profession as attorneys refuse to delegate appropriate tasks for fear their employees will be committing unauthorized practice of law. On the other hand, if paralegals are allowed to do more than they should, there looms the possibility of attorney discipline, malpractice claims, and even criminal sanctions against the paralegal. Therefore, it is imperative that we face the problem and develop clearer standards so that the role for paralegals can continue to expand, thereby offering attorneys the opportunity to provide quality legal service at a reasonable cost (1991, pp. 49-50).

Note that this author refers in the last part of her remarks to the type of paralegal that works directly under the supervision of an attorney. As later described, the data I report consist overwhelmingly of UPL complaints made against the independent paralegals, document preparers, and other nonlawyer practitioners who elect not to work under an attorney's supervision.

6.   The authors of that study, which reports data from interviews of almost 800 Chicago lawyers in 1975, state that "we know of no reason to believe that its [Chicago's] bar is unrepresentative in fundamental ways of those in other large American cities with diversified economies (1982, p. 18).

7.   Recently, these researchers found that the real income of sole practitioners has continued to decline, that demand for corporate law services has continued to grow faster than personal law services, that the bar is even less cohesive than it was in 1975, and that the two hemispheres of legal practice which they found in 1982 are now changing into "small clusters," "hemispheres" no longer a term that accurately captures the current reality (Heinz et al. 1997).

8.   Carlin (1962, pp. 207-208) provides the following analysis of the work of what he refers to as the "individual practitioners," or what Heinz and Laumann (1982) call the "personal plight" lawyers:

[They are] rarely called upon to exercise a high level of professional skill. . . . Most matters that reach the individual practitioner—the small residential closing, the simple uncontested divorce, drawing up a will or probating a small estate, routine filings for a small business, negotiating a personal injury claim, or collecting on a debt—do not require very much technical knowledge, and what technical problems there are, are generally simplified by the use of standardized forms and procedures. . . . As a result, the legal work of the individual practitioner lawyer is, in most instances, reduced to a fairly routine, clerical-bookkeeping job—the very kind of job which many nonlawyers and lay organizations are as well, if not better, equipped to handle than the lawyer. The considered success that such lay groups have had in taking away this type of business from the lawyers has led to some bitter controversies.

9.   Originally, attorneys were not necessarily trained in law. They were "largely traders, factors [agents or brokers], and speculators and laymen of clear penmanship and easy volubility, whom parties employed to appear and talk for them in the courts. The few persons who acted as professional attorneys were at first mostly pettifoggers [a lawyer whose methods are petty underhanded, or disreputable], or minor court officers such as deputy sheriffs, clerks, justices, who stirred up litigation for the sake of petty court fees" (Warren 1966, p. 5).

10.   A more detailed discussion of the historic ebb and flow of UPL enforcement may be found in Christensen (1980).

11.   Rhodes's (1981) study was criticized by Munro (1990), who argues that Rhodes' study suffers from, among other things, certain methodological flaws, including: (1) sample size too small, (2) insufficient complaint data, and (3) a focus on complaints filed by clients themselves, when bar association or even attorney complaints may have been made on behalf of an injured client (1990, p. 236). The data reported here remedy these problems.

12.   One writer described the controversy as follows:

"The ABA approval process has been controversial from its inception. Those who oppose total attorney control of paralegalism feel that the bar associations are inappropriate mechanisms to regulate training institutions. Since a major objective of attorneys is to increase their profits by employing paralegals, critics argue that it is a conflict of interest for attorneys to control the field totally. When regulation decisions must be made on matters such as the approval of schools, whose interest would the attorneys be protecting in making these decisions? The interest of the paralegals? The interest of the public? Or the profit interest of the attorney-regulators? (Statsky 1993, p. 32).

13.    Johnstone and Wenglinsky (1985, p. 84) noted in their study of paralegals under attorney supervision, "Repeatedly in our interviews paralegals expressed dissatisfaction with the repetitive, boring, and uncreative tasks regularly if not exclusively assigned to them." They referred to their work as "idiot work," "mindless work," "robot work," and "work that could be performed by trained apes."

14.    Abbott points out that the central public argument against workplace assimilation is that "subordinates lack the theoretical education necessary to understand and use what they know by assimilation. This is often a fiction, since the theoretical education in the dominant profession is often irrelevant to practice. . . Dominant professions seem to be successful in hiding from the public the excessive assimilation of professional knowledge in the workplace" (1988, p. 68).

15.    The Arizona Supreme Court later relaxed its previous position regarding UPL. In *Hunt v. Maricopa County* (1987), the plaintiff was a court employee involved in an employment grievance who was represented in an administrative hearing by a nonlawyer union official. The court rejected the county's argument that this constituted UPL by the union representative. The court noted an existing statute, since repealed, authorizing such representation in employment hearings such as this. It then decided that UPL had not occurred, since authorizing such activity was the clear intent of the legislation and because the claim involved had a value of less than $1,000. Interestingly, the court took notice of the existing legal market; it held that the low value of the claim made it unlikely that the plaintiff would have been able to obtain counsel. This justified its departure from established UPL law, and pointed to one reason why the public's needs are unmet.

16.    Abbott examined about 800 UPL complaints from urban and rural England and the United States from the period 1882 to 1950. His American data are drawn primarily from legal publications and annual reports of the New York County, New York City, and Pennsylvania bar associations (the latter being the only locations where "records" of the bar were studied). As he states, "Each record contains three simple items of information: the occupation accused of poaching, the area of work invaded, and the year of occurrence" (1988, p. 257). The data reported here provide greater level of detail regarding the work performed by nonlawyer practitioners.

17.    Complaints by members of other professions (i.e., physicians and stockbrokers) were counted separately from those by "individuals" when the UPL allegations arose in the context of their professional roles.

18.    For divorce cases, he writes to both husbands and wives. He advises men that, "Although I am not an attorney and do not give legal advice, I assist those who cannot afford an attorney to represent themselves." This is followed by the offer (1) to file an answer on the individual's behalf, (2) to prepare a "child custody plan," and (3) to prepare court papers leading to "a temporary Court order establishing an *immediate* visitation schedule" (emphasis in original).

For wives, Mr. Jones' letter states that he will "assist women who cannot afford an attorney to represent them." He offers a variety of services "if we determine that mediation is not immediately possible." These include (1) preparing an answer to the husband's petition, (2) if no support is being paid, preparation of a petition for "*Pendente Lite* [temporary] Relief and a Petition for an Order to Show Cause re: child support, and division of community income so that you can begin receiving support *immediately*" (emphasis in original). He goes on to promise that, "If your husband is represented by an attorney, we will be able to cause him to expend so much money to be represented by an attorney, that he will want to mediate the issues of your dissolution very quickly"!

In contract cases, Mr. Jones advises defendants, "Although I am not an attorney and do not give legal advice, I assist those people who either cannot afford an attorney, or refuse to pay the high cost of hiring an attorney, and wish to represent themselves." He offers (1) to prepare an answer to the complaint, "based upon your direction as to which allegations you wish to admit or deny and which defenses you choose to allege, if any," (2) to prepare "any other documents required during the pending litigation;" and (3) to "attempt to mediate a resolution to your dispute," noting that "I have had good success reaching satisfactory resolutions to disputes."

19.    Interestingly, the same committee, prior to the latter report, had circulated a draft of a new definition of the "practice of law." It stated:

One who acts in a representative capacity in protecting, enforcing or defining the legal rights and duties of another is engaged in the practice of law. It also includes counseling or advising another in connection with their legal rights and duties. One is deemed to be practicing law whenever he furnishes to another advise or services which require the exercise of legal judgment. The practice of law creates a professional relationship of competence and trust based upon the giving of legal advice.

The question remained, even after this clarifying definition, as to what constitutes "services which require the exercise of legal judgment" and "legal advice"? This definition, like all others established by courts, is tautological.

20.    The attorney-supervised paralegals would have also benefited from a very controversial provision of the proposed rules which would have permitted legal assistants employed by attorneys to appear in certain limited-jurisdiction (i.e., minor) matters, provided they met certain educational requirements.

# REFERENCES

Akers, R. 1968. "The Professional Association and the Legal Regulation of Practice." *Law & Society Review* 2: 463-482.

*Akron Bar Assoc. v. Greene*, 673 N.E.2d 1307 (Ohio 1997).

American Bar Association. 1983. *Annotated Code of Professional Responsibility*. Chicago, IL: American Bar Association.

American Bar Association. 1994a. *Just Solutions: A Program Guide to Innovative Justice System Improvements*. Chicago, IL: American Bar Association.

American Bar Association. 1994b. *Responding to the Needs of the Self-Represented Divorce Litigant*. Chicago, IL: American Bar Association.

American Bar Association. 1995. *Nonlawyer Activity in Law-Related Situations: A Report with Recommendations*. Chicago, IL: American Bar Association.

Anleu, S. 1992. "Jurisdictional Conflicts and the Professions: The Legal Regulation of the Medical Practice." Paper presented to the Annual Meeting of the Law & Society Association, Chicago.

*Arizona Attorney*. 1995. "Legislative Round-Up." *Arizona Attorney* 35 (July): 35.

*Attorney Grievance Commission v. Hallmon*, 681 A.2d 510 (Md. 1996).

*Brotherhood of R.R. Trainmen v. Virginia* ex rel. Virginia State Bar, 377 U.S. 1 (1964).

Cannon, T. 1996. *Ethics and Responsibility for Legal Assistants*, 2nd edition. Boston, MA: Little, Brown.

Carlin, J. 1962. *Lawyers on their Own: A Study of Individual Practitioners in Chicago*. New Brunswick, NJ: Rutgers University Press.

Carr-Saunders, A., and P. Wilson. 1933. *The Professions*. Oxford, UK: The Clarendon Press.

Christensen, B. 1980. "The Unauthorized Practice of Law: Do Good Fences Really Make Good Neighbors—Or Even Good Sense?" *American Bar Foundation Journal*, 159-216.

*Conway-Bogue Realty Investment Co. v. Denver Bar Association*, 312 P.2d 998 (Colo. 1957).

Dingwall, R. 1995. "Herbert Spencer and the Professions: Occupational Ecology Reconsidered." *Sociological Theory* 13: 14-24.

Durkheim, E. [1895] 1958. *The Rules of Sociological Method*, 8th edition, edited by G.E.G. Catlin. Glencoe, IL: Free Press.

Farinacci, J. 1991. "Paralegalism: The Undefined Profession." *Journal of Paralegal Education and Practice* 8: 147-165.

Friedson, E. 1973. *Professionals and their Prospects*. New York: Sage.

Friedson, E. 1994. *Professionalism Reborn: Theory, Prophecy, and Policy*. Chicago, IL: University of Chicago Press.

*Goldfarb v. Virginia State Bar*, 421 U.S. 773 (1975).

Goldschmidt, J. 1994. "A Brief History of Unauthorized Practice of Law." Pp. xi-xx in *1994 Survey and Related Materials on the Unauthorized Practice of Law/Nonlawyer Practice*. Chicago, IL: American Bar Association.

Goldschmidt, J., and I. Pilchin. 1996. *User-friendly Justice: Making Courts More Accessible, Easier to Understand, and Simpler to Use*. Chicago, IL: American Judicature Society.

Greacen, J. 1994. "Arizona's Agony Over the Unauthorized Practice of Law, or Why We All Need to Care About the Public's Opinions of Lawyers." *Judges' Journal* 15 (Spring): 38.

Green, D., J. Snell, R. Corgiat, and T. Paramanith. 1990. "The Professionalization of the Legal Assistant: Identity, Maturation States, and Goal Attainment." *Journal of Paralegal Education and Practice* 7: 35-42.

Halpern, S. 1992. "Dynamics of Professional Control: Internal Coalitions and Crossprofessional Boundaries." *American Journal of Sociology* 97: 994-1021.

Haug, M. 1975. "The Deprofessionalization of Everyone?" *Sociological Focus* 8: 197-211.

Heinz, J., R. Nelson, E. Laumann, and E. Michelson. 1997. "Chicago Lawyers: Hemispheres, Tectonic Plate Movements, and Continental Drift." Paper presented at the Annual Meeting of the Law & Society Association, St. Louis.

Hughes, E. 1958. *Men and Their Work*. New York: The Free Press.

Hughes, E. [1965] 1994. "Professions." Pp. 37-49 in *Everett C. Hughes: On Work, Race, and the Sociological Imagination*, edited by L.A. Coser. Chicago, IL: University of Chicago Press.

Humphries, S., C. Weixel, D. Rising, and Z. Cohen. 1992. "Paralegal Representation Before the Social Security Administration." *Journal of Paralegal Education and Practice* 9: 20-58.

*Hunt v. Maricopa County*, 619 P.2d 1036 (Ariz. 1980).

*Ingham County Bar Association v. Walter Neller Co.*, 69 N.W.2d 713 (Mich. 1955).

*In re* Burson, 909 S.W.2d 768 (Tenn. 1995).

*Jacobellis v. Ohio*, 378 U.S. 184 (1964).

Johnson, T. 1972. *Professions and Power*. London: Macmillan.

Johnstone, Q., and D. Hopson, Jr. 1967. *Lawyers and their Work: An Analysis of the Legal Profession in the United States and England*. Indianapolis, IN: Bobbs-Merrill Co.

Johnstone, Q., and M. Wenglinsky. 1985. *Paralegals: Progress and Prospects of a Satellite Occupation*. Westport, CT: Greenwood Press.

Kellermann, P. 1990. "Professions and Expert Labor." *Innovation* 3: 185-194.

Kimball, B. 1992. *The "True Professional Ideal" in America: A History*. Cambridge, MA: Blackwell.

Klegon, D. 1978. "The Sociology of Professions: An Emerging Perspective." *Sociology of Work and Occupations* 5: 259-283.

Kreutzer, R. 1994. "New Rules Take Wrong Turn." *Arizona Attorney* 34(March): 20-23, 51.

Kritzer, H. 1992. "Toward a Theory of Representation in Adversary Proceedings." Paper presented at the Annual Meeting of the Law & Society Association, Philadelphia, PA.

Kritzer, H. 1998. *Legal Advocay: Lawyers and Nonlawyers at Work*. Ann Arbor, MI: University of Michigan Press.

Kronus, C. 1976. "The Evolution of Occupational Power: An Historical Study of Task Boundaries Between Physicians and Pharmacists." *Sociology of Work and Occupations* 3: 3-37.

Larson, M. 1977. *The Rise of Professionalism: A Sociological Analysis*. Berkeley: University of California Press.

Levine, J. 1994. "A Better Plan than Licensed Non-Lawyer Legal Technicians." *Maricopa Lawyer.* 12(1, February): 2.

Lockwood, D. [1958] 1992. *The Blackcoasted Worker*. London: Allen and Unwin.

Macdonald, K. 1995. *The Sociology of the Professions*. London: Sage Publications.

Merton, R. [1958] 1982. "Functions of the Professional Association." Pp. 199-209 in *Social Research and the Practicing Professions*, edited by A. Rosenblatt and T.F. Gieryn. New York: University Press of America.

Miller, R., and M. Urisko. 1995. *West's Paralegal Today: The Legal Team at Work*. St. Paul, MN: West Publishing.

Monahan, S. 1994. "Organization and Jurisdictional Competition: The Relation Between Formalized Lay Roles and Shared Work in Churches." Paper presented at a meeting of the American Sociological Association. *Sociology Abstracts* 42.

Munro, M. 1990. "Deregulation of the Practice of Law: Panacea or Placebo?" *Hastings Law Review* 42: 203-48.

*NAACP v. Button*, 371 U.S. 413 (1963).

Nemeth, C. 1995. *The Paralegal Resource Manual*. Cincinnati, OH: Anderson Publishing.

*Paralegal Institute, Inc. v. American Bar Association*, 475 F.Supp. 1123, *aff'd* 622 F. 575 (1st Cir. 1980).

Parsons, T. 1968. "Professions." Pp. 536-537 in *International Encyclopedia of the Social Sciences*, Volume 12, edited by D.L. Shils. London: MacMillan.

Pavalko, R. 1988. *Sociology of Occupations and Professions*, 2nd edition. Chicago, IL: F.E. Peacock.

Pilchen, I., and S. Ratcliff. 1993. *Conducting State Court Futures Activities*. Chicago, IL: American Judicature Society.

Podgers, J. 1994. "Legal Profession Faces Rising Tide of Non-Lawyer Practice." *Arizona Attorney* (March): 24-30.

Rhode, D. 1981. "Policing the Professional Monopoly: A Constitutional and Empirical Analysis of Unauthorized Practice Prohibitions." *Stanford Law Review* 34: 1-112.

Rhode, D. 1990. "The Delivery of Legal Services by Non-Lawyers." *Georgetown Journal of Legal Ethics* 4: 209-233.

*R.J. Edwards, Inc. v. Hert*, 504 P.2d 407, 416 (Okla. 1972).

Rothman, R. 1984. "Deprofessionalization: The Case of Law in America." *Work and Occupations* 11: 183-206.

Sales, B., C. Beck, and R.K. Haan. 1993. *Self-representation in Divorce Cases*. Chicago, IL: American Bar Association.

Sallen, P. 1993. "State Bar Board Okays Concept of Non-Lawyer Practitioners." *Maricopa Lawyer* 12(December): 12, 42.

Samborn, R. 1993. "So What Is a Lawyer, Anyway? The Fight Over Non-Lawyers and Legal Services Comes to a Head." *National Law Journal* 15(June 21).

Schrader, G. 1991. "Legal Assistant Professional, Educational and Career-Related Organizations." *Journal of Paralegal Education and Practice* 8: 63-85.

Shely, L. 1994. "Where We Are Today." *Arizona Attorney* 34(February): 11.

Simmons, J. 1994. "Non-Lawyer Practice Rules: No Turning Back." *Arizona Attorney* 19(March): 32-33.

State Bar of Arizona. 1994. *Non-Lawyer Practice Rules* (January 25).

*State Bar of Arizona v. Arizona Land Title and Trust Co.*, 366 P.2d 1 (1961)

State Bar of California. 1988. *Report of the Public Protection Committee* (April 22). Los Angeles, CA: State Bar of California.

State Bar of California. 1990. *Report of the State Bar of California Commission on Legal Technicians* (July). Los Angeles, CA: State Bar of California.

*State Bar of New Mexico v. Guardian Abstract and Title Co.*, 575 P.2d 943, 948 (NM 1978).

Statsky, W. 1993. *Paralegal Ethics and Regulation*. St. Paul, MN: West Publishing.

Torren, N. 1975. "Deprofessionalization and its Sources." *Sociology of Work and Occupations* 2: 323-337.

*UMW Dist. 12 v. Illinois State Bar Ass'n*, 389 U.S. 217 (1967).

U.S. Department of Labor. 1994. *Occupational Outlook Handbook: 1994-95 Edition*. Washington, DC: Superintendent of Documents.

Warren, C. [1911] 1939. *A History of the American Bar*. New York: Howard Fertig.

Weber, M. 1978. *Economy and Society*. Berkeley, CA: University of California Press.

# CAREER COMMITMENTS:
## WOMEN AND MEN LAW SCHOOL GRADUATES

Kandace Pearson Schrimsher

## ABSTRACT

This chapter is based upon a larger study of the careers of women and men 11 years after they graduated from Case Western Reserve University School of Law. The study examined the similarities and differences of law school, employment, and family experiences. Specifically, this chapter focuses on the career commitment of the women and men graduates. Career commitment is measured by continued employment in the legal profession and the side bets (investments) made in maintaining career commitment.

A qualitative and quantitative analysis of eight specific side bets found that: (1) women and men had similar law school experiences, and class rank, not gender, influenced opportunities and experiences; (2) women were more committed than men to the legal profession, as measured by job mobility, which was increasing for both groups; (3) although the vast majority of graduates experienced stress in their work, the conflict experienced between work and family relationships was different among women and men; (4) the career satisfaction of prestige of position, prestige in community, and advancement opportunities differed between women and men; and (5) the women accepted restrictions on their careers if they came from their husbands. Most women were not getting the support they needed from their husbands or

Current Research on Occupations and Professions, Volume 10, pages 193-215.

the workplace. What they did get was not as relevant to their careers as the type of spouse/partner career support received by their male counterparts.

## THE LEGAL PROFESSION

The astonishing growth in the number of lawyers in the United States since World War II has been accompanied by an unprecedental increase in demand for legal services from both business clients and individuals. By 1990, the legal profession had become a $91 billion a year service industry, employing more than 940,000 people, and surpassing the medical profession in the number of licensed professionals (MacCrate 1992).

The rapid growth in the number of lawyers has affected the manner in which law is practiced and how law firms are structured and organized. It has allowed greater specialization in law practice and an increased division of labor among lawyers. New areas of law and regulation, for the most part designed by lawyers, have created whole new fields of legal services.

Along with the rapid growth of lawyers has been an increased demand for all kinds of legal services, particularly from the business community. For the past 40 years, the economic base that supports legal services has expanded greatly, being matched by the steadily increasing number of clients willing and able to pay for legal services (MacCrate 1992).

In 1991, there were 805,872 lawyers in the United States, of which 80.2% were men and only 19.8% were women (Curran and Carson 1994). Practicing lawyers are employed in several types of settings, including private practice (solo and firm), private industry, and government employment (federal and state). Private practice is the most common organization of employment for lawyers. In fact, in 1991, 73% of lawyers were employed in private practice, of which 45% worked solo and 55% in firms (Curran and Carson 1994).

Private practice may consist of firms ranging in size anywhere from one to over 100 lawyers. Generally, these firms may also include legal researchers and support staff. Lawyers employed in private practice are salaried professionals with an opportunity for profit sharing if and when they get promoted from associate to partner status.

Internal structures of law firms generally consist of administrations ordered by a series of committees and a system of ranks, each with distinctive obligations and privileges. The single most significant distinction in a law firm is that of rank. "Partners" own the firm and, therefore, divide its considerable profits at the end of the year. "Associates," on the other hand, do not share in the profits of the firm; they are its employees, working for a generous but nevertheless fixed salary (Spangler 1986, p. 35). The enormous differences between associate salaries and partnership salaries (sometimes more than 10:1) symbolize both the subordination of the former and the rewards that follow from accepting it gracefully (Abel 1989, p. 222).

Firms admit associates to partnership after seven to 10 years of service to the firm. In all firms, the timetable for full partnership is similar: after a decade, attorneys expect to be full and permanent members of the partnership. Under the current system, then, salaried employment (associate status) is a temporary status designed to lead into either partnership (ownership) or exit from the firm. The possibility of a partnership secures the associates' loyalty, although the probability of being made partner is declining (Spangler 1986, p. 36).

Another position within a firm is that "of counsel." Traditionally, this position title applied to those individuals who were retired partners of their law firm. However, in the last decade or so, this position has been redefined. Currently, the status "of counsel" implies more of a lateral category; a nonequity salaried employer of the firm. Individuals who occupy this position are not considered associates because they are not on a partner-track.

To a significant extent, firm size and practice setting have had a direct relationship to the types of clients served, the type of law practiced, and the financial rewards of practice. Community-oriented solo and small firm practitioners work predominantly for individuals, whereas lawyers in larger firms work predominately for business clients.

In general, the financial rewards for legal work for individuals (except for personal injury claims) have been less than the rewards for representing business. Various studies of the legal profession have indicated a strong relationship among the size of a firm and the source of its income: as firm size increases, the percentage of fees from business clients also increases while the percentage of fees from individuals decreases. Hence, the larger the firm, the greater the concentration of work from business clients and, thus, the larger the average income of the firms' lawyers.

Generally, working in the legal profession requires an involvement in cases, paperwork, and community. Moreover, in most firm environments, earnings are to some extent dependent on initiative—ambition, drive, and motivation. Hence, the time demands of the work increase dramatically. A full day at the office is not enough for an aspiring lawyer. The norms of the legal profession equate success and excellence with hard work, measured in part by long hours. Putting in extra hours at night and working on weekends are not only common but generally expected. Furthermore, besides having the pressure to work hard, there is also the expectation of spending all extra time on legal affairs, such as taking classes (i.e., continuing legal education), bar association projects and meetings, and community services such as pro bono work.

The legal profession, as well as most professions in general, are what Coser (1974) describes as "greedy institutions." Not only do firms and corporations impose time demands on employees, but the employees themselves seek other options to get access to more challenging and interesting work and/or seek to participate or increase participation in organizational problem solving. Thus, in these types of settings, women and men have projects with goals that require additional work and reflection beyond the time permitted at the office (Kanter 1977). Moreover, it is the organization's intention that time records and billing demands keep lawyers competitive and overproducing.

## CONCEPTS OF CAREER COMMITMENT AND SIDE BETS

In order to understand fully what is meant by career patterns, the concept of career must first be defined. In broad terms, Hughes (1958) defined the concept of career as "the more or less orderly and predictable course of one's work life" (1958, p. 12). According to Wilensky (1961), a career is "a succession of related jobs, arranged in a hierarchy of prestige, through which persons move in an ordered (more or less predictable) sequence (1961, p. 523).

Moreover, the work of one person is usually related to the work of another or a larger social unit. According to Lopata, Barnwaldt, and Miller (1986), this means that one's work is usually woven into social roles, defined as sets of interdependent social relations between a social person and a social circle involving duties and rights (Znaniecki 1965, as cited in Lopata et al. 1986; see also Lopata 1994). In other words, jobs are social roles containing work and other aspects of social relations with all members of a social circle, where the social circle consists of everyone with whom the person interacts in order to carry forth a role.

Throughout this discussion, Becker's (1960) concepts of commitment and side bets are used along with the side bets proposed in Lopata's (1992) article, "Career Commitments of American Women: The Issue of Side Bets." Although Lopata presented 10 side bets, only those side bets that are relevant to the occupational/professional commitment of law school graduates are examined and discussed here. As Lopata did in her study, I broaden the concept of side bets to include relevant aspects of the total life space (professional and personal) which women and men graduates can bring into line because these can affect, with ease or with difficulty, the types of career choices made and the level of commitment desired and/or maintained.

When discussing the concept of commitment, Becker (1960) referred to two distinct types: (1) commitment to the organization, and (2) commitment to the occupation. I examined the investments and commitments the women and men graduates have to the legal profession, as opposed to their commitment to the organization of employment. Hence, for the purpose of this study, commitment to the profession will be measured by continued employment in the legal profession.

### Side Bet Theory

Becker (1960) contended that commitment occurs through a process he called "placing side bets," or types of investments. Becker asserted that the greater the number of side bets, the greater the degree of commitment of the individual to a course of action. It follows that these investments strengthen one's commitment to employment and career goals, both directly and indirectly, making it beneficial to continue such commitments. In Becker's analysis, commitment to the organization and commitment to the occupation are the result of a series of conscious and unconscious side bets, or investments. A committed person has acted in such a way as to involve her or his other interests, originally unrelated to the action the person is engaging in, to be directly related to that action (Stebbins 1968, p. 527). For example, Stebbins (1970), in his study of job commitment, found that age, education, marriage, children, and salary were all associated with a strong degree of commitment as well as a strong development of professional identity. So it follows that these investments strengthen one's commitment to employment and career goals, both directly and indirectly, making it beneficial to continue such commitments.

A consistent line of activity will often be based on more than one kind of side bet; several kinds of "things" valuable to the person may be staked into a particular line of activity. Ritzer and Trice (1969) asserted that there needs to be an analysis of values with which side bets can be made, by examining what kinds of things the individual desires and those types of losses feared. Hence, in order to understand commitment fully, one must discover the system of values upon which the side bets are made.

Lopata's (1992) study examined the side bets of American women in the form of: education, occupational preparation, employer selection, full-time involvement, role-conflict avoidance, incorporation of career into self-concept, and the building of a relatively congruent construction of reality. Following Becker (1960), Lopata contended that choosing an occupation or employing organization provides the individual with rewards which can be hard to give up; thus, leaving it may result either in penalties or in costs that the person can be increasingly unwilling to face. Thus, the side bets tie the person to that line of action in many ways that she or he may not be aware of until she or he contemplates the decision to leave this occupation or organization. On the other hand, a person may consciously increase the ease of following the committed line of action by purposely building "side bets" into her or his life (Lopata 1992, p. 4). The focus of this analysis is on the latter, that most of the women and men graduates have purposely built side bets into their lives.

Several major studies of the changing commitments of women to family and work roles indicate possible forms of side bets. For example, a person who may perceive her or himself as committed to an occupation or a career line which demands high commitment is thus recognized as such by her or himself as well as by her or his closest "significant others," such as mentors, friends, and colleagues, and by the "generalized others," such as members of the profession.

Traditional research indicates that the greater the complexity of the occupation, as well as its status in the structure, the higher the commitment of the persons in it (Kohn and Schooler 1973, 1983; Lopata et al. 1985a, 1985b, as cited in Lopata

1992). This paper utilizes eight of Lopata's hypothesized 10 side bets. The eight side bets discussed here are:

1. Preparation for the profession through schooling;
2. Selection of an organization which can be expected to support commitment to professional goals;
3. Involvement in a job which enables one to work in that profession and/or pursue a career line and provide job complexity;
4. Positive feedback from above involvement;
5. Association with people of similar commitments, colleagues, and friends;
6. Support from spouse/ partner;
7. Support in the role of parent to make possible commitments; and
8. Developing a relatively congruent construction of reality at the sociopsychological and behavioral aspects of commitment.

These side bets are examined to identify similarities and differences in side bets among women and men and what effect, if any, these differences and/or similarities have on their measured commitment (continued employment) to the legal profession.

Lopata (1992) emphasized that these side bets have been expected for most men in modern America, due to the "vestige" of the two-sphere "world" in which a male's professional commitment to an occupation/employee role is taken for granted. This society was developed through a strong focus upon the economic institution, to be carried forth directly by men (Weber 1953, as cited in Lopata 1992). Until very recently, members of this half of the American population have needed no justification for organizing their lives around occupations. Socialization throughout the life course provided men with a solid base of a congruent construction of reality. The occupational system continues for the most part with "greedy" commitment demands, making it difficult for women who also must meet the demands of equally "greedy" families (Coser 1974; Coser and Coser 1974). Thus, career commitment by women and men in two-career families requires investment in many side bets, both internally, in terms of mutual support and self-concept, or identities; and externally, from the environment, especially from people in the various role circles in which they are involved (Lopata 1992, p. 11). However, studies indicate that women and men do not receive equal internal and external support in their careers.

## STUDY DESIGN AND METHODS

The findings of this study are based on the data of an alumni survey of the women and men 1981 graduates of Case Western Reserve University School of Law, Cleveland, Ohio, conducted during the summer of 1992. Approximately 45% of the graduates participated in the study—35% women and 65% men (which was also the actual gender composition of the class of 1981). The data were complied from a questionnaire that was 22 pages long and comprised mostly of multiple choice questions that required participants to circle a number which best described their answers, as well as open-ended, and rank-ordered questions. The CWRU Alum Survey included the following sections: (1) Personal Background Information; (2) Law School Experiences; (3) Employment Histories; (4) Current Employment; (5) Work Projections; (6) Additional Background Information; and, (7) Housework and Child Care Responsibilities.

Other studies of law school graduates referred to throughout this chapter include: the New Mexico Study (Teitelbaum et al. 1991), University of Michigan Study (Chambers 1990), the Minnesota Study (Mattessich and Heilman 1990), the Stanford Project (1988), the Harvard Project (Vogt 1986), and Liefland (1986).

Specifically, the New Mexico Study examined alumni of the University of New Mexico, classes of 1975-1986. The University of Michigan Study examined the class of 1985 (five years out of law school at the time of the survey). The Minnesota study was commissioned by the Minnesota Women Lawyers Task Force on the Status of Women in the Profession. That study examined the careers of law school graduates of 1975, 1978, 1982, and 1985 from three Minnesota law schools: University of Minnesota (35%), William Mitchell (44%), and Hamline University (21%). The Stanford Project was an empirical study which examined gender, legal education and the legal profession with law school graduates and current students. The Harvard Project examined the classes of 1959, 1969, 1974, and 1981 from seven northeastern law schools: Boston College, Boston University, Columbia University, University of Connecticut, Harvard, Northeastern, and Suffolk University. This study had a range of small, medium and large schools, representing schools with regional or national prestige. Finally, Liefland's study examined the career patterns of male and female lawyers of the classes of 1976, 1977, and 1978 from four prominent law schools: University of California at Berkeley, Columbia University, University of Pennsylvania, and New York University.

# EXAMINATION OF SIDE BETS

The findings of this study suggest that 11 years after graduating from law school, most CWRU graduates have maintained a high level of commitment to the legal profession, as evidenced by the high number, 84% of the women and 81% of the men, who were practicing law. When eight of Lopata's (1992) proposed side-bets were examined, many similarities and differences among the women and men graduates were revealed.

## Side Bet 1: Preparation Through Schooling

Side Bet 1 examines commitment to professional goals by the graduates of the same law school by means of education. Clearly, all of the women and men made an investment in their career commitment through attaining a law degree. They invested their time, energy, and money in obtaining an undergraduate degree, prior to their three year commitment to earn a law degree.

## Side Bet 2: Selection of an Organization of Employment

Side Bet 2 is the selection of an organization which can be expected to support commitment to professional goals. This investment was made by the majority of both women and men and is reflected in the high rate of job mobility among the graduates. During the first 11 years after graduating from law school, most graduates had held at least two different jobs. Moreover, 47% of the women and 50% of the men had held three or more jobs. Twenty-one percent of the women and 24% of the men had had yet another job after job three. This finding suggests that when professional goals are not being met or are perceived as being compromised in a job or an organization, most CWRU women and men will seek alternative employment perceived as more conducive to the achievement of professional goals and commitments. Therefore, the high rate of job mobility among the women and men indicates an unsatisfactory match between lawyer and organization at the beginning, and for some throughout, their legal careers.

At the time of the survey, the majority of women and men practiced law in private firms (see Table 1). These findings were similar to several other studies, with the exception of first job after law school (no differences were found among

***Table 1.*** Current Organization Of Employment

|  | Female (%) | N | Male (%) | N |
|---|---|---|---|---|
| Private firm/solo | 43 | 1 | 55 | 36 |
| Federal government | 3 | 1 | 5 | 3 |
| State/local government | 14 | 5 | 3 | 2 |
| Quasi-government (World Bank) | 0 | 0 | 2 | 1 |
| Legal services/public defender | 3 | 1 | 2 | 1 |
| Public interest | 0 | 0 | 2 | 1 |
| Fortune 500 industry/service | 14 | 5 | 8 | 5 |
| Other industry/business | 9 | 3 | 3 | 2 |
| Banking/finance | 6 | 2 | 5 | 3 |
| Accounting firm | 0 | 0 | 6 | 4 |
| Other service | 9 | 3 | 9 | 7 |
|  |  | (N = 65) |  | (N = 35) |

**Notes:** $F = .07$; Sig. $= .796$

**Table 2.** Current Position Of Employment

|  | Female (%) | N | Male (%) | N |
|---|---|---|---|---|
| Practice law | 82% | 28 | 72% | 49 |
| Trial/appellate judge | 6 | 2 | 0 | 0 |
| Other legal position | 6 | 2 | 3 | 2 |
| Non-law employment | 6 | 2 | 25 | 16 |
|  |  | (N = 34) |  | (N = 67) |

**Notes:** $F = .43$; $Sig. = .513$

CWRU women and men), where gender differences were found among organizations and positions of employment (Mattessich and Heilman 1990; Liefland 1986; Vogt 1986; Curran and Carson 1985; White 1965).

As indicated in Table 2, at the time of the CWRU survey, the men graduates had greater employment diversity among firm sizes, while the women were concentrated in the very small or the large firms. Moreover, many more men than women were employed outside of the legal profession. The majority of women employed in the legal profession practiced law; two were judges and two women were in other types of legal work. The majority of women who practiced law were employed in solo or small firms of four or less and large firms of 100 or more (see Table 3). Similar to the women, the majority of men practiced law, and two men did other legal work. However, the majority of men who practiced law were employed in firms of five to 15 lawyers (36%), one-fourth were in large firms of 100 or more, and 21% were in small firms of four or less.

The high concentration of women employed at solo and large firms may be because of the perceived advantages associated with each setting. Solo practice offers an environment with a sense of independence for setting one's own hours and the freedom to turn away and accept clients; also, solo practitioners need to answer to no one but themselves in pursuing cases that appeal to them (Spangler 1985). Large firms, on the other hand, have made the greatest strides toward equality among men and women as they have elsewhere in the legal profession and in

**Table 3.** Firm Size of Practicing Lawyers

|  | Female (%) | N | Male (%) | N |
|---|---|---|---|---|
| 4 or less | 47 | 7 | 21 | 9 |
| 5-15 | 7 | 1 | 36 | 15 |
| 16-29 | 0 | 0 | 2 | 1 |
| 30-49 | 0 | 0 | 5 | 2 |
| 50-100 | 7 | 1 | 12 | 5 |
| More than 100 | 40 | 6 | 24 | 10 |
|  |  | (N = 15) |  | (N = 42) |

**Notes:** $F = .03$; $Sig. = .852$.

other professions. There is equality in recruitment; equality in pay in the initial years; access to specialization in all areas of the law; and institutionalization of policies such as maternity leave, unpaid leave, and availability of part-time work tracks (Epstein et al. 1995).

These findings and differences must be explored further to determine the underlying factors prevalent in organizations of employment. For example, are these choices made by the women and men? Are they the result of unsuccessful employment in a different firm size and/or organization? Or are they the only alternative options available for personal responsibilities? Are women and men making their employment decision based on their own needs, or are they based on the needs of others?

At the time of the study, more than twice as many men than women had achieved partner status at their law firms. Among those women who were in firms with status-level positions, 33% were partners, 12% associates, 7% were of counsel, and 7% were employed in other firm positions (see Table 4). The majority of men were partners in their firms (69%); a further 8% were associates, 3% were of counsel, and 8% were employed in other firm positions. It should be noted that the differences among the firm positions of the women and men may be attributed to the fact that a high proportion of women were employed in solo practice, where generally there are no distinct associate and partner positions.

Another explanation of the differences of status-levels among the women and men may be that some of the women are taking or have taken nontraditional career paths. For example, in her study of women lawyers, Fritz (1990) found that many law firms provide maternity leave and offer flexible arrangements to those associates, virtually all of whom are women, who wish to work part-time or take extended leaves to care for their children. But those women who accept part-time arrangements (at a law firm, this can mean working 30 hours a week) generally forfeit the opportunity to become partners. In my study, 58% of the women and 16% of the men, at some point in their careers, had worked part-time and/or stopped employment; parental leaves had been taken by 42% of these women and only one man.

***Table 4.*** Current Law Firm Status

|  | Female (%) | N | Male (%) | N |
|---|---|---|---|---|
| Solo | 40 | 6 | 13 | 5 |
| Partner | 33 | 5 | 69 | 27 |
| Associate | 13 | 2 | 8 | 3 |
| Of counsel | 7 | 1 | 3 | 1 |
| Other | 7 | 1 | 8 | 3 |
|  |  | (N = 15)) |  | (N = 39) |

**Notes:**   $F = .26$; $Sig. = .610$.

The Law Society's Working Party on Women's Careers found that with women's career advancements, "the principal difficulty that women have to face is the reconciliation of their social responsibilities for children with the needs of a career" ("Women in the Profession" 1988, p. 11). The traditional male career model of success would suggest that a person occupy an associate status for five to seven years and then be promoted to partner status. However, the timing of associate status coincides with women's peak fertility, while mens' fertility lasts much longer. The demands of the firm tend to be the highest at the associate status, a time when women may prefer to work less. The CWRU women who were at the associate level status may have "slowed" or stopped their career path for family/ spouse obligations or commitments (Pearson 1990; Spencer and Podomore 1987; White 1984). Therefore, those women who took "nontraditional" career paths would be more likely to achieve partner status much later in their careers than those persons who followed the traditional male career model of the legal profession. Or, perhaps some women may have redefined their priorities and no longer aspire to partner status and the male career model of success.

### Side Bet 3: Pursual of Career Goals and Job Complexity

Side Bet 3 reveals that the majority of both women and men were involved in jobs which enabled them to work in the legal profession and/or pursue their career-line and provide job complexity. Again, the investment of career commitment is reflected in the high rate of mobility within the legal profession among both women and men graduates.

The factors liked the most among both women and men about their current careers include: intellectual stimulation, challenges, and problem solving—all characteristics of job complexity. Work-related stress was the aspect they liked the least. The majority of both women and men attribute their work-related stress to deadlines, which require immediate output, making quality work difficult. Also, many women attribute their work related stress to their attempt to balance work and family demands, and problems with superiors. The men, on the other hand, attribute their stress primarily to client demands, problems with superiors and balancing work and family demands. Even though family was more frequently a source of stress for the women, among the men family demands were the fourth most common source of stress. Perhaps the differences in the acknowledgement and identification of stress due to family demands may be influenced by the fact that 66% of the men (as opposed to all of the women) have spouses who are also employed in the paid labor force. Hence, these men, as opposed to their male counterparts whose wives are not employed, may experience, to some extent, the potential career disadvantages that result from the time demands most women in the paid laborforce experience in their attempt to juggle both work and family responsibilities and commitments.

At the time of the survey, most of the graduates had children who were quite young. However, in their study of women lawyers, Spencer and Podomore (1987) found that child care concerns and responsibilities are not just indicative to the early years. A senior barrister in their study pointed out that her teenage children made a lot of emotional demands "that you can't delegate" (Spencer and Podomore 1987, p. 55). Moreover, in her research on changing families in America, Stephanie Coontz (1997) reflects on her own family life and contends, "as a mother and a researcher, I've found that it's the parents of teenagers who need the most flexibility."

It could be argued that those women who aspire to achieve higher positions, or who occupy higher positions within their organization, have the potential to shoulder heavier "out of work" demands, particularly family responsibilities. Research suggests that women, despite their full-time employment status, continue to do the bulk of the family and household work (Hertz 1986; Hochschild 1989, 1997; Kanter 1989). Hence, like the women lawyers in Epstein's study (1981) and the women professionals in Hochschild's study (1997), the CWRU women graduates appear to have additional time demands, or what Hochschild (1997) terms "time binds," created by their multiple roles of professional, mother and spouse.

### Side Bet 4: Positive Feedback From Employment Involvement

Side Bet 4, a positive evaluation of both instrumental and secondary involvement, is evident in the graduates' evaluation of their career satisfaction. Regardless of the high level of stress experienced by the graduates, in the workplace and/ or at home, the vast majority of both women and men had overall high levels of career satisfaction (see also Chambers 1990; Mattessich and Heilman 1990; Stanford Project 1988).

The majority of the women had the highest level of career satisfaction with solving problems, intellectual challenges, intellectual stimulation, and degree of independence. The majority of men had high levels of career satisfaction with: intellectual challenges, prestige of position, degree of independence, intellectual stimulation and solving problems.

**Table 5.**   Gender Differences: Career Satisfaction

|  | Very Important | | Somewhat Important | |  |  |
| --- | --- | --- | --- | --- | --- | --- |
|  | Female (%) | Male (%) | Female (%) | Male (%) | F | Sig. |
| Prestige of position | 18 | 51 | 50 | 37 | 9.60 | .00 |
| Prestige in community | 21 | 52 | 48 | 32 | 5.84 | .01 |
| Advancement in community | 24 | 39 | 39 | 34 | 3.09 | .08 |

**Notes:**   The sample size of the women was 33, except for Prestige of Position, which was 34. The sample size of the men was 63, except for Advancement Opportunity, which was 61.

As indicated in Table 5, the significant statistical differences of sources of career satisfaction among the women and men was prestige of position, prestige in community, and advancement opportunities. The men had a much higher degree of career satisfaction from the prestige of their position than did the women. These dramatic differences among women and men may be indicative of the fact that more than twice as many men than women had achieved partner status in their law firm. High-status positions generally tend to be associated with high salaries and power and are occupied primarily by men. Therefore, it may be presumed that prestige in the community would be higher for those persons with high-level positions, salaries, and power, since generally these aspects tend to be admired in American society and are often directly correlated with each other, hence leading to higher job satisfaction.

The differences in sources of career satisfaction among women and men needs to be further investigated in order to discover whether the dissatisfaction is also related to job position, salary, organization of employment, or public perception of the significance of job positions.

## Side Bet 5: Association with People of Similar Commitments

Side Bet 5, association with persons of similar commitments, also reveals differences among women and men. The American Bar Association (ABA) is the national professional organization of the legal profession. There are additional bar associations at the state, county and city levels and also in different areas of specialty (special bars). Memberships in professional organizations allow persons to seek recognition from outside groups and professional groups. Therefore, membership affiliations are one indication of professional commitment and loyalty to the profession.

Gouldner (1957, 1958) studied the professional commitment and organizational loyalty of faculty members of a small liberal arts college. He distinguished between two types of members: "the cosmopolitan" and "the local." The cosmopolitan member has little loyalty to the local organization, a strong commitment to specialized skills, and a strong identification with reference groups representing professional specialty. The local member, on the other hand, displays a strong loyalty to the local organization and the profession, a weak commitment to specialized skills, and a strong identification with reference groups located within the organization.

When professional membership affiliations were examined using Gouldner's (1957, 1958) concepts of "local" and "cosmopolitan" professional role orientation, the findings of this study suggest that the women graduates took more of a "local" role orientation, while the men were more "cosmopolitan." The majority of the women were not members of the American Bar Association, while the majority of the men were.

Those women and men who did not participate in professional organizations identified time and/or financial constraints as the primary reasons for nonmembership. They did not have enough time to devote to being members, and/or the membership rate was too expensive, preventing them from joining if their firm did not pay for it, or causing them to pick and choose those organizations they believed to be the most beneficial to join. However, it should be noted that women were not allowed to join the American Bar Association until 1910, and currently there is a significant underrepresentation of women in the governance of the American Bar Association coupled with the commonly held perception that the American Bar Association caters to middle-sized and larger male firms. Moreover, the low rate of membership among the women may be attributed to the expenses and time commitments of bar involvement which must be approved by firm partners, a position that, as revealed in this study and several others, is more often occupied by men than women.

## Side Bet 6: Support From Spouse/Partner

Although it has been documented that the role of spouse/partner is regarded as being somewhat competitive to careers (especially for women), like most other studies, the majority of both women and men were married at the time of the CWRU survey (see also Chambers 1990; Mattessich and Heilman1990). However, all of the women and only 66% of the men were involved in marriages where either both spouses had careers or one had a career and the other was employed full-time in the paid labor force (see Table 6).

Eighty-seven percent of the women had spouses/partners who were employed professionals (of whom 32% were lawyers). The remaining 13% of the husbands were also employed full-time. Among the men, however, only 46% had a spouses/partners who were employed professionals (of whom 15% were lawyers). Thirty-one percent had wives who were homemakers, 21% had wives who were employed full-time, and 2% had wives who were college students. These findings were similar to the Michigan study (1990), where the women were typically linked with partners who earned about as much or more than they did. By contrast, the

***Table 6.***   Spouse/Partner Employment

|  | Female (%) | N | Male (%) | N |
|---|---|---|---|---|
| Professional | 55 | 17 | 31 | 17 |
| Lawyer | 32 | 10 | 15 | 8 |
| Other employment | 13 | 4 | 21 | 11 |
| Student | 0 | 0 | 2 | 1 |
| Homemaker | 0 | 0 | 31 | 17 |
|  |  | (N = 31)) |  | (N = 54) |

**Notes:**   $F = 2.92$; *Sig.* $= .091$.

great majority of men with partners were linked to someone who earned much less than they did or did not have a job in the paid workforce.

This study hypothesized two possibilities of partner support: (1) the majority of women are married to professionals who are likely to be in "greedy" organizations and thus have their own competitive career goals, and (2) since fewer than half of the men are married to professionals or women who are employed full-time, and close to one-third are married to "homemakers," more of the men are likely to have spouses who support their own competitiveness, professional goals and commitments.

It could be argued that those women and men who have spouses/partners with similar professional interests and goals may receive a greater degree of encouragement and understanding of their own work responsibilities and commitments. On the other hand, those persons involved in dual-career marriages or relationships, or one-career/two-income marriages or relationships, would tend to encounter different types of strains and tensions and different types of career support than a graduate who was the primary career person in the marriage, with a spouse who had little, if any, employment in the paid labor force.

The ways in which a spouse/partner gave career support differed drastically among women and men. The majority of women received "general career support," followed by consulting with work and career, and helping with child care and household responsibilities. On the other hand, the majority of men received spouse/partner career support in terms of tolerance of the time demands of their work, taking sole responsibility for child care and household responsibilities, and tolerance of their frequent and extensive travel.

Among the men graduates, partner career support was defined more in terms of a tolerance of job responsibilities. For example, one man wrote, "[My wife] does most of the child care, manages my wardrobe, [and] consults on matters in her field." Another indicated, "[My wife] puts up with long hours [and] lots of travel.... She does a great job with our kids—reducing the stress on me." Still another man reported, "She stays home with the kids, which makes my work priorities easier to handle. She's also excellent with social matters." Another wrote, "[My wife] bore the burden of spending extra time with the children." Hence, most men, but few women, get full time help from their spouse/partner.

The women tend to define spouse career support more in terms of general career support and help with child care and/or household responsibilities. The men, on the other hand, orient their spouse/partner support in terms of their own professional responsibilities—a characteristic related directly to career performance and commitment. This finding supports the "second shift" ideology (Hochschild 1989) that working women, in addition to their professional obligations and commitments, continue to be responsible for the lion's share of both child and household obligations and commitments. However, it should be noted that some of the men recognize the fact that their wives take *sole* responsibility for child care and household obligations, therefore giving them the opportunity to be more deeply involved with their professional obligations, commitments, and goals—something most of their female counterparts are not likely to ever experience.

Among the women, the husbands pose a major problem since the women's careers are not getting the support they need, and also the support provided by the husbands is not as relevant to their career commitment and goals as what the men graduates receive from their wives. In their study examining the position of women lawyers in eight New York City law firms, Epstein and her colleagues (1995) found that support from one's spouse was one of the greatest factors that their participants noted as an important influence in the career paths of women.

When spouse/partner career hindrances were examined, again differences among women and men were revealed. Many of the women claimed that simply choosing to have a family limited their choices, as did receiving little if any help with child care and household responsibilities. For most of the men, a spouse/partner hindered their career by complaining of the long hours they have to work, and in the tension they experience as the result of having a spouse who was also employed full-time.

Like Hochschild (1989), who found in her study that family responsibilities were a major theme in the problems of the women, this theme also prevailed in my study. For example, among the women of CWRU, responses such as these were all too common: "He does nothing to ease the responsibilities at home and with our child. By leaving those responsibilities to me, my career goals have had to be deferred." Another woman reported, "He does not do his share to run the household or child care responsibility." Yet another wrote, "Too many household chores fall on my shoulders."

Among the men, an intolerance of job responsibilities was the most frequently mentioned aspect of spousal career hinderance. For example, one man indicated, "[My wife] has, on occasion, been less than enthusiastic about . . . [my] long hours." Another wrote, "[My wife has] inflexible hours with her job." And yet another man indicated, "She has a full-time career also, so we have to juggle kids, household chores, etc." Changes in the attitudes of men begin to appear among those lawyers with children who have wives who are also employed full-time. Hence, these men, to some degree, experience the conflict and stress of juggling family and career obligations and commitments.

Traditionally, male lawyers have not had to participate in or solve such dilemmas. Their professional ambitions and the work and time necessary to realize these ambitions have been treated by American society as an unquestioned condition of professional life. In their study of the position of women lawyers, Epstein and her colleagues (1995) found that most of the male attorneys interviewed in their eight-firm study acknowledged that the time demands of their work can produce pressures on marital relations. However, such strains are generally mitigated by a wife's primary interest in her husband's career advancement. This is something their female counterparts may rarely experience.

## Side Bet 7: Support in the Role of Parent

Side Bet 7 examines support in the role of parent to make professional commitments possible. Although it has definitely been documented that the role of parent competes with careers of women (and as this study reveals, to some extent, men), the majority of both women and men graduates had two to three infant, preschool, or grade-school age children. Most graduates found that their work conflicted with their ability to devote enough attention to their children (see also Hochschild 1997). However, as indicated in Table 7, the women were twice as likely than the men to experience this conflict "often" or "very often."

For example, one woman wrote, "[The conflict is] time demands—sometimes when I travel, I won't see my kids for three days. I have missed parent/teacher conferences. Sometimes I feel I've not spent enough time with my children." Another woman reported, "[I'm] away from home for weeks at a time (including weekends). I've missed one full month of my baby's so far six-month life." And another woman wrote:

> I work all day, so the important times—lunch, right after school—for good communication, are missed. It's hard to be involved in school. And evenings are only three hours long, including dinner time. If I devote any time to me or my husband, to get emotionally prepared to go to work the next day, there is little time left for a child.

Therefore, if a woman professional decides to have children, she is faced with the dilemma of how much time should she devote to her children and her spouse/partner and how much to her career. This dilemma involves more than simply the allocation of time, as Coser and Rokoff (1971) pointed out. The issue of "family" and "career" implies conflict between deeply held social values, too: "Professional women are expected to be committed to their work 'just like a man,' at the same time they are normatively required to give priority to their family (Coser and Rokoff 1971, p. 535). Moreover, it has been documented in many studies that the conflict experienced by many women in the paid labor force, as indicated in the responses of the CWRU women, is often associated with guilt (see Pearson 1990;

***Table 7.*** Conflict Experienced between Work and Children

| | Female (%) | N | Male (%) | N |
|---|---|---|---|---|
| Rarely | 15 | 4 | 18 | 9 |
| Sometimes | 42 | 11 | 57 | 28 |
| Often | 23 | 6 | 16 | 8 |
| Very often | 19 | 5 | 8 | 4 |
| | | (N = 26)) | | (N = 49) |

**Notes:**  $F = 2.23$; *Sig.* = .139.

Hochschild 1997). Also, along with the guilt, some of the women appear to project a sense of sadness.

Among some of the men, lack of time to spend with their children also contributes to their stress. For example, one man wrote, "I am not able to spend as much time with [my children] as I would like, or attend doctor appointments, or games or programs. I am never home for dinner unless we eat out on a weekend." Another man reported, "I am not able to devote as much time to my daughter as I would like. Often I don't get home until after her bedtime." Still another man wrote, "It is extremely important for a parent to spend time (not just 'quality time,' but absolute quantitative time also) with children, and my work inhibits that occasionally."

## Side Bet 8: Developing a Congruent Construction of Reality

Side Bet 8, examined the development of a relatively congruent construction of reality around career identity. Lopata (1992) hypothesized the higher the level of occupational/professional commitment, the more the person is likely to pull together a congruent image of the self, the job, and the environment. She contended that the very process of making conscious career commitments can push a person toward increased congruence. Therefore, projected career goals and the perceived possibility of their goal attainment could be reflective of the degree of occupational commitment and congruency.

It appears that the vast majority of the CWRU women and men graduates have pulled together a congruent image of the job. However, does this congruency of their professional goals and commitments come at a cost to their personal selves? According to the graduates, it most certainly does. When the extent that their work conflicts with their personal interests was examined, 45% of the women and 34% of the men experienced conflict "often" or "very often" and 33% of the women and 53% of the men "sometimes" experienced conflict. As discussed throughout this paper, "time demands" are the primary source of conflict.

One woman, in response to her work conflicting with personal interests wrote, "Are you kidding! I work, I raise two kids and I maintain our household—how much time can be left?" Another woman reported, "When there is simply not enough time, personal interests get sacrificed before family interests." And, one of the few women in this study who was not employed in the paid labor force wrote, "My 'work' is primarily raising the kids, with the exception of volunteer work on political campaigns or as director on a board. However, I've found that kids will eat up every available minute, often to the detriment of the parent's personal interests." Keeping this in mind, one can only imagine how those women with full-time careers and employment married to men with similar working commitments, experience time demands.

Time demands are also experienced by the men, as evident in this response, "There is just very little time to relax enough and do the personal things I like, and not take time from [my] wife and children." Another man indicated, "I can't read or play as much as I would like. I've had no time to learn golf and little time to travel."

In Hochschild's (1997) study of work and family "time binds" among the women and men employed at Americo, she asserted that the more women and men do what they do in exchange for money and the more their work in the public realm is valued and/or honored, the more, almost by definition, private life is devalued and its boundaries shrink (1997, pp. 198-199).

Nevertheless, despite the conflict the graduates experience, whether it was with job changes, stress, or conflict with spouse or parental roles, most of the graduates had high levels of career satisfaction and continued to be employed in the legal profession. This holds true even when the vast majority of women and men experienced a high level of conflict with work and personal interests. Regardless, nearly half of the women and men were "very satisfied" with their family life today, and nearly one-third of both women and men were simply "satisfied" (see also Chambers 1987; Mattessich and Heilman 1990).

Perhaps the legal profession, and professions in general, which have incredible time demands and pressures—along with the socialization process at both the law school level and the professional level—drive and maintain the image of congruency. As a result, such congruency, as Lopata (1992) asserted, creates a strong commitment to the profession. This may be the only way to maintain the self, job, and the environment in order to maintain professional goals and commitments.

Hence, although the graduates cope daily with the stress and conflict of meshing their professional and personal lives, overall they are satisfied with the results and continue to be committed to their professional goals, evidenced by the fact that they generally do spend their time mainly on the job (see also Hochschild 1997).

## CONCLUSIONS

Although the women and men graduates did equally well in law school, differences and similarities emerge in career and family experiences once outside of law school. An examination of the eight side bets revealed gender differences among the graduates which include: (1) men occupy different career positions, practice law in more diverse settings and were more than twice as likely to have achieved partnership status in their firms; (2) dramatic differences emerged among three aspects of career satisfaction, with women being more dissatisfied with the prestige from their position, prestige in community and opportunities for advancement; (3) men were more involved in professional associations than women; and (4) men experienced spousal career support that is relevant to their career ambitions, goals and advancements. The women, however, received more "tolerance," a positive attitude or supportive attitude toward their career commitments and goals. Moreover, when men did experience career hindrances from their wives, it was in the form of an "intolerance" of their long hours and frequent and extensive travel, a hinderance which is an attitude but not a direct barrier to career performance or advancement. The women, on the other hand, had hindrances that were the result of having the primary burden of child care and household responsibilities, hindrances which decrease the amount of time available for career commitments and obligations—a direct barrier to career performance and achievement.

The similarities among the women and men graduates appeared when men had children and wives who were employed full-time. These similarities include: (1) men who indicated having experienced work related stress due to their attempt to balance work and family life; (2) men who reported their career hindrances were a direct result of having a wife who also had a career or was employed full-time; and (3) men who revealed that they experience stress because their professional obligations and commitments do not allow them to spend as much time with their families as they would have liked.

When examining career commitment of the women and men graduates, it may be concluded after looking at side bets that American society, the legal profession, and careers are not set up in such a way that women with spouse and child responsibilities can continue total professional commitment. So, although women can put all of their eggs in one basket for law school, afterward they have to withdraw some of them because of children and lack of cooperation and division of labor with a spouse. It appears, for the most part, that although the spouse is aware of and sympathetic to her needs, she is the one who adjusts her work schedule and shoulders the primary responsibility of the children.

This study indicates that women cannot maintain side bets as well and as easily as the men can. Many significant gender differences were rooted in child care/ household responsibilities. This is an important issue, directly or indirectly, for most of the CWRU women and men graduates, especially since the vast majority have young children (infant, preschool, and early grade school age). Primary responsibility for family obligations and commitments has career consequences. As indicated in this study, most men tend to construct their lives around their jobs, while most women tend to construct their jobs around their personal lives.

The majority of the men do not experience the demands of child care/household responsibilities as being detrimental to their career opportunities and choices, but most women do. This could be because nearly one-third of the men have wives who are "homemakers," not currently employed in the paid labor force. As a result, the majority of the CWRU men are the sole career person and sole or primary source of income for their families. This puts less constraints on their time for professional responsibilities and obligations, which could put them at a professional advantage over most of their female counterparts in attaining positions of employment and salary. In turn, this places men at an advantage in maintaining the eight side bets.

Moreover, many women accepted restrictions on their careers if they came from their husbands. Not only are they not receiving support from their spouses, but they accept this. And for those women who receive career support, the vast majority are not getting the type of support they need, and what they get is not as relevant as the spousal career support received by their male counterparts. Therefore, this study suggests that career differences among women and men are not attributable to their law school experiences per se but to the lack of cooperation they receive from their spouses, and the workplace, for their family responsibilities and obligations.

While overall the graduates are relatively satisfied with their jobs, it is necessary to ask whether or not they will continue to be satisfied—as they appear to want to have *both* their family lives and their professional lives—if the profession continues to follow the traditional male career model of success, that is, for a man with few, if any, responsibilities. This model appears to dissatisfy not only women but some men as well. Perhaps women and men may become increasingly dissatisfied if the legal profession fails to confront the idea that both women and men lawyers need to lead satisfying personal lives, and participate in their communities, in addition to practicing law and remaining committed to their professional goals.

However, it should not be overlooked that CWRU women and men have similar high levels of career satisfaction. Perhaps women are, in their own ways, attempting to redefine the traditional male career model of success, making it more conducive to fit their unique needs. Moreover, with the increase of men in solo firms, one can not help but wonder if they, too, are attempting to redefine the traditional male career model of success by defining success more on an individual level, as opposed to what is expected in the legal profession and in American society.

The overrepresentation of women in solo and large law firms reported by CWRU women and men deserves further attention. These differences suggest that while larger numbers of women have entered the practice of law and will continue to be employed in legal positions, the goal of fully assimilating women into the profession has not yet been met. The findings of this study suggest that a goal more conducive to full participation and maximum productivity of both women and men may be to integrate the unique needs identified in this study. After all, for many families, a mother in the paid labor force is not an option; it is necessary for economic survival. As revealed in this study, problems that have fallen solely on the shoulders of most women—the struggle of attempting to be accountable for both work and family responsibilities—are now being experienced by those men who also have a spouse employed in the paid labor force.

In conclusion, as Lopata (1992) contended, most men in modern America are expected to develop and maintain professional side bets. It has only been in recent years that men have needed to justify organizing their lives around their occupations. As evidenced in the literature, and as further indicated in the experiences of the CWRU women and men law school graduates, the occupational system continues to be a "greedy institution" with "greedy" commitments, demands, and expectations. Moreover, for many of the women, and for those men involved in dual-career or one career/two-paycheck marriages, the family is an equally

"greedy institution" with equally "greedy" demands, commitments, and expectations of its own. Therefore, the degree of career commitment for most of the
CWRU women and men is dependent upon the investment each makes in side bets
in terms of self-concept and identities, and external contributions from the environment. Such investments help to maintain a continued involvement with their
professional goals, which in turn dictate the strength of their professional commitment.

In other words, the differences in the careers of women and men are, to a great
extent, a consequence of differences in whom they marry, child responsibility, and
the type and degree of support women and men give each other in their career
commitments and family/household responsibilities and obligations.

## REFERENCES

Abel, R.L. 1989. *American Lawyers*. New York. Oxford University Press.
American Bar Association Commission on Women in the Profession. 1985. *Women in Law: Statistical
    Data*. Chicago, IL: American Bar Foundation.
American Bar Association Commission on Women in the Profession. 1988. *Summary of Hearings ABA
    Mid-year Meetings*. Philadelphia, PA, February 6-7.
American Bar Association Commission on Women in the Profession. 1988. *Report to the House of
    Delegates*. Chicago, IL, June.
Becker, H.S. 1960. "Notes on the Concept of Commitment." *American Journal of Sociology* 66: 32-40.
Becker, H.S., and J. Carper. 1956. "The Development of Identification with an Occupation." *American
    Journal of Sociology* 61: 289-298.
Chambers, D. 1987. "Tough Enough: The Work and Family Experiences of Recent Women Graduates
    of the University of Michigan Law School." Draft report, University of Michigan, Ann Arbor.
Chambers, D. 1990. "Accommodation and Satisfaction: Women and Men Lawyers and the Balance of
    Work and Family." *Law & Social Inquiry* 14: 251-287.
Coontz, S. 1997. *The Way We Really Are: Coming to Terms with America's Changing Families*. New
    York: Basic Books.
Coser, L., and R.L. Coser. 1974. "The Housewife and her 'Greedy Family.'" Pp. 89-100 in *Greedy
    Institutions*, edited by L. Coser. New York: Free Press.
Coser, R.L. 1974. *Greedy Institutions*. New York: Free Press.
Coser, R.L., and G. Rokoff. 1971. "Women in the Occupational World: Social Disruption and Con-
    flict." *Social Problems* 18(4, Spring): 535-554.
Curran, B., and C.N. Carson. 1994. *The 1994 Lawyers Statistical Report*. Chicago, IL: American Bar
    Foundation.
Epstein, C.F. 1971. "Law Partners and Marital Partners." *Human Relations* (December 24): 549-564.
Epstein, C.F. 1983. *Women in Law*. New York: Anchor.
Epstein, C.F., R. Sauté, B. Oglensky, and M. Gover. 1995. "Glass Ceilings and Open Doors: Women's
    Advancement in the Legal Profession." *Fordham Law Review* LXIV(2, November): 291-449.
Fritz, N.R. 1988. "In Focus: A Close-up Look at What's New in the HR Picture: Legal Equals?" *Per-
    sonnel* 65: 45.
Gouldner, A. 1957. "Cosmopolitans and Locals: Toward an Analysis of Latent Social Roles—I."
    *Administrative Sciences Quarterly* 2: 281-306.
Gouldner, A. 1958. "Cosmopolitans and Locals: Toward an Analysis of Latent Social Roles—II."
    *Administrative Sciences Quarterly* 2: 444-480.

Hertz, R. 1986. *More Equal than Others: Women and Men in Dual-Career Marriages.* Berkeley: University of California Press.

Hochschild, A.R. 1989. *The Second Shift.* New York: Viking.

Hochschild, A.R. 1997. *The Time Bind.* New York: Metropolitan Books.

Hughes, E.C. 1963. Professions. *Daedalus* 92: 665-668.

Kanter, R.M. 1978. "Reflections on Women and the Legal Profession: A Sociological Perspective." *Harvard Women's Law Journal* 1: 1-7.

Liefland, L. 1986. "Career Patterns of Male and Female Lawyers." *Buffalo Law Review* 35: 601-631.

Lopata, H.Z. 1992. "Career Commitments of American Women: The Issue of Side Bets." *The Sociological Quarterly* 34: 257-277.

Lopata, H.Z. 1994. *Circles and Settings: Role Changes of American Women.* New York: State University of New York Press.

Lopata, H.Z., C. Miller, and D. Barnewolt. 1986. *City Women in America: Work, Jobs, Occupations, Careers.* New York: Preager.

MacCrate, R., ed. 1992. *Legal Education and Professional Development: An Educational Continuum,* student edition. New York: West Publishing.

Mattessich, P., and C. Heilman. 1990. "The Career Paths of Minnesota Law School Graduates: Does Gender Make a Difference?" *Law & Inequality* IX: 59-114.

Pearson, K.M. 1988. "Women and Men in the Legal Profession: Are the Scales of Justice Unbalanced?" Unpublished paper, University of Wisconsin, Milwaukee.

Pearson, K.M. 1990. "Illusions and Realities: An Examination of Women and Men Attorneys in Small Law Firms." Unpublished Master's thesis, University of Wisconsin, Milwaukee.

Ritzer, G., and H.M. Trice. 1969. "An Empirical Study of Howard Becker's Side-Bet Theory." *Social Forces* 47: 459-479.

Schrimsher, K.P. 1996. *Professional Socialization, Career Patterns & Commitments: A Gender Analysis of Law School Graduates.* Ann Arbor, MI: UMI Dissertation Services.

Spangler, E. 1986. *Lawyers for Hire.* New Haven, CT: Yale University Press.

Spencer, A., and D. Podomore. 1987. "Women Lawyers: Marginal Members in a Male Dominated Profession." In *A Man's World: Essays on Women in Male Dominated Professions,* edited by A. Spencer and D. Podomore. London: Tavistock Publishers.

Stanford Project. 1988. "Gender, Legal Education, and the Legal Profession: An Empirical Study of Stanford Law Students and Graduates." *Stanford Law Review* 40: 1209.

Stebbins, R. 1970. "Career: The Subjective Approach." *Sociological Quarterly* 11: 32-49.

Stebbins, R. 1971. "The Subjective Career as a Basis for Redefining Role Conflict." *Pacific Sociological Review* 14: 383-402.

Teitelbaum, L., A.S. Lopez, and J. Gender. 1991. "Legal Education and Legal Careers." *Journal of Legal Education* (September/December): 443-481.

Vogt, L. 1986. "From Law School to Career: Where Do Graduates Go and What Do They Do?" Report prepared for Harvard Law School Program on the Legal Profession.

White, J.J. 1967. "Women in Law." *Michigan Law Review* 65: 1051-1122.

Wilinskey, H. 1964. "Careers, Lifestyles, and Social Interaction." *Social Sciences Journal* 12: 553-558.

"Women in the Profession: Accountants Should Share Lawyers' Fears." 1980. *Accountancy* 101: 11.

# CAREERS OF EARLY
# AFRICAN–AMERICAN
# WOMEN EDUCATORS

Kathleen F. Slevin and C. Ray Wingrove

## ABSTRACT

Using feminist and life course frameworks, this chapter examines how historical forces, as well as race, class and gender, influenced the career paths and work experiences of 30 African American women educators who are now retired. Through their narratives we explore the challenges faced by these women as they developed careers and worked in both segregated and desegregated education systems. The effects of race and gender discrimination are apparent, yet we come to appreciate the extent to which these educators survived and resisted injustice by calling on the strengths of their race and gender. Their reminiscences illustrate the many ways that their professional identities were linked to "uplift work" in which they strove to create a better world for the next generation of African Americans.

Current Research on Occupations and Professions, Volume 10, pages 217-237.
Copyright © 1998 by JAI Press Inc.
All rights of reproduction in any form reserved.
ISBN: 0-7623-0034-5

*I ran into a very bad situation with a* [male] *principal and my doctor advised me to get out of that situation because I developed an ulcer. So, I had requested a transfer. And, to be perfectly honest, I* [told] *the Director of Personnel . . . that I'd take a job in West hell. Because I'm in East hell now and West can't be much worse.* (Retired elementary teacher and school librarian)

*        *        *        *        *

*And . . . at that time the rule of thumb was* [that] *you had to come out of the classroom* [without pay] *before you wore maternity clothes. So, at the end of your third month you had to come out. And you couldn't come back until the child was eight months old.* (Retired elementary school teacher)

*        *        *        *        *

*. . . had I been white, my work experiences would have been different. I would have excelled further in status, position, and everything else.* (Retired specialist in special education)

*        *        *        *        *

*I prepared my work, and I enjoyed doing it. Of course, some of the children some-time would make you feel like killing them. Nothing bothered me more than to have a child with a good brain* [who did] *not use it. I was real concerned about that. And then I was concerned about the children's home life.* (Retired elementary school teacher)

Thirty retired African-American women allowed us to come into their homes and shared reminiscences such as these from their careers in education. Their stories transported us to the earlier decades of this century as they reminded us how their lives were shaped by coming of age in a time when legal segregation affected their lives so powerfully. They discussed why they had channeled their hopes for a better life into the field of education, and they described the steep climb toward their goals. They told us about the shortcomings as well as the benefits which they received from their work, and they spoke at length about what it was like to work in both segregated and integrated work worlds.[1] As they reflected on their lives, we learned that they had achieved success despite the odds against them. They prevailed, notwithstanding barriers stemming from the interlocking social systems of race, social class, and gender. They overcame obstacles which most would consider formidable. Indeed, the push toward achievement had been uphill most of the time. Yet, as one woman aptly expressed it, "We made stepping stones from stumbling blocks."

# THE INTERVIEWS

We conducted our interviews in the summer and fall of 1993 and the spring of 1994. Using a "snowball" sampling technique, we established a list of women who met our criteria for inclusion in the study. We began the process with the help of two prominent African-American women, each in a different southern city, who were in positions to identify other retired professional African-American women. Each was told the nature of our investigation and was asked to supply us with names of older African-American women who had been employed in the professional or business world.

All the interviews were conducted in the homes of the women, and interviews averaged three hours, with some lasting as long as five. With the exception of three women who were uncomfortable being recorded, all interviews were taped.[2] At the close of each interview, we asked for additional names of retired, professional or business women who might agree to be interviewed. We continued interviewing until our sample size reached 50, 30 of whom are the educators discussed in this chapter.

## Our Sample: A Brief Profile

The 30 educators we interviewed ranged in age from 54 to 87, but only two were in their 50s and six in their 80s. The average age was 71. Only one woman had been born and reared outside the South.

Eighteen women hold master's degrees and seven have earned doctorates. Seven retired as college professors or administrators and the remaining 23 had filled a variety of teaching/supervisory positions in elementary or secondary education. Most of them married, and 17 still live with their spouses. Nine of the 11 widowed live alone as does the one divorced woman. The remaining three share homes with an adult child or companion.

A large majority of the women live quite comfortably today. Eighteen of them enjoy family incomes exceeding $3500 per month. Their incomes, however, are less than they feel they would have had if they had been white and male. Lifetimes of salaries and opportunities limited by race and gender are reflected in correspondingly lower retirement incomes. They realize that in this way, the conspiracy of race and gender continues to haunt them in their retirement years.

# THEORETICAL ORIENTATIONS
# GUIDING OUR DATA ANALYSIS

We were guided in our search for recurrent themes in these women's lives by the perspectives of the life course and feminist frameworks. A life course perspective is described by sociologists Stoller and Gibson as one that explores "the ways in which people's social location, the historical period in which they live, and their personal biography shape their experience" (Stoller and Gibson 1994, p. xxvii). Such a perspective demands that we be sensitive to the way in which the time period in which people live relegates them to prescribed social roles which in turn limit their opportunities and narrow their chances of achievement. People who live at the same time and who are of similar ages experience their life stages and transitions within the same historical period. Nevertheless, the same historical event can have very different meanings for individuals' lives depending upon their social locations. Thus, the feminist perspective blends very well with that of the life course as it insists that we be mindful of the importance of race, social class, and gender and their intersections with the impact of historical events and individual biographies.

The life course perspective also points out the importance of cohort analysis in the understanding of human responses to historical events (Stoller and Gibson 1994, pp. 7-8). However, even though our sample includes several cohorts, we have chosen not to engage in cohort analysis. As we examined our data, we discovered the extent to which the lives of these African-American women defied such analysis. A pervasive similarity characterized so many of their descriptions of critical life experiences that the absence of cohort differences was often palpable. In a fundamental way, their lives were powerfully shaped by structural constraints that were ubiquitous and consistent. All grew up in a society characterized by segregation and all received the bulk of their education in such a system. The effects of this system of racism, intertwined with systems of class and gender inequality, led to a pervasive and homogenous oppression for most of their lives. Hence, we concluded that cohort analysis for our sample would not be productive.

We were also mindful, as we looked for the common threads in these career women's lives, to avoid the use of the male model as our point of comparison. Feminists Diane Beeson (1975) and more recently Diane Gibson (1996) remind us of the pitfalls associated with relying upon the experiences of men as the pivotal standard for understanding the lives of women.

## CREATING A COLLECTIVE BIOGRAPHY

In order to create a collective biography of these women's lives, we used the analytical inductive method central to grounded theory. In this way, we uncovered patterns and common analytical themes (see Glaser and Strauss 1967). We employed a process referred to by sociologist Margaret R. Somer as "causal emplotment"—a process involving social narrativity in which connecting stories is the essence that gives meaning to individual instances. She says, "narrativity precludes sense-making of a singular isolated phenomenon. Narrativity demands that we discern the meaning of any single event only in temporal and spacial relationship to other events . . . the chief characteristic of narrative is that it renders understanding only by *connecting* (however unstable) parts to a constructed *configuration* or a *social network* (however incoherent or unrecognizable)" (Somer 1992, p. 601, italics in the original).

Throughout our discussion, we highlight the patterns and themes that were shared by the majority of the educators interviewed. Where relevant, we note exceptions to these common patterns. Our overall focus is not on the single experience of one woman but on the *patterns* of experiences that are *corroborated*. Thus, the themes we present do not mirror or match specific questions asked of the women. Instead, they emerge from our interpretation of *all* of the data collected. We use narratives from individual life stories in order to illustrate collective themes—creating a certain unavoidable but creative tension between the personal and the anonymous.[3]

## Remembering the Past

Asking someone to remember her earlier life invites her inescapably to use the lens of lived experience to reconstruct the past. For example, when discussing racial inequality and race relations in their early childhood, the retirees frequently followed a description of the discrimination faced by African Americans in those days with the words, "That's the way things were, we didn't know any different *then*" (author's emphasis added). Similarly, as they looked back on the discrimination that they faced because they were women, a number of them commented on the fact that awareness of gender inequities was not part of their consciousness in their earlier lives. "We were not into the gender thing then," were the words one woman used to explain that many aspects of gender inequality were invisible to them during their youth.

Since reconstruction of the past seems inevitable, the question of accuracy arises—especially when we called upon respondents to recall events which spanned at least six decades. However, because our central focus concerns the exploration of the lives of a particular group of women from *their* perspectives, and because we are committed to allowing *their* voices to interpret *their* lives, we assume that the women we interviewed told their stories as they saw fit. In fact, it is precisely *their* interpretations of *their* life experiences that form the essence of this study. Our task is to analyze and to give voice to their subjective experiences. In this regard, we agree with the insight of feminist historian Mary Jo Maynes: "the line between lived experience and memory of experience, while impossible to draw, is for many purposes not relevant. Since people act not on the basis of unassimilated facts of an existence but on the basis of the sense they make of a particular experience within the context of a lifetime, for many analytical purposes recorded memories are the pertinent evidence" (Mayes 1992).

We begin by exploring the ways that race, class, and gender constrained the employment and career opportunities of the women educators we interviewed.

Their stories of thwarted opportunities and challenges in segregated as well as integrated environments help us to fathom the extent to which racism *and* sexism prevailed in their careers. Yet, despite the discrimination they experienced in the world of work, it is clear that these women approached their lives as survivors and not as victims. They are tenacious women who faced adversity without being consumed by it, who turned stumbling blocks into stepping stones.

Race and Gender: Their Impact on Career Choices and Work Experiences

In step with the social geography of the time, all but one of the women interviewed *began* their careers in elementary education. Even though some women spoke of youthful dreams for professions in law or medicine, they were aware of the limited career choices open to them as college-educated African-American women. One woman described the taken-for-granted nature of gender restrictions with these words: "Well, girls didn't go in for much law or dentistry or medicine. It was just an accepted fact that you didn't do it." "There weren't too many fields that were open to us then," was another phrase that we heard repeatedly during these discussions. Teaching, nursing, and social work were almost exclusively the career opportunities available to college-educated women and especially to women of color.[4] The explanation of one woman's decision to be a teacher puts a human face on these realities, "I didn't want to be a nurse, I was really interested in becoming a doctor but . . . if I couldn't be the real thing [I'd become a teacher]."

As they began their careers, their accounts illustrate a keen awareness of the restrictions imposed by race. In contrast, gender constraints were largely taken for granted and were relatively invisible to them.[5] As one woman explained, "Sexism just didn't play a role [in how we saw things] at that particular time," and another mused, "Back then we were not into the gender thing."

The pervasive extent of the invisibility of gender restrictions is illustrated as these women recollect their career choices. One woman, among the very top students in a very competitive college preparatory class in a Northern high school, recalled: "I was preparing to be a teacher. If I had been a man I probably would have been a lawyer or a doctor." She remembered that as a young woman she was very conscious of how her race created impediments, but it had never dawned on her that she had been limited in her choices also because she was a woman. Today, current awareness of sexism and women's inequality unavoidably influences how she and a number of the other women reconstruct their pasts.

Twenty-two of the 30 educators we interviewed worked the majority of their careers in settings which were predominantly African American. Even though such work arrangements offered them respite from many of the daily personal assaults of a racist society, they did not remove the reality of racial inequities. Salaries were depressed for black American professionals, as they were for the black American population in general. One finds more than a strong hint of the consequences of such restrictions in the comment of a retired college professor who spoke of how, "when I saw the black attorneys, they were all in these frayed white collars, all frayed because they were so poor." In one instance, a former teacher told us that she started teaching for $2500 per year in a county where white women started at $3500. Another recalls how in her early days of teaching, she worked in the evenings as a domestic worker for two white women who were also teachers:

> Even after I was in the school system, I worked in the evenings. Because my younger sisters and brothers had to have car fare. And I went to work for these two teachers. . . . And they used to ask me questions . . . "where do you get the energy to come and work for us in the evenings? And we do the same thing that you do and we're so pooped out." But, you see, you do what you have to do. And that little change that they paid me, I gave it to my sisters and brothers . . . for bus fare.

Here we get a strong sense of how race influenced the status of this teacher. Because she was African American and paid less than whites, her profession did not provide her an adequate income to meet her family obligations. Yet, as she pointed out, the white women teachers could afford the "little change" it took to hire a black woman to do their housework.

These women also experienced blatant discrimination in matters of promotion, and geographical location made little difference. For example, a woman who worked in the North in elementary education told how she found herself in settings where she was one of relatively few African-American women. In the 1940s, she applied for the position of acting vice-principal in her school but the job was given to a white woman. She confronted the white superintendent and asked him why she had not been appointed, "Is it because I'm colored?" After a long silence on his part he responded, "Well, it could be." In a comment that reinforced her own sense of injustice, the white woman who was appointed later said to her, "Alice, let me tell you something. The only reason you didn't get this job is because you're colored. I'm telling you now. That's the only reason."

Despite low salaries, the female-dominated profession of teaching provided something African-American professional women had to have—a steady and reliable income. Yet, we again find that gender was entangled with race to magnify their salary inequities. Men were paid more than women, and white women were paid more than black women. The reminiscences of one retiree illuminates the ways that race and gender discrimination were intertwined within the black American community. Recalling the sacrifices it took to save the money needed to attend college in order to get her teaching credentials, she spoke of how men who had considerably less education that she could earn more money. She remembered her husband's remark as he reacted to her first pay check as a teacher, "custodians [are] making more money than you."

Although racial and gender inequities were often intertwined within the black American work environment, African-American scholars suggest that black women who worked in predominantly African-American workplaces found gender to be a more personally salient factor than race (Collins 1990, p. 86). Our conversations with the women who worked in such environments support these findings. It is clear, however, that retrospection heavily influenced the ways that most of these women recognized gender hierarchy and gender discrimination in their work environments. Retrospection aside, gender inequities that had been either invisible and taken for granted at younger ages—particularly within the context of primary groups and informal relationships—became more striking within formal groups and secondary relationships. Working on a day-to-day basis with men, especially where men were the majority, is key to understanding their experiences. For instance, the minority who claimed never to have experienced gender discrimination were those who worked in predominantly female environments.

Those who were most impassioned about the effects of gender discrimination were the seven women who worked as professors and administrators in black American universities and colleges which were predominantly male. They spoke about how they frequently found themselves overworked, underpaid, and seldom appreciated by the male administrators with whom they worked. For women who had gone to extraordinary lengths to acquire the doctoral degrees that placed them among the academy's elite, their second-class citizenship often proved particularly galling. Among the numerous contradictions these women faced, the one that stands out especially is how their gender affected their experiences with "race uplift" work.[6] While both professional African-American men and women shared a commitment to the uplift of their race, the *consequences* of this commitment were gendered. Women's commitment was often used in pernicious ways by male superiors in order to keep them in subordinate positions and to discourage them from complaining too much or too publicly about such inequities. One retired university professor who chaired her department for many years and directed her division for several more, spoke of having experienced gender discrimination very often. Nevertheless, fighting against it was difficult—so difficult that she "internalized a lot and got an ulcer." In discussing her career, she said:

> [I was discriminated against as a woman] in pay, in promotions. [There was also] a double standard for evaluating my performance. One of my female associates studied the differences between women and men and was chastised. When I became Director I got $3,000 to bring me to the *base* for administrators at that rank. . . . My contributions were not given equal evaluation or recognition as men. Evaluation[s] determined raises and promotions. If I made a speech or attended meetings or held [professional] offices, I was not given the same credit. I got a "3" and men got a "4."

This woman went on to tell how she always taught 15 hours per semester in addition to the work load required by her administrative duties. As a result of her heavy work load, she experienced a further drawback that was mentioned by several other women professors—"lack of opportunity to do research or write." Retired women professors spoke too of the "academic perks," such as summer support, that were routinely given to male colleagues and not to them. They put up with such inequities not only because that was "the way things were" but also because of their commitment to education and to doing their part to improve conditions for their race.

Despite the fact that most held doctoral degrees, women who held positions in university administration usually worked most of their careers as administrative assistants where, in the words of one, "the connotation in discussion with others was [that you were] secretarial and not administrative at all. Although [the latter] was the kind of work that you were doing." Promotions or opportunities to move up the administrative ladder were routinely blocked. One retired administrator bitterly complained, "when openings came . . . a second-rate man would be considered rather than the first-rate [woman] who knew what was going on." And then when "second-rate" men got the promotions, "[they] would have to come to you anyway for the information." Another woman remembers her blocked aspirations:

> Several times there were openings in key central administrative positions and I indicated an interest in them and it was quite obvious that I was capable of doing the job. And the answer that I got was that it was a man's job and not a woman's job. I think it was a gender problem. That, plus the fact that I'm not too sure that they wanted to lose me from the position that I held.

As this comment suggests, she felt that *because she was a woman* there was an underlying assumption that she should sacrifice her ambition to the common good. Another story from a woman who was well-qualified to assume a major university administrative post, reinforces how women were blocked for promotions and expected to accept it. Of the university in consideration, she said, "[It] is not good to women, [it] historically has not had women in top administrative positions." As she describes her plight, her words highlight how survival also dictated that these women "pick their battles":

> And then a new Director was hired, a male with less education, less experience [than I had]. . . . And I reacted, I objected. And the president of the university was adamant that he wasn't going to let me tell him what to do, that he was going to hire the man. Now this is when I really ran into the sexism thing.

Given the power and adamancy of the president, resistance could only go so far. Survival in this environment required that she accept her fate.

Women who worked in the mostly female world of elementary or middle schools more frequently gave accounts that suggest that they were not touched on a daily basis by gender inequities. Nevertheless, such environments were not without tensions between the women teachers and the few men who were almost always their principals. A retired elementary teacher was very pointed in her criticism of her male principals. When asked to discuss the negative aspects of her work life she summed them up in the following pithy words, "jack-assed principals!"

The 14 women who remained in the classroom throughout their careers often felt that they themselves could have made excellent principals, but in most cases they did not even apply for such positions. They were satisfied with the contribution they were making in the classroom and they spoke of recognizing that their contributions directly helped improve black American children's lives and opportunities. One woman, defending her choice to remain in the classroom, said, "I was satisfied with who I was and what I did." In addition, they were keenly aware that administrative positions were almost always reserved for men. As one woman's words succinctly summarized it, "Had I been a man, in earlier years, perhaps I might have moved into [the] principalship. . . . I probably would have."

Not surprisingly, those women who taught at the high school level where the number of men teachers increased considerably were generally more aware of gender inequities. A former high school coach and chair of the physical education department recalled how shortly after integration, she helped organize black *and* white women coaches in her region to file a class-action suit for sex discrimination. In words that illustrate how the demise of segregation allowed at least some women to move beyond survival and to resist legally the inequities that all women faced, she explained, "we felt we were underpaid and we really felt that we were doing more work than the men. And we felt that we just weren't being treated fairly." She went on to recount how some of the women in other districts, "were threatened by their superintendents, and they were given $500 for coaching several activities, so they stayed out of it." "But we plowed on," was how she described the fight that led to the eventual law suit that garnered more equal treatment for all women coaches.

## TEACHING: DOING RACE AND GENDER WORK

The formative work experiences of all but the youngest of the women educators we interviewed were rife with the inequities spawned by a racially segregated system of education. With poor facilities as a given, their everyday work lives were dominated by two inescapable conditions—large classes and woefully inadequate supplies. The words of one woman who began her teaching career in 1948 describes the conditions: "When I first started teaching. . . . I stood up and looked at 42 children. Now that was a difference between the segregated schools and the integrated schools . . . we never had less than 30 children." The same woman, who taught for 41 years, went on to recall what it was like to teach fewer children once her school was integrated in the late 1960s. "I had 25 children. . . . I'd never seen just 25 . . . and I said, 'This is a picnic'." When it came to supplies for the children, teachers remembered how they bought them out of their meager salaries. Responding to a question about her sense of how things would have been different for her had she been white, a former teacher said, "I would've had more supplies to deal with and [I] wouldn't have used my own money for supplies. We did that for years. Bought our own . . . pens, paper and anything else that we wanted the children to have." To do this was both an act of survival and an act of resistance to a white administration that denied them the basic necessities.

Although salaries were low, classes large, and work conditions poor, there were intangible benefits which made survival worthwhile. Respect for teachers in the African-American community was high. Teachers were still revered, not only because they were professionals at a time and in a place where most were not but also because they were the professionals who could provide black American children with the education they needed to make dreams of a better life come true. In a very real sense, teachers were vehicles of class mobility. The words of one former teacher capture this idea, "When I first started teaching . . . the parents were glad for us to . . . try to teach their children the things that would make them better . . . if you were a teacher . . . you were highly respected." In a paradoxical way, black American teachers also enjoyed a measure of freedom in the classroom that was unavailable to most whites. For instance, while black schools were subject to white policy on matters such as budgets, plant facilities, and general curriculum, they escaped white scrutiny of their everyday operations. This meant that black teachers had a certain autonomy that allowed them, for example, to teach black history. Undetected by white eyes, they could give their students that special attention which they felt black children needed in order to have any chance at all in their lives. These aspects of life as professionals provided them with a strong sense of satisfaction and reward and allowed them to carry on the tradition of survival and resistance. The genesis of such fulfillment was certainly intimately tied to the oppression of the broader African American community. As Patricia Hill Collins notes, "This culture was a culture of resistance, essential to the struggle for group survival" (Collins 1990, p. 147).

A major tactic in the fight for survival of the next generation involved "uplift work" or "race work" which has already been identified. Here gender came into play in yet another way. Some of the women interviewed thought, for example, that certain aspects of race work were unique to women, "Men didn't have the sense of mission [we women had]." For these women, a key dimension of their jobs as teachers was to prepare the next generation to *survive* in a racist society. Another was to provide the children with the ammunition needed to *resist* the negative messages from the dominant society. A third involved the drive to provide young people with the educational tools that would allow them to *prosper*, to be upwardly mobile, to move toward a better life than their parents, "to make something of themselves." Some teachers spoke of their duty not only to confront but also to dispel the negative stereotypes that were held by the dominant society regarding African Americans' intelligence. One woman touched on this directly when she talked about how hard she worked to make her students excel: "We dogged those children and got the best out of them . . . when I say dogged, I mean

we dogged them. Stayed on them." Referring specifically to the dawning of desegregation, she continued, "I knew that our black kids were going to put the lie to
what white folks had said, because we had graduated many, many good students."
Another woman alluded to the same issue when she said, " It wasn't right to see
talent wasted, potential wasted. We cared. We knew the road and we insisted that
they work up to their potential. We drove them." The allusion to a common road
is an important one because it speaks to a common foundation that characterized
the black American experience and created the conditions that ensured that gender
and class took a secondary status to race.

These teachers' concerns for their students' chances in the broader community
reflected their social class and gendered perspectives as well as their racial identity. Reared in communities where they themselves were treated as children of the
community, where the family extended into the community and the community
embraced the family, where teachers were required to visit the homes of their students several times a year, it is hardly surprisingly that these women teachers
became surrogate mothers to their own students. For example, if children from disadvantaged members of the community were not taught the basics of hygiene and
personal care at home, teachers took it upon themselves to teach them at school as
the following comments illustrate:

> When I first started working . . . conditions were extremely poor, children would
> come to school unkept. . . . Finally, I decided [that] I would get a washcloth and soap
> . . . we had a little wash-pan that we're supposed to wash the board with . . . and we
> would get water and the kids would go in the bathroom and wash themselves up.

*   *   *   *   *

> I bought clothes for many [students], shoes and underwear [too]. I'll never forget one
> girl came one day and we were having a play. And, of course, when [she] bend over
> [you could see that] her pants [were] just like that floor there. So I went to the office,
> asked the principal if I could send somebody up to a place called Troy's, a discount
> store, and get something to put on her.

Others spoke of how they taught children good grooming, telling students, "Your
hair needs combing," or "You need to change clothes after a bath." As a former
teacher summarizes, "Black teachers taught everything—-etiquette, dress and
manners," or put differently by another, "[Black] teachers taught a kid, not a subject." In fact, a number of former teachers spoke directly about the human element
of their work, that is, how they taught children, not subjects. This in no way
intends to imply that academic content or rigor was irrelevant. The underlying
message is amplified in another description, "Black teachers knew from whence
the black child came. They knew the black experience and . . . had a vested interest
in seeing that child made progress . . . and that child . . . knew it."

It was important to these teachers that students knew they cared. The philosophy was that cared for children responded better in the classroom. A former teacher explained, "Children know if you care. You know, you don't have to say. . . . 'I care about you' . . . it's your facial expressions, your touch . . . the way you talk . . . they know that." What is strongly hinted at, but not explicitly spelled out, in all these comments is the fact that all black Americans—whatever their class, gender, or position in life—shared a feeling of unity, a sense of togetherness or community born from a common oppression. This meant that in the specific case of teachers, they served as a bridge between an oppressed group and the meritocratic social structure.

Over and over again, these women claimed that they approached their jobs differently from men. They felt that African-American women teachers took on different types of responsibility than their male counterparts.[7] One woman who taught especially poor children not only remembers how she and other women teachers paid attention to concerns beyond the academic, but she also suggests that such attention separated women from men teachers. She said, "A child coming to school hungry has behavior problems. They're all rude and crude. I don't know whether a man would have tried to find out . . . what was wrong with the child." As these words suggest, getting to know the "whole" child was an integral part of segregated education and it was a key to understanding the gendered component of how these women defined their work as teachers. The positive effects of a holistic approach to teaching children was something they had experienced in their own young lives, and they were keenly aware of "giving back," of doing the same for their own students. They knew no other way. The ethos of sharing, of "lifting as you climb," of combining the unpaid with the paid labor of women, inevitably infused how they approached their work. Repeatedly, we heard women speak about the importance of family, and we heard women use familial language to describe their work. These women approached the children they taught holistically; they saw them as part of their racial family. "Black schools were like family," was the way one retiree summed it up.

## INTEGRATION: THE ELUSIVENESS OF EQUALITY

The arrival of legal integration provides an interesting forum from which to explore how race continued to have significance in the work experiences of these women. This was particularly the case in elementary and secondary education, where the majority (23) worked throughout their careers. Teachers, whatever their race, confronted a new order of daily business as a result of the government mandate to integrate public schools. Legal integration forced the creation of a new system to replace what had previously been two systems with separate schools, separate professional meetings for the races, and different pay scales for black and white teachers. What white teachers had always taken for granted was now extended to many black Americans. Suddenly, they found themselves with new books and available school supplies. They no longer had to use their own money to get their students the basic necessities. A former teacher who remembered, "I never had a brand new book," contrasts the situation in desegregation where, "Most of the time, if I asked [the principal] for something, I got it that evening or the next morning. Scared me so bad, I didn't know what in the world to do! [Laughs]"

Other obvious material advantages surrounded the dismantling of legal racism such as better pay and better facilities, but desegregation also brought in its wake a variety of less positive consequences. Many of these women who had lived and worked in a predominantly black world now faced the challenges that accompanied being in a work environment where they were not the majority. A former teacher described the earliest days of desegregation in her school, where 70% of the teachers were white. She recalled how important it was that "we knew all the black teachers" because "I think it helped tremendously." Already well-known to each other because they came from close-knit communities, these women made it clear from their comments that the bonds among them strengthened as they faced the challenges of desegregation together. When asked to talk about relationships with their new white colleagues, one woman chose cautious and telling words to describe them: "I don't want to say that they [white teachers] were terrible." Such a comment suggests much that is left unsaid. These words, plus the fact that she immediately shifted the conversation to a former principal who had tried hard to create a positive atmosphere, "to make the tone for the school," suggested that she wanted to avoid discussing something that was negative. Indeed, by shifting the conversation, she tried to put a positive face on what clearly had been a less than positive situation.

Other stories told by these women point to the fact that the post-1960s era did not herald the end of discrimination and unjust treatment for African-American women. It simply assumed different forms.[8] In fact, a former guidance counselor claimed that her troubles *began* with the arrival of integration. "I didn't really experience [discrimination] until integration came. I really didn't . . . [but] when they integrated the schools, I was head of the guidance department and they sent some [white] folk from across town and I was in charge of the department and they resented it. And they let me know it!" In this case, the arrival of legal equality between the races may have influenced the balance of racial power and challenged the hierarchial nature of white privilege, but it did not stop whites, who now had to report to a black woman, from registering their disapproval of her authority. The stories of other women reinforce this same notion—even when they were in charge, their authority rarely went unchallenged by whites, at least initially.

What these women educators learned about working in desegregated schools was that race took on a *different* significance than it had in segregation. The dominance of race, and race discrimination, far from disappearing from their lives, now appeared in new and different forms that required new and different strategies for survival and resistance. While the rules shifted and the players appeared to come to the table as equals, the game changed little for them. These women spoke vehemently about the price African Americans paid, and continue to pay, to be part of a system where equality remains largely elusive. They recalled at length what they saw as contradictions of a system which was meant to improve the condition of black American life but yet had so many negative consequences for them. Black educators found themselves the targets of new forms of discrimination. For example, the closing of black schools created employment hardships for many. Statistics from that era indicate that large numbers of black teachers and principals lost their positions.

Those who were hired by the white establishment faced other new challenges. Classroom teachers recalled how, at least initially, some white parents "were very hard on black teachers." One former teacher claimed, "White parents did not want their children taught by a black teacher." Another explained it this way: "I think they were apprehensive simply because . . . they had read [that] black teachers were not as good as white teachers." Others remembered how some white parents not only did not respect black teachers but "made [things] very difficult for them." Grading was often a source of contention. A number of former teachers recalled how white parents routinely questioned any grade below a "B," implying, as one former teacher put it, "that it was my fault because [their child didn't get an 'A']."

Another example further illustrates the types of disrespect black teachers suffered at the hands of some white parents. A former elementary teacher in a large Southern city talked about the pervasiveness of racist attitudes on the part of whites. She spoke at some length about the "poignancy of racism." Her school served very poor neighborhoods, and here is how she remembers a specific incident of racism:

> I was teaching the whites from Maryhill and you are talking about some folk who are poor as Job's turkey, where incest is just rampant, where lice and filth is the order of the day . . . and I had this little boy who was in my class and, anyway, he had lice in his head. And the mother came to school and said, "Well, he got it from that old black teacher that's teaching there." . . . I made it a point to prove [that] black folk are not the ones that really carry lice and filth and dirt. . . . I went down to the public health department and got data to prove to her that black folk are less inclined to get lice in their heads than whites.

Apart from illustrating blatant racism, this narrative demonstrates how race and class intersect to the disadvantage of the teacher who is formally in authority. It shows the assumed privilege that being white gave a lower-class mother and, in turn, the lack of privilege that accompanied this African-American middle-class teacher. By contrast, it seems inconceivable that a white teacher would ever be so accused, or indeed, if accused, would feel any obligation to defend her entire race from such an inappropriate slur. Having left a system where the teacher's authority was nearly sacred in the eyes of students and their parents, for teachers like her, we can only imagine how difficult it was to find herself in a situation where some students and their parents assumed that she was either unfair, less than competent, or unhygienic.

Another disturbing consequence of desegregation discussed with both sadness and anger by a number of retired teachers concerned the treatment of black American children in desegregated schools. Desegregation created a new world where these children were no longer surrounded by teachers who knew them, who knew their parents, and who understood their communities. Although most black American teachers survived this separation, many black children did not. In a very real sense, desegregation meant that black teachers lost control of black children, and many of these women felt sure that, as a consequence, "something was lost." Rarely did these women suggest that white teachers *intentionally* mistreated their black students. Instead, it was their uniform assessment that white teachers did not understand black children, that they lived in what one woman called a "little antiseptic world," and hence were ill-equipped to teach black children. Again, the notion of teaching a subject rather than a child captures the essence of the difference these women saw in how white teachers approached a black child. Social class also played a distinct role in this judgment because all were clear that *their own* children, those who were now among the most elite black Americans, managed to prosper quite well in a desegregated system because they were able to provide them with the necessary tools of survival and resistance. Many spoke, however, of the devastating effects desegregated schools had on the "average" black American child. Each had a different way of describing it and, of course, all recognized that the issue was an enormously complex one that involved many other factors. Two lengthy quotes paint for us the picture of why these women see the white world and white teachers as problematic:

> I think with integration, with white teachers, I think [that] they're sincere, I really do. But I think they come with certain myths, one of these is that [black children] cannot learn, or they can only learn so much. And this is the way they operate . . . this is just from where they have grown up. They are in their own little white world. They are not going to live next to any black people, they're not going to know any. All they do is read a magazine, or read a book and . . . it's a subjective thing. It's not objective. It's very subjective. Well, that's what they grew up with all around [them]. Then, when they get in a room full of little black children, you know all of these things come to play. And they really don't know what to do.

*   *   *   *   *

> Oh, if I could turn back the clock, I would segregate the schools again, and I would get all of those good, fine black teachers. And I'd get them in and they'd work for our youth. Because when they work[ed] with them, they got respect. Kids respected them. I think this is what happened to the schools. We have now an inte-

grated school system where there have been whites who have felt that [black kids] are not going to learn, they're hoodlums, so that's what they expect, and that's what they get.

In this second narrative we hear the longing for a return, not to inequality but to the shared sense of mission that characterized black segregated education.

While most women tempered their criticism of white teachers, one former educator, a woman who taught for over 40 years in elementary school and who is still very active as a community leader, was less charitable. Indeed, she was adamant that, both in the classroom and beyond, whites in general did not try to understand black people. Accommodation and understanding have always been one-way, she argued. As the following words make clear, she wants to register the taken-for-granted nature of the white privilege that accrued to the researcher's race, "We've had to understand you. We've had to figure out your world and what made you tick as white people but you've never done that for us. You've never tried to figure out what is special about us, what is our uniqueness." Her message is that black Americans are by necessity bicultural. Their survival has been predicated upon understanding, "figuring out," the dominant culture. By contrast, she argued, whites as the majority group have not made an effort to understand or appreciate African Americans.

Although negative encounters with white administrators and teachers clearly outweighed the positive, we also heard stories of whites who made things better, who treated all their colleagues as equals, who went the extra mile, who became friends. Overall, we sensed, however, that such stories were highlighted because they stood out in some way as notable, as exceptions. It is also our sense that these conversations, more than any others in our time together, were ones very much influenced by the racial differences between researcher and retiree. We both felt, for example, that the women we interviewed dug deep to find positive examples of white behavior because they did not want us to think they were being unfair. Yet as they spoke, they inevitably returned to how difficult things had been for them because they were black Americans. Perhaps one former teacher, who spent most of her life in segregated schools, summed up what others repeatedly inferred about the more subtle aspects of racism that they encountered each day. She recalled how in her integrated school, "one of the administrators was very obviously a person who had trouble dealing with blacks. And even though he tried to hide the fact, it eventually came out. Not verbally, but . . . it usually does [come out]. They can't hide it but so long."

As already suggested, some African-American women who worked in a largely white work world found that opportunities opened up for them after the mid-1960s. This was especially true in the case of several women in our study who were in a position to take advantage of a predominantly white work world which now felt legal pressure to diversify its work force and to hire and to promote women of color. For example, a woman who earlier had been denied a job as acting vice-principal was able to leverage to her benefit federal affirmative action programs that forced employers to "go outside the mainstream." Her ascent in the occupational hierarchy began with a federally funded program that provided her with the opportunity to teach in a college for a year as a teacher/lecturer and field supervisor for student teaching. She then returned to her home base and, "just started climbing faster and faster." How far and how fast did she climb? Upon returning to her school, "I became what you call a project coordinator, that was a federal job . . . but still in the school. Then the following year, I became a teacher to assist the principal. The following year, I became the vice-principal. And the following year, I became a principal."

## POSTSCRIPT: THE CONTINUING
## SIGNIFICANCE OF RACE AND GENDER

The decade of the 1960s was a watershed that created different work environments for minorities and women. The government become actively involved in ensuring fairer treatment for black Americans and for women—both in the world of work and in general. As a result, African Americans and women not only became more aware of race and gender issues but they also became more vocal about discrimination.

When we look at the work history of our youngest interviewee, who began her career in the 1960s, we see some differences between her and the majority of the women, who had started their work lives in prior decades when race and gender oppression were so blatant. Unlike the majority of older educators, this woman faced fewer structural constraints because obvious economic oppression for minorities had been reduced. For instance, this respondent—unlike the women in the older cohorts—made no mention of blocked promotions and salary inequities. As a former middle school principal in a predominantly white school, she moved from teaching middle school through a series of other educational positions in fairly rapid succession and claimed, "I kind of think being black has worked as an advantage for me, especially with integration because they needed . . . blacks . . . I don't think my blackness worked against me." In an interesting way, her choice of words hints at the different world that younger African-American women experienced. In contrast to the older educators, who experienced clear disadvantages because of their race and gender, the younger women faced different career prospects. The shift that had occurred is alluded to in the sentiment that white organizations "needed blacks." To be a black American ceased being an absolute negative and, instead, sometimes became a qualified positive. What the comments and experiences of this women illustrate, however, is that race and gender remained crucial to her social location.

In contrast to many of the older women interviewed, the woman above, as well as others like her who are included in the entire study, were also cognizant of the interactions of race and gender in their careers. Illustrating awareness of gender, the former school principal remembers how she came to accept her job, "When the superintendent asked me about taking a principalship, I really loved the school I was in and I didn't want to do that. And then I thought, If I don't go where they ask me to, then I won't open up room for another female." As well as awareness of the positive, she was also aware of the costs she bore because of her race and gender. For instance, she recalled how as school principal she faced difficulties because of the perception that "women weren't principals." She felt that "it was more of a gender thing than a black thing." Her words tell us what it felt like to be in charge:

> I feel that everywhere I went I had to prove myself. I had to work very hard to show that I was competent, to show that I could do the job, to show that I could handle stress. Because when things go wrong people look to see if you're going to cry because you're a woman, if you're going to fold up because you're a woman. And there were many times when I was in my office and I probably wanted to cry but when I came out, nobody knew it [Laughs].

As well as the obvious gendered component to this narrative, there is also the racial. In fact, these negative experiences were likely a reflection of how race *and* gender in combination created discriminatory consequences.

## CONCLUSION

Devotion to work and to doing the very best job possible is an important motif in the work stories of these African American women educators. As we have seen throughout the chapter, such devotion was influenced by a variety of forces. Their work biographies reflect the power of race, class and gender, as well as the influences of historical and sociocultural forces. Through their narratives, we see illustrations of how these forces shaped the lives of African-American women during legal segregation. Their choices of professions and careers, as well as their everyday work experiences as teachers, were hampered by inequity. Their reminiscences of segregated systems provide us a glimpse of the complexities of human responses to unjust institutions—we see both the negative consequences of discrimination and the positive effects that were born from resistance to injustice. Despite the racism and sexism that confronted these women, we see how they used the strengths of their race and gender, as well as their class positions, to engage in uplift work. We learn, for example, how their resistance to injustice took different forms—sometimes it took the form of providing their own school supplies in the absence of adequate administrative support. At other times, we come to appreciate the extent to which they were both vehicles of mobility for disadvantaged children and surrogate mothers for all their students.

The challenges that accompanied African-American educators in the desegregation era are also illuminated in the narratives of these women. The comparisons and contrasts of both segregated and desegregated systems help us understand how structural elements shape work experiences. We learn, too, how continuingly elusive is equality for African Americans—whatever their professions or privilege. We hear how racial inequity and racism does not disappear but, instead, takes on different meanings in desegregated institutions. Thus, over and over again, we hear in their narratives the themes of survival and resistance. While their stories illuminate multiple "stumbling blocks," they also allow us to appreciate the tenacity with which these women approached such impediments. Despite all the obstacles, they managed to build the necessary "stepping stones" to success. The lessons learned in earlier life—to be the best, to pick their battles, to have self-confidence—were all ones that they called upon in order to successfully navigate a lifetime of employment under trying conditions.

## ACKNOWLEDGMENT

This chapter is based on data from a larger study of retired African-American professional women reported in a forthcoming book to be published by New York University Press.

The authors wish to acknowledge the Provost's Office at the College of William and Mary and the Research Awards Committee at the University of Richmond for their financial aid in support of this study.

## NOTES

1. Most of the women in the study referred to themselves as black. Consequently, throughout this chapter we use "African American," black American," and black" interchangeably.

2. All transcribing was done by an African-American woman.

3. We grappled with the issue of naming each woman (through the use of pseudonyms in order to protect confidentiality and anonymity) as we quoted her or mentioned her experiences, but it became so cumbersome that we abandoned it. In all cases where the names of people or places are used, they are fictionalized in order to protect identities. Furthermore, in the telling of a woman's story, we occasionally changed details in a way that did not change the point of her story but that ensured that she could not be identified.

4. Sociologist Elizabeth Higginbotham describes black American professional women as "colonized professionals" because they have disproportionately offered their professional services within the confines of the African American community (Higginbotham 1984). They also have been concentrated in the female-dominated professions of teaching, nursing and social work.

5. Using the continuum of sociologists Gerson and Peiss (1985), these women were at one end of the gender consciousness continuum. Certainly, they were aware of gender ("gender awareness")—in other words, they could provide "a non-critical description of the existing system of gender relations" (p. 324). However, unlike their racial consciousness, they had not arrived at a level of gender consciousness that Gerson and Peiss describe as feminist/antifeminist consciousness. This point represents the other end of the consciousness continuum and involves an articulated ideology of challenge or defense of the existing system of gender relations.

6. See the discussion of the importance of race uplift work in the lives of African-American women in Collins (1990, pp. 147-151).

7. For further discussion of the special role played by African-American women in their communities, see Gilkes (1982).

8. For an in-depth analysis of how members of the black middle class continue to be haunted by racism, see Feagin (1991).

# REFERENCES

Beeson, D. 1975. "Women in Studies of Aging: A Critique and Suggestion." *Social Problems* 23: 52-59.

Collins, P.H. 1990. *Black Feminist Thought*. Boston, MA: Unwin Hyman.

Feagin, J.R. 1991. "The Continuing Significance of Race: Antiblack Discrimination in Public Places." *American Sociological Review* 56(February): 101-116.

Gerson, J.M., and K. Peiss. 1985. "Boundaries, Negotiation, Consciousness: Reconceptualizing Gender Relations." *Social Problems* 32(4, April): 317-331.

Gibson, D. 1996. "Broken Down By Age And Gender: 'The Problem of Old Women' Redefined." *Gender and Society* 10(4): 433-448.

Gilkes, C.T. 1982. "Successful Rebellious Professionals: The Black Woman's Professional Identity and Community Commitment." *Psychology of Women Quarterly* 6(3, Spring): 289-311.

Glaser, B.G., and A. Strauss. 1967. *The Discovery of Grounded Theory*. Chicago, IL: Aldine.

Higginbotham, E. 1984. *Work and Survival for Black Women*. Research Paper 1. Memphis, TN: Center for Research on Women, Memphis State University.

Mayes, M.J. 1992. "Autobiography and Class Formation in Nineteenth-Century Europe: Methodological Considerations." *Social Science History* 16(3, (Fall): 517-536.

Somer, M.R. 1992. "Narrativity, Narrative Identity, and Social Action: Rethinking English Working-Class Formation." *Social Science History* 16(4, Winter): 591-630.

Stoller, E.P., and R.C. Gibson. 1994. *Worlds of Difference: Inequality in the Aging Experience*. Thousand Oaks, CA: Pine Forge Press.

# A CRITICAL EVALUATION OF THE EXPERIENCES OF WOMEN AND MINORITY FACULTY:
## SOME IMPLICATIONS FOR OCCUPATIONAL RESEARCH

Toni M. Calasanti and Janice Witt Smith

## ABSTRACT

In this chapter, we describe aspects of the work experiences of white women and women and men of color who are members of the professoriate. A review of the literature provides a backdrop for our own study of the subjective experiences of faculty in six doctoral-granting institutions, a location which is taken to signify "success" in this occupation. To ascertain the extent to which women and minority group members feel well-integrated, we examine how race/ethnicity (being white, African American, or Asian American) and gender influence experiences of institutional and social isolation as well as the expression of affective organizational commitment. Our data reveal three findings relevant to research on faculty and on occupations in general. First, the same occupation, even in the same organization, can be and is experienced differently by race/ethnicity and gender. Second, important differences among groups defined by race *and* gender emphasize the need to

Current Research on Occupations and Professions, Volume 10, pages 239-258.
Copyright © 1998 by JAI Press Inc.
All rights of reproduction in any form reserved.
ISBN: 0-7623-0034-5

derive data in a way that allows us to examine the intersections of race/ethnicity and gender. Dividing groups on the basis of race *or* gender, or comparing minority groups only to the majority, obfuscates critical differences among these groups. Finally, our research confirms the recent critiques of tokenism, demonstrating the importance of exploring the power relations that render occupational experiences variable for different "token" groups.

# INTRODUCTION

As an occupation, the professoriate represents a particularly interesting case for examining the treatment and experiences of white women and women and men of color. Widely regarded as having the characteristics considered pertinent to the definition of a profession (Hodson and Sullivan 1995), an extensive literature documents the importance of a sense of autonomy and intrinsic rewards to success in academic careers. And as is the case in other professions, rules of entry into this occupation involve specific and fairly well-defined sorts of training. A command of the requisite knowledge is taken to be signified by the possession of an advanced degree, usually a doctorate, in various disciplines (Aisenberg and Harrington 1993). For white women and women and men of color to have gained entry, then, into this and similarly defined professions means that they have already achieved a certain degree of career success. As Aguirr, Martinez, and Hernandez (1993, p. 382) note, "minority faculty, for the most part, have embarked on 'extraordinary careers' in that they have managed to transcend the social expectations ascribed to them by U.S. society." From the standpoint of research into work and occupations, one pertinent question, then, is: what are the experiences of "successful" minority[1] group members in a white, male-dominated workplace? Having proven that they can fulfill the requirements of graduate training and also having obtained jobs in higher-level universities, what are the experiences of minority group members who have, in fact, "made it"?

From the standpoint of academe, understanding the experiences of white women and women and men of color in faculty positions is also critical. Many argue that the professoriate ought to reflect the increasing diversity of college students themselves: for nearly two decades, more than half of all degrees from associate through master's level have been awarded to women, and the participation of people of color in college education has also risen (Glazer, Bensimon, and Townsend 1993). Others are concerned about the projected shortfall of faculty after the turn of the century, especially in science and engineering (Olsen, Maple, and Stage 1995; Tack and Patitu 1992). While they have been consistently underrepresented in natural sciences and engineering, the numbers of white women and women and men of color completing doctoral degrees in all disciplines have continued their steady rise in the 1990s. In 1995, women earned a record number of doctorates (16,333), a figure nearly 10 times higher than 30 years before. Similarly, U.S. racial/ethnic minorities earned a record 3,489 doctorates in 1995. Notably, this represents a reversal in the downward trend among African Americans in the late 1970s and 1980s, the only group to have not demonstrated the generally upward growth. In 1995, 4.7% of doctoral degrees were awarded to blacks, a figure which topped their previous high in 1977 of 4.4% (Henderson, Clarke, and Reynolds 1996).

Finally, although the number of white men receiving doctoral degrees continues to rise, the proportion entering the professoriate after earning graduate degrees in general, and degrees in science and engineering in particular, has been steadily decreasing (Henderson et al. 1996; Blackburn and Lawrence 1995). Doctoral recipients from all groups increasingly are seeing other occupations as more viable and better-paying career choices than faculty positions (Tack and Patitu 1992). Still, among 1995 recipients of doctoral degrees, women were far more likely than men (46.2% vs. 33.3%) to say that they planned on entering educational institutions (Henderson et al. 1996).

Whether for these reasons or simply as a matter of social justice, universities and researchers have become more concerned with increasing the diversity of those who hold faculty positions. At the same time, "enthusiasm for increasing the number of women and minorities on . . . campuses has outstripped our understanding of the experience of these traditionally underrepresented groups in academe" (Olsen et al. 1995, p. 276).

Our first goal in this chapter, then, is to address this gap in our knowledge concerning the experiences of white women and minority men and women in the professoriate. We do not attempt an exhaustive review but instead seek to describe selected aspects of the work experiences of some of the "most successful" women and minority faculty. This focus is in part based on the assumption that if the "best-placed" are not well-integrated and progressing, it is even less likely that others will be. These are the ones who have, for the most part, "bought" the white, male model; they have "played the game" and played it well, up to the point of entry into the occupation.

Drawing from the literature and recent data, we begin by describing the overall status of minorities within the occupation in relation to objective indicators, including their representation among faculty; rank, including tenure and promotion; types of institutions in which they are located; and salaries. For the most part, available data are broken down by either gender or race/ethnicity but not by both; as a result, our overview must often do the same. Similarly, our discussion of racial-ethnic groups is also necessarily limited by present research; we examine African Americans and, when possible, Asian Americans, Hispanics, and American Indians. Next, we examine how the job is experienced, as this, together with career progression, would influence the likelihood of recruitment and retention (assuming, of course, that other options exist). The aspects that we focus upon are those typically identified as being important in both the general workplace literature and literature on faculty in higher education, namely, those experiences taken to be associated with tokenism.

Generally missing from the literature, however, is a theoretical consideration of race/ethnicity and gender as intersecting social locations in relation to these objective indicators and subjective experiences. Our second goal, then, is to explicitly consider the significance of race/ethnicity and gender on faculty experiences with an eye toward the possible implications for occupational research in general. Specifically, we build on the concept of tokenism and subsequent critiques as a way of beginning an exploration of how the simultaneous experience of race/ethnicity and gender influence the subjective aspects of work.

The remainder of the chapter, then, is devoted to presenting data collected among faculty at six doctoral-granting institutions in a southeastern state, on feelings of isolation and affective organizational commitment for groups defined by both their gender and race/ethnicity. We draw on the recent work of others (Yoder and Aniakudo 1997; Pierce 1995; Martin 1994) whose studies led them to conclude that the same occupation was experienced differently by each race/gender group. If we also find that the intersections of race/ethnicity and gender influence how an occupation is experienced, this lends greater weight to the argument that our theories must go beyond treating tokenism as a demographic phenomenon and must instead consider variations in workplace experiences in the context of power relations. Empirically, it would also speak to the necessity of examining the unique experiences of groups defined by their membership in both gender and racial/ethnic categories and then comparing across groups in order to understand occupational experiences and how these are structured by race and gender.

## THE POSITION OF WOMEN AND MINORITIES

### Representation

While the entry requirements for employment in academia are straightforward, rules governing actual employment and advancement are not as clear and can vary considerably according to such things as time, place, and specific institutional norms (Aisenberg and Harrington 1993). For example, governmental policies concerning monies spent on higher education as well as employment policies have had tremendous impact on the growth or constriction of disciplines and on the employment opportunities of various oppressed groups. The National Defense Education Act of 1958 served to spur growth in mathematics and the sciences, while the Civil Rights Act of 1964, and later, affirmative action legislation, heightened the employment opportunities for women and people of color (Dwyer, Flynn and Inman 1991).

Indeed, a brief look at historical trends in faculty demographics by gender and race/ethnicity is quite revealing. Finkelstein's (1984) examination of faculty from

1930-1981 revealed that, in 1980-1981, women accounted for just over one-quarter (26.3%) of all full-time faculty. While this represented a 7% increase from the previous decade, it also signaled a decline of 1.7% from 1929-1930, when women constituted 28% of this same group (Finkelstein 1984, p. 181). Continued increases in women's employment in the 1980s have resulted in their finally reaching their 1930 level of representation (Dwyer, Flynn, and Inman 1991).

Overall, racial/ethnic minority group members account for 11% of full-time faculty. Of these, Asian Americans have the highest representation, at 4%, followed by blacks, at 3% (Russell 1991). Problems with data collection have resulted in the proportion of Hispanics being estimated at anywhere from .5% to 3%, but data from the National Center for Educational Statistics placed their representation at 2% (Blackburn and Lawrence 1995). Finally, American Indians comprise approximately 1% (Russell 1991).[2]

## Institutional Location

Faculty experiences are greatly shaped by the type of institution in which they are employed. For example, as we previously noted, the greatest resources are located in the most elite institutions. However, white women and women and men of color have consistently found achieving entry into the highest level of faculty employment outside their grasp (Olsen et al. 1995). The distribution of women faculty has not changed much, as they are still concentrated in the community colleges and liberal arts colleges and underrepresented in elite research and doctoral-granting institutions (Dwyer, Flynn, and Inman 1991). In 1980-1981, women were disproportionately located in community colleges (35.5%) and underrepresented at research universities (17.6%) (Finkelstein 1984), numbers which showed little change by the end of the decade (Blackburn and Lawrence 1995). As of 1995, women were slowly increasing their representation at the more elite, doctoral-granting institutions,[3] yet their representation among faculty remained inversely related to institutional type. White women constituted 37.7% of faculty at public, two-year institutions; similar disproportionate representation was evident for women of all racial/ethnic types, with the exception of Asian Americans (Finnegan, Webster, and Gamson 1996). Overall, then, women's representation among faculty was increasing at all institutions but remained smallest in the most desirable locations—research universities—which offer the greatest resources. By contrast, by 1995, men's representation at each level was positively related to institutional level (Finnegan et al. 1996).

Context must be considered when examining the position of faculty of color, since numbers alone will not tell the complete story. Indeed, these data are an excellent example of why combining all racial/ethnic minority groups into one, "nonwhite" category is problematic (Blackburn and Lawrence 1995). For example, in 1960, prior to the Civil Rights movement, it was estimated that African Americans comprised about 3% of faculty. By 1975, this figure remained virtually unchanged. Yet, while their representation remained stable, their distribution did not (Finkelstein 1984). In 1960, African-American faculty—as many as 80%— were overwhelmingly located in historically black colleges and universities (HBCUs). Importantly, HBCUs not only tend to have far fewer resources for research, including computer resources and library facilities, but also heavier teaching and service requirements. In the 1970s, blacks' representation at HBCUs decreased to about half, while their share of positions in predominantly white institutions (PWIs) rose 8%, including a 12% increase at the highest levels (Finkelstein 1984; Locke 1997). As of 1995, African Americans of both genders, however, were still disproportionately located at liberal arts and comprehensive colleges (Finnegan et al. 1996). This is a reflection of their employment at HBCUs; about half of HBCUs are private, liberal arts colleges that employ large numbers of blacks (Blackburn and Lawrence 1995).

Significant differences among minority groups are apparent in their likelihood of being located in particular types of institutions. The experiences of African Americans stand in stark contrast to those of Asian Americans of both genders, who are disproportionately represented at elite institutions (Finnegan et al. 1996). American Indians, whose numbers are quite small, do not appear to be concentrated in any type of institution, while data on Hispanics indicate a somewhat bimodal distribution as they tend to be located in both research institutions and community colleges (Blackburn and Lawrence 1995).

## Fields

While women have tended to be concentrated in "traditionally female fields"—especially the performing arts, foreign languages, health-related professions, and English—they have begun to make greater inroads into other disciplines (Finkelstein 1984). While the percentage of women is still highest in English, where women accounted for 33.6% of faculty positions in 1988, the greatest gains since 1969 have been in the fields of sociology and history. Some improvement has occurred in the field of mathematics and the natural sciences, but women's representation is still quite low (Blackburn and Lawrence 1995). As recently as 1995, women comprised 32.5% of doctorates awarded in all sciences and engineering; however, of this amount, the vast majority were in the life sciences. And while 22.3% of doctoral degrees in mathematics and 11.6% in engineering went to women, women in both disciplines were more likely than men to have committed to being employed in teaching (56.1% versus 40.4% in mathematics, and 16.4% versus 11.8% in engineering) (Henderson et al. 1996, pp. 112-114).

Data on racial/ethnic minorities reveal wide variations across populations. As many as one-half of African Americans have been located in education—historically, one of the few fields in which African-American women could participate and one in which they have a long and distinguished legacy (McKay

1997)—and social sciences (Finkelstein 1984). Hispanics are somewhat more likely to be in humanities, agriculture/home economics, fine arts, and education; the numbers of American Indians are small but appear to be somewhat dispersed. By contrast, an overwhelming preponderance of Asian Americans are located in engineering, followed by business and the natural and health sciences (Blackburn and Lawrence 1995).

## Rank, Promotion, and Tenure

Gains in employment have not necessarily translated into career advancement, especially at elite institutions (Clark and Corcoran 1993). Women and minorities are most likely to be concentrated in lower faculty ranks and in non-tenure-track positions (Olsen et al. 1995).

Turning first to women:

> Despite an increased proportion of females receiving doctoral degrees and pursuing postdoctoral training, they represent a disproportionate share of the unemployed, part-time status, special program, or adjunct employees (Dwyer, Flynn, and Inman 1991, p. 214).

Similarly, men are more likely to be tenured (Russell 1991). With the exception of particular disciplines, women have not made much progress in being promoted to associate and full professor, and overall, they are still promoted more slowly than men (Dwyer, Flynn, and Inman 1991). Even when the sample is restricted to those with doctorates or first professional degrees, women are far less likely than men to be full professors, and far more likely to be assistant professors, instructors, or lecturers (Russell 1991).

Wide variation is again exhibited among racial/ethnic groups. Asian-Americans' rate of promotion to full professorships far exceeds (two times as likely) the national average of 25%. In this regard, it is interesting to note that this is not the case in relation to promotion to associate. Somewhat analogously, Hispanics are promoted to full professor at just above the national average, but again, not at the associate level (Blackburn and Lawrence 1995). Interpreting such numbers alone is risky, but it may be the case that those who are promoted to associate level have to be "that much better" than their white counterparts, a fact that is reflected in their subsequent higher rates of promotion to full.

By contrast, African Americans lagged far behind the national rate, at 15% (Blackburn and Lawrence 1995). Not only are their rates of promotion to full professor significantly worse, but blacks also appear to be concentrated in the lower ranks. This is particularly the case when they are employed at PWIs, where their likelihood of holding part-time or non-tenure-track positions far outstrips that of their white counterparts (Finkelstein 1984).

Salary

Finally, similar to other occupations in the United States, a significant salary gap between women and men still exists among the professoriate (Dwyer, Flynn, and Inman 1991). This differential is often attributed to men's greater prevalence in research and doctoral institutions as well as their greater likelihood to be of higher rank. However, the gender discrepancy is strong even within rank and within program areas. In a 1988 national survey, especially large differences were documented between men and women of the same rank in professional program areas. Here, total income for male full professors averaged $81,702, compared to an average of $57,146 received by similarly-ranked women. Full professors in the arts and sciences also exhibited · differences; male full professors received $57,098; women, $48,966 (Russell 1991). Similarly, a recent institutional review of salaries at a Research I institution found a consistent wage gap among men and women of similar rank and background (Olsen et al. 1995).

Data on salary for racial/ethnic groups are sparse. But in the 1980s, those few studies that existed indicated that black faculty receive compensation at rates similar to their white counterparts (Finkelstein 1984). A recent examination of one state system also found that minority faculty were paid comparably to white men with similar backgrounds (Olsen et al. 1995).

## EXPERIENCES OF WOMEN OF COLOR IN THE ACADEMY

The critical importance of examining the experiences of women of color has been aptly expressed by Spelman (1988), who noted that the phrase "women and minorities" provides no location for someone who is both, thereby essentially erasing the experiences of minority women. However, few of the data on faculty speak to the unique position of women of color in academe. Or perhaps it is more accurate to refer to their unique positions as there are not only similarities but also differences among minority women along racial/ethnic lines. To date, however, research attention has only begun to focus on the experiences of black women faculty; Asian-American, Native American, and Hispanic women, as well as other groups of women, have been virtually ignored (Glazer et al. 1993). Pinpointing their differences, aside from the data concerning their representation in types of institutions and tenure and promotion figures, is speculative at best.

Because of what some have termed the "dual status" of women of color—their membership in two minority groups (Wyche and Graves 1992)—their experiences are shaped by the interactions of racism and sexism (Graves 1990; Moses 1989). Speaking of African-American women faculty, Graves (1990, p. 7) maintains that they face a situation of "double jeopardy" that results in black women facing "a different job description" than whites of either gender. While white women may be expected to serve on committees dealing with gender issues, for example, minority women are further called upon to participate on any committees that could potentially deal with issues of race/ethnicity. Similarly, black women are expected to be role models and to mentor African-American students of both genders and also contribute to programs geared toward women students. But ultimately, such service is not weighed heavily, if at all, in tenure and promotion decisions (Hine 1997; Locke 1997; Kawewe 1997; Graves 1990). And similar to research that has found that the competence of professional women and minorities is more likely to be challenged by both co-workers and clients (Pierce 1995; MacCorquodale and Jensen 1993), minority women's scholarship is often marginalized by white academics (Brewer 1997). As Locke (1997, p. 342) explains:

> Oftentimes white institutions and scholars feel that they have cornered the epistemological market. They tend to view research on race, gender, and ethnicity as not being "real" scholarship, particularly when it is presented from an Afrocentric perspective, and they have the same perception of journals that publish this research.

Similarly, students question her credibility in the classroom. Overall, women of color make up less than 1% of all full professors (Essein 1997). Data from the University of Michigan indicate that, from 1978-1990, the numbers of women of color, at any rank, did not really increase, despite affirmative action. In 1978, only 20 of 501 (4%) assistant professors were women of color; by 1990, this proportion had actually declined, to only 29 of 690 faculty at this rank. A similar decrease occurred at the associate level (only 12 of 529 associate professors were women of color in 1990), while the increase from three to six full professors is almost meaningless. It appears that women of color are not being promoted as often as white men, or are leaving, or both (Blackburn and Lawrence 1995). Indeed, in 1991, only 6.6% of tenured faculty were African-American; 88.2% were white (Locke 1997). Part of the explanation, Kawewe (1997) maintains, is that while affirmative action opened up opportunities, it does not monitor retention, a fact which adversely affects African-American women. Others have also observed that, overall, affirmative action has had an important impact on women being employed in higher education, but not on their career advancement, salary, or movement into administration (Dwyer, Flynn, and Inman 1991). And while women earn less than men, African-American women's median salaries are even lower than those of their white counterparts (Phillip 1993). The following recent expression of pessimism is as important for what is says of African-American women faculty's experiences as for the sources perceived to be at fault:

> Earlier I was confident that we could collectively find a solution to some of the problems of race, class, and sex in the academy; . . . Change will not come until those responsible for the current conditions decide to end the impact that racism, classism, and sexism among white faculty and administrators have on the lives of others. My pessimism springs from feeling that nothing we have seen over the last twenty-five years indicates such action . . . domination,

empowerment, legitimacy, and authority remain in the places they have always been (McKay 1997, p. 17).

Hispanics of both genders represent only a small percentage of all full-time faculty; their numbers get even smaller as the rank goes up, and smaller still for Hispanic women versus Hispanic men (.2% versus .9% at the rank of professor) (Nieves-Squires 1993). Nieves-Squires (1993) suggests that critical cultural differences shape Hispanic women faculty experiences in particular ways by creating cultural conflicts. For example, Hispanic culture encourages tolerance of differences, as well as cooperation and cohesion. The atomistic culture of academe, with its emphasis on challenge and debate, runs counter to these values. Similarly, differences in interpersonal relationships can be problematic, as cultural norms allow for much greater physical proximity and signs of affection among friends of the same gender. While the behavior of both Hispanic men and women may be construed as homosexual as a result, women's behavior might also be seen as inviting intimacy by male members of the dominant culture. Finally, the stereotype of Hispanic women as being only concerned with home and family can also shape their faculty experiences (Nieves-Squires 1993).

Baraka (1997) also notes cultural differences between African Americans and dominant whites that cause conflict and miscommunication in the academy, including "differences in communication styles, in the role of emotion in human behavior, in beliefs about equality and progress, and in practices around beliefs about affiliation versus individuality" (Baraka 1997, p. 236). Further, whites confuse collegiality and friendship, assuming a broader definition of the latter; the result can be a violation of personal and social space (Baraka 1997).

At least two points relevant to our present analysis emerge from the above discussion. First, implicit in the discussion of both Hispanic and African-American cultural differences is the fact that such variations become meaningful when one group—Euro-Americans—is dominant, thereby setting the occupational norms to be followed. Second, cultural variations between Hispanic and African-American women indicate one source of critical differences in their occupational experiences. Collapsing the experiences of women and minorities into one or the other group not only "erases" women of color, it also obscures crucial variations, thereby disguising sources of and possible solutions to inequities. Thus, in the discussion which follows, when possible we explore gender and race simultaneously.

## WHITE WOMEN AND MINORITY WOMEN AND MEN FACULTY: ARE THEY "TOKENS?"

As the above review indicates, while some changes have occurred, parity has not been reached by white women or men and women of color. Depending on the outcome of interest, a variety of explanations for continued inequities have been offered. Generally speaking, the majority can be viewed as *individual-level* explanations, regardless of their focus. For example, approaches that emphasize discrimination point to employers' and co-workers' attitudes and practices (McElrath 1992); in this instance, equality would be achieved by changing the dominant group. Other explanations focus on those who receive unequal treatment, and range from human capital and personal choice differences, to problematic socialization, to "role overload" (Finkelstein 1984). In examining gender disparities in compensation, for example, included among these latter approaches are explanations that assume differences in research productivity or career disruption result in women reaping fewer rewards (McElrath 1992).

However, a recent national study on one academic field revealed that interrupting one's career did have a substantial effect on tenure—but only for women. Given that no productivity differences by gender were found, the author suggests that perhaps when women interrupt their careers, they are then perceived as being "less serious" scholars and, therefore, less worthy of tenure. The same study also found that the gendered salary differences within all ranks except that of assistant professor persisted when controls for productivity, length of service, and other relevant factors were introduced (McElrath 1992).

A somewhat more *structural* strand of research has drawn on Kanter's (1976, 1977) findings concerning "tokens." This approach, which connects objective factors in the workplace to both subjective and objective career outcomes, is one of the most often used explanations for the experiences of faculty women (Dwyer, Flynn, and Inman 1991) and people of color.

Based on her study of women and men managers, Kanter asserted that three structural variables shape the workplace experiences of men and women. These factors include the opportunity structure, or possibilities of occupational mobility; the organizational power structure, which includes both formal and informal networks; and an occupation's sex ratio. The fact that women express less organizational commitment or have lower job aspirations, she maintained, is due to their relationship to each of these structural factors. Were men and women equally subject to these variables, they would display similar behaviors.

It is in relation to the last factor, the demographic composition of an occupation, that Kanter's concept of tokenism emerged. Essentially, Kanter felt that group members became tokens when their representation in an occupation was highly skewed—confined to less than 15% of the whole. As tokens, the women in Kanter's study experienced heightened visiblity, which led to pressure to perform well. Isolation from both informal social and professional networks was also a consequence; this stemmed, in part, from "boundary heightening," wherein the differences between the majority group (men) and the token group (women) were exaggerated. Finally, token women were stereotyped and were encouraged to act in ways "appropriate" to their gender.

Despite the presence of other perspectives, Kanter's approach constitutes the primary and virtually only structural explanation used by researchers in examining gender and racial/ethnic disparities in academe. In particular, tokenism is depicted as an organizational factor that leads to greater stress levels for women and minority group members. As a review of the literature on women faculty expresses this theory,

> [A] skewed distribution of social types based upon minority status within the group generates certain damaging perceptions of the token members by the majority members. Circular causation is set in motion. The perceptions determine the nature of the interaction between the token and the majority members of the group which, in turn, results in pressures placed upon the tokens. In turn, token responses to these pressures fit the stereotypical perceptions of the majority (Dwyer, Flynn, and Inman 1991, p. 212).

Indeed, many of the experiences of faculty women of color speak to the negative effects of heightened visibility. The heavy service load is one such result, but other types of performance pressures exist as well. As a black woman vice-president of academic affairs put it:

> There is a lot of pressure to perform well that I don't think you find for a White person in the same situation. A Black must represent the whole race. You want to do a good job and not let the race down. There is tremendous pressure for an individual. . . . Not only do you want to do well for yourself—whatever the position dictates—but you are doing it as a first or only Black person (quoted in Benjamin 1991, pp. 126-127).

Similarly, exclusion from both formal and informal networks, which not only serve as an important source of career knowledge but also social support, has been reported by women and minority faculty (Blackburn and Lawrence 1995; Moses 1989).

While Kanter's work has provided a fruitful approach, a variety of critiques of her original formulation have been offered, each ultimately asserting that more than sheer numbers are involved in shaping token's experiences (Yoder 1991), and that the nature of these experiences is highly variable. For example, men who are numerical tokens in female-dominated occupations do receive different treatment—they do *better* than their female counterparts and are often fast-tracked onto a "glass escalator" to higher-level positions (Williams 1992). The fact that the effects of underrepresentation are neither gender- nor race-neutral (Yoder and Aniakudo 1997; Martin 1994; Nkomo 1992) indicates that the "experience of tokenism is very different" for members of different groups. Further, what is critical in determining the content of these experiences "is the social status of the token's group—not their numerical rarity" (Williams 1992, p. 263). In other words, a group's power in the wider society defines the token experiences of its members.

## A STUDY OF SUBJECTIVE EXPERIENCES

We build upon the research tradition on tokenism and subsequent critiques to examine selected aspects of the experiences of a sample of faculty women and people of color. Through exploratory research at six doctoral and research institutions, we assess the extent to which men and women of different racial/ethnic groups experience the workplace in similar and different ways. Specifically, we examine the experiences of social and institutional isolation that different gender and racial/ethnic groups may experience in the workplace. Given the importance of retention, we also briefly explore their affective organizational commitment. To examine the effects of race and gender apart from rank or institution, it is useful to focus on faculty from more homogenous institutional settings. And selecting doctoral-granting institutions allows researchers to focus on those who are presumably chosen for employment because of their research potential and performance (Olsen et al. 1995).

From a standpoint of examining occupational inequities, faculty at doctoral-granting institutions represent those who have achieved the most prestigious positions; they have followed the "rules of the game" and have succeeded. In fact, access to these "elite" occupations, which offer rewards, autonomy, power, and influence not generally offered to nonprofessionals, has "defined" middle-class success (Sokoloff 1992). While it might be assumed that such "achievers" will not experience the effects of tokenism, research has not borne this out. In other occupations, for example, Nkomo and Cox (1990) examined the experiences of African-American MBAs and found that while they had attained some objective measures of success within their organizations, these MBAs did not feel as if they "belonged." Their entree into the organization and even their involvement in informal interracial social activities did not change this perception. Similarly, Ibarra (1992) found that minorities who engage in informal social interaction with their white counterparts reaped little career or social psychological benefit. We argue, then, that even among this successful group, important differences in experiences likely exist between white men and other gender and racial/ethnic groups.

Finally, we also examine isolation and organizational affect in relation to race and gender simultaneously. The importance of this is suggested by recent empirical work in other occupations as well. For example, both Martin (1994) and Yoder and Aniakudo (1997) found that while similarities existed, the experiences of white and African-American women in white male-dominated fields differed in critical ways. As a result, these authors maintain, the occupation *itself* was not the same for each group, a claim which echoes the voices of the African-American women faculty above.

Implicit in such occupational studies is Robinson and McIlwee's (1991) assertion that occupational experiences are greatly determined by the work culture in which they occur. Further, their research among men and women engineers found that conformity with the expectations of this culture—that is, competence—is not as important as *how* this conformity is expressed. As both the work culture and its appropriate expression are defined by members of the dominant group, women and minorities may well be perceived to be less competent and be treated as such. In academe, the marginalization of their scholarship is perhaps a manifestation of the fact that productivity, for example, must be expressed in particular ways, defined by white men.

## Research Design

What follows is based on mail survey data garnered from full-time tenure-track faculty at six doctoral-granting institutions in a single state. Of the 765 usable surveys, only 1.2% identified themselves as Hispanics, too small a group to include in statistical tests. Of the remainder, 5.6% were African-American, 6.6% were Asian-American, and 86.8% were white. By gender, two-thirds of the sample were men, and one-third were women.

The two types of isolation, social and institutional, included in this study are analogous to Kanter's discussion of the isolating effects of tokenism in term of both formal and informal networks. *Institutional isolation* was defined as the perception that: (1) one lacks knowledge about, access to, interaction with and/or utilization of organizational sources of power, prestige, support, and information critical to one's success; and (2) regardless of one's position, training, or educational background, others important to career success ignore one's opinion unless it is validated by member(s) of the dominant group. Thus, individuals who feel institutionally isolated feel they are excluded from decision-making processes, have little input into matters which impact them, and are kept out of the "inner circle."

*Social isolation* refers to a feeling of exclusion that can be manifested in a variety of ways: (1) feeling that one is often singled out, on display, tolerated but not accepted due to racial and/or gender identity; (2) feeling that co-worker relationships are superficial because others cannot relate to one's experiences, because one has to be a "bridge" and translator of one's experiences to other-race and other-gender individuals, and because one is taken to be "representative" of one's racial/gender group; and (3) feeling that one is alone, without a support network.

These related but analytically distinct forms of isolation were measured using a scale previously developed and tested (see Smith et al. 1994; Smith and Markham 1997). Differences by gender were ascertained using simple $t$-tests. To assess the extent to which each race or race/gender group felt isolated, mean differences were compared using a general linear model analysis of variance (GLM ANOVA).

In addition to isolation, faculty members' feelings of organizational commitment were assessed. An 8-item, 5-point Likert-type scale developed by Meyer and Allen (1984) was used to assess *affective commitment*: individuals' positive feelings toward their institution and resultant desire to stay in that workplace. GLM ANOVA was used to ascertain group differences on this measure.

## Results

By gender, men and women do not differ in their feelings of institutional isolation. The picture is more complex by race. While institutional isolation scores for

African Americans and Asian Americans were very similar (2.56 versus 2.61), only the latter reached statistical significance ($p < .05$), indicating that Asian-American faculty feel more institutionally isolated than their white peers (2.35). The closeness of African Americans' scores to those of Asian Americans suggests that it is practically and theoretically significant and deserves to be explored further in subsequent research.

Within subgroups defined by both race and gender, both Asian-American and African-American women (2.75 and 2.69, respectively) felt more institutionally isolated than any other subgroup based on the GLM ANOVA analysis ($p < .01$). White men felt the least institutionally isolated (2.33)

A statistically significant difference in the level of social isolation was also evident by both race and gender. Turning first to race, African Americans felt more socially isolated than either Asian Americans or whites (2.36 versus 2.07 and 1.83, respectively; $p < .05$). In turn, Asian Americans felt significantly more socially isolated than whites.

Despite popular cultural beliefs regarding women's social nature, in this study, faculty women, on average, were significantly more likely to report feeling socially isolated than faculty men (2.03 versus 1.79; $p < .01$). This is consistent, however, with the literature on tokenism in general and in academe in particular. Finally, in terms of race and gender group differences, African-American women experienced greater social isolation than Asian-American men or white men and women (2.59 versus 2.16, 1.75, and 1.97, respectively; $p < .01$). In this regard, it appears that race is a critical defining factor in social isolation, with membership in a dominant racial ethnic group providing greater social resources even among women.

Do these varied experiences of social and institutional isolation influence affective commitment, as tokenism would predict? Despite the intergroup variations found on the two isolation measures, and contrary to expectations concerning tokens, no differences emerged on organizational commitment. That is, the feelings of exclusion or lack of support that significantly differentiated groups did not lead to similar variations in their desire to stay with their academic institution.

Rather than discuss each of these findings in a vacuum, we consider some of their implications in relation to the literatures on higher education and on occupations in general.

# DISCUSSION

In this chapter, we have highlighted aspects of the experiences of women and minorities within the professoriate with two related goals in mind. One goal was to bring together literature on this subject in order to paint a clearer picture of the contemporary status of women and minorities within this occupation. But we also hoped to use this literature to set the context for examining faculty experiences in a way that might advance future theory and research not only among those within this occupation but also in the field of occupational research in general.

The data on rank, salary, place of employment, and the like demonstrated continued inequities among faculty by gender and race/ethnicity. To be sure, changes have occurred, but these have been uneven not only by group but also by institution and rank. Further, there are suggestions that such progress has been stymied. These indications are especially troublesome in light of recent state and court decisions concerning affirmative action as well as the restructuring of higher education.

In addition, these objective indicators, in combination with the subjective aspects of the occupation, gleaned both from the literature and our data suggest at least three things of relevance to occupational studies in general and of faculty in particular. First, occupations cannot be viewed as merely the sum of tasks involved (Lopata and Figert 1996). This is not an original idea, to be sure. But we are also suggesting, in line with others such as Robinson and McIlwee (1991), Martin (1994), Pierce (1995), and Yoder and Aniakudo (1997), that occupations are experienced differently—that they *are* different for various groups defined by socially salient characteristics. Both the literature and our results indicate that being a faculty member is not the same across the board, even within the same institution. Indeed, even the "objective" tasks that make up the occupation may be experienced differently, as well as the balance of tasks expected/required and rewarded. For example, the greater likelihood of students challenging the credibilty of African-American women no doubt makes the task of "teaching" different for them than, say, for white men or even white women. Similarly, the greater service demanded of women of color results in their often carrying a different mix of tasks, or a heavier burden. In essence, we are not only pointing to "occupational differences;" we are also suggesting that it might be more fruitful to conceive of this as a situation of *different occupations*: occupations that are obviously related and share enough similarites to be discussed in tandem, but also different enough to warrant separate treatment. The view that occupations are not simply a set of tasks but also a set of social relations (Lopata and Figert 1996), combined with research on workplace culture, helps us understand how these differences might arise. Still missing, however, is an explicit understanding of how power relations intervene in and shape this process.

This is related to our second point: that these differences are based on a combination of race and gender (at the very least). As Spelman (1988), Moses (1989), Brewer (1997), and countless others have maintained, to separate our research into gender *or* race categories denies the reality of the intersections of these. And these intersections often spell highly diverse experiences among those believed to have belonged to "one common group," such as African-American men and women, or Asian-American and Hispanic women. But because most research does not separate out each minority group and virtually none look at race/gender groups, such differences are generally hidden. The importance of the data reported above is that they suggest that men and women wtihin each of the minority faculty groups experience isolation, and that they do so *differently*. These disparities are significant for understanding not only variations among groups but also the experiences of those within race/gender groups that would otherwise be hidden.

For example, the finding of greater institutional isolation among Asian Americans is critical as it contradicts the "model minority" belief held by the majority. To be sure, national statistics show that Asian-American faculty fare better than any other minority group in terms of institutional placement, and better than any racial/ethnic group in relation to promotion to full professor. However, it may be that the term "model minority" is used more to punish other minority groups than to reward Asian Americans (Cho 1995). Even more, this label may also create more pressure for Asian Americans to be successful, setting up the expectation that they will succeed without providing the corresponding support, and also not acknowledging their efforts as their success is, simply, expected. Both situations might be conducive to institutional isolation.

Finally, it is apparent that theory and research on faculty and in occupations must continue to go beyond tokenism to explore and understand the structural bases for diverse occupational experiences. The work of Acker (1990) and Nkomo (1992) has provided a beginning for this endeavor, laying the ground for understanding how organizations are gendered and racialized. But the impact of these scholars is embryonic in research on faculty (but see Park 1996). And while we have pointed to several recent works that have critiqued tokenism for excluding considerations of power relations, less attention has been paid to the ways in which the influence of such power relations are shaped by intersecting systems of privilege/oppression. That is, both the voices of African-American women faculty and recent research on other occupational groups suggest that the experiences of tokens is not the same by virtue of their rarity. But this is true not only of majority (white men) tokens versus minority group tokens, as Williams (1992) aptly demonstrates, but *among* minority tokens as well. The effects of being a "token" cannot and should not be assumed to be the same *across* minority groups. Further, combined racial/ethnic and gender variations also have not been much theorized. Relatedly, even less has been done in terms of cultural differences that may influence the "fit" between workplace culture and *varying* racial/ethnic cultures, as well as gender variations therein. While it might be possible to generalize to the extent of saying that groups with less power (i.e., those who are not white, middle-class, heterosexual men) will experience the negative aspects of being a "token," *how* this will be experienced is another issue entirely. In terms of faculty experiences, women and people of color may all experience being tokens; they might even all experience being isolated. But the ways in which the isolation occurs and is felt, the types of isolation, and the consequences are variable. For example, being a "model minority" appears, on the one hand, not to be detrimental to employment and promotion as Asian Americans are disproportionately located at Research 1 institutions. At the same time, even they experience isolation—and perhaps a lack of credit for their achievements, as it is expected that they will do well. A similar lack of recognition for achievement may be experienced by a white woman or an African-American man or woman, but for a different reason: the assumption is that they are in "good" jobs only because of affirmative action.

We are arguing, then, for greater consideration of similarities and differences among racial/ethnic and gender groups, and not simply between those in the majority versus all "others." Again, this can only occur when research no longer engages in implicit or explicit comparisons between white men and everyone else.

## ACKNOWLEDGMENTS

The authors would like to thank Alan Bayer and Elizabeth Creamer for generously sharing their resources and expertise. In addition, we are also grateful to Kerstin Popp and Meeta Mehrotra for their research and editorial support.

## NOTES

1. In this chapter, we use the term "minority" in it broad sense, pointing to relatively powerless groups, and thus including white women and women and men of color.
2. Numbers do not add to 11% due to rounding.
3. By race/ethnicity, white women were at 22.4%; American Indian women were .1%; African-American women were 1.6%; Hispanic women were .6%. The one exception, which is discussed more below, is Asian-American women who, while only constituting 1.9% of faculty, achieved their highest representation at that institutional type.

## REFERENCES

Acker, J. 1990. "Hierarchies, Jobs, and Bodies: A Theory of Gendered Organizations." *Gender & Society* 4: 139-158.
Aguirre, A., Jr., R. Martinez, and A. Hernandez. 1993. "Majority and Minority Faculty Perceptions in Academe." *Research in Higher Education* 34: 371-385.
Aisenberg, N., and M. Harrington. 1993. "Rules of the Game." Pp. 387-398 in *Women in Higher Education: A Feminist Perspective*, edited by J.S. Glazer, E.M. Bensimon, and B.K. Townsend. Needham Heights, MA: Ginn Press.
Baraka, J.N. 1997. "Collegiality in the Academy: Where Does the Black Woman Fit?" Pp. 235-245 in *Black Women in the Academy: Promises and Perils*, edited by L. Benjamin. Gainesville: University Press of Florida.
Benjamin, L. 1991. *The Black Elite: Facing the Color Line in the Twilight of the Twentieth Century.* Chicago, IL: Nelson-Hall.
Benjamin, L., ed. 1997. *Black Women in the Academy: Promises and Perils.* Gainesville: University Press of Florida.
Blackburn, R.T., and J.H. Lawrence. 1995. *Faculty at Work: Motivation, Expectation, Satisfaction.* Baltimore, MD: The Johns Hopkins University Press.
Brewer, R.M. 1997. "Giving Name and Voice: Black Women Scholars, Research, and Knowledge Transformation." Pp. 68-80 in *Black Women in the Academy: Promises and Perils*, edited by L. Benjamin. Gainesville, FL: University Press of Florida.

Cho, S.K. 1995. "Korean Americans vs. African Americans: Conflict and Construction." Pp. 461-470 in *Race, Class, and Gender: An Anthology*, 2nd edition, edited by M.L. Andersen and P.H. Collins. Boston, MA: Allyn and Bacon.

Clark, S.M., and M. Corcoran. 1993. "Perspectives on the Professional Socialization of Women Faculty: A Case of Accumulative Disadvantage?" Pp. 399-414 in *Women in Higher Education: A Feminist Perspective*, edited by J.S. Glazer, E.M. Bensimon, and B.K. Townsend. Needham Heights, MA: Ginn Press.

Dwyer, M.M., A.A. Flynn, and P.S. Inman. 1991. "Differential Progress of Women Faculty: Status 1980-1990." Pp. 173-215 in *Higher Education: Handbook of Theory and Research*, Vol. VII. New York: Agathon Press.

Essien, F. 1997. "Black Women in the Sciences: Challenges along the Pipeline and in the Academy." Pp. 91-102 in *Black Women in the Academy: Promises and Perils*, edited by L. Benjamin. Gainesville: University Press of Florida.

Finkelstein, M.J. 1984. *The American Academic Profession: A Synthesis of Social Scientific Inquiry Since World War II*. Columbus, OH: Ohio State University Press.

Finnegan, D.E., D. Webster, Z.F. Gamson. 1996. *Faculty and Faculty Issues in Colleges and Universities*, 2nd edition. Needham Heights, MA: Simon & Schuster Custom Publishing.

Glazer, J.S., E.M. Bensimon, and B.K. Townsend. 1993. "Prologue." Pp. ix-xvii in *Women in Higher Education: A Feminist Perspective*, edited by J.S. Glazer, E.M. Bensimon, and B.K. Townsend, Needham Heights, MA: Ginn Press.

Graves, S.B. 1990. "A Case of Double Jeopardy? Black Women in Higher Education." *Initiatives* 53: 3-8.

Henderson, P.H., J.E. Clarke, and M.A. Reynolds. 1996. *Summary Report 1995: Doctorate Recipients from United States Universities*. Washington, DC: National Academy Press. (The report gives the results of data collected in the "Survey of Earned Doctorates," sponsored by five federal agencies: NSF, NIH, NEH, U.S. Dept. of Ed., and USDA and conducted by the NRC.)

Hodson, R., and T.A. Sullivan. 1995. *The Social Organization of Work*, 2nd edition. Belmont, CA: Wadsworth Publishing.

Hine, D.C. 1997. "The Future of Black Women in the Academy: Reflections on Struggle." Pp. 327-339 in *Black Women in the Academy: Promises and Perils*, edited by L. Benjamin. Gainesville: University Press of Florida.

Ibarra, H. 1992. "Personal Networks of Women and Minorities in Management: A Conceptual Framework." *Academy of Management Review* 18: 56-87.

Kanter, R.M. 1976. "The Impact of Hierarchical Structures on the Work Behavior of Women and Men." *Social Problems* 23: 415-430.

Kanter, R.M. 1977. *Men and Women of the Corporation*. New York: Basic Books.

Kawewe, S.M. 1997. "Black Women in Diverse Academic Settings: Gender and Racial Crimes of Commission and Omission in Academia." Pp. 263-269 in *Black Women in the Academy: Promises and Perils*, edited by L. Benjamin. Gainesville: University Press of Florida.

Locke, M.E. 1997. "Striking the Delicate Balances: The Future of African American Women in the Academy." Pp. 340-346 in *Black Women in the Academy: Promises and Perils*, edited by L. Benjamin. Gainesville: University Press of Florida.

Lopata, H.Z., and A.E. Figert. 1996. "Introduction." Pp. 1-8 in *Current Research in Occupations and Professions, Volume 9: Getting Down to Business*, edited by H.Z. Lopata and A.E. Figert. Greenwich, CT: JAI Press.

MacCorquodale, P.L., and G. Jensen. 1993. "Women in the Law: Partners or Tokens?" *Gender & Society* 7: 582-593.

Martin, S.E. 1994. "'Outsider Within' the Station House: The Impact of Race and Gender on Black Women Police." *Social Problems* 41: 383-399.

McElrath, K. 1992. "Gender, Career Disruption, and Academic Rewards." *Journal of Higher Education* 63: 269-280.

McKay, N.M. 1997. "A Troubled Peace: Black Women in the Halls of the White Academy." Pp. 11-22 in *Black Women in the Academy: Promises and Perils*, edited by L. Benjamin. Gainesville: University Press of Florida.

Meyer, J.P., and N.J. Allen. 1984. "Treating the 'Side-bet Theory' of Organizational Commitment: Some Methodological Considerations." *Journal of Applied Psychology* 69: 372-378.

Moses, Y.T. 1989. *Black Women in Academe: Issues and Strategies*, Washington, DC: Project on the Status and Education of Women, Association of American Colleges.

Nieves-Squires, S. 1993. "Hispanic Women: Making Their Presence on Campus Less Tenuous." Pp. 205-222 in *Women in Higher Education: A Feminist Perspective*, edited by J.S. Glazer, E.M. Bensimon, and B.K. Townsend, Needham Heights, MA: Ginn Press.

Nkomo, S. 1992. "The Emporer Has No Clothes: Rewriting Race in Organizations." *Academy of Management Review* 17: 487-513.

Nkomo, S., and T.H. Cox, Jr. 1990. "Factors affecting the Upward Mobility of Black Managers in Private Sector Organizations." *The Review of Black Political Economy* 18: 39-57.

Olsen, D., S.A. Maple, and F.K. Stage. 1995. "Women and Minority Faculty Job Satisfaction: Professional Role Interest, Professional Satisfaction, and Institutional Fit." *Journal of Higher Education* 66: 267-293.

Park, S.M. 1996. "Research, Teaching, and Service: Why Shouldn't Women's Work Count?" *Journal of Higher Education* 67: 46-84.

Phillip, M. 1993. "Feminism in Black and White." *Black Issues in Higher Education* 10: 12-17.

Pierce, J.L. 1995. *Gender Trials*. Berkeley: Univeristy of California Press.

Robinson, G., and J.S. McIlwee. 1991. "Men, Women, and the Culture of Engineering." *The Sociological Quarterly* 32: 403-421.

Russell, S.H. 1991. "The Status of Women and Minorities in Higher Education: Findings from the 1988 National Survey of Postsecondary Faculty." *College and University Personnel Association* 42: 1-11.

Smith, J.W., and S.E. Markham. 1997. "Dual Construct of Isolation: Institutional and Social Forms." Paper presented at the Annual Meetings of the Academy of Management, Boston, August.

Smith, J.W., S.E. Markham, R. Madigan, and S. Gustafson. 1994. "Exploration of Isolation as Two Independent Constructs: Social Isolation and Institutional Isolation." Paper presented at the Center for Creative Leadership Conference on Work Team Performance in the Context of Diversity, Greensboro, October.

Sokoloff, N.J. 1992. *Black Women and White Women in the Professions.* New York: Routledge.

Spelman, E. 1988. *Inessential Women: Problems of Exclusion in Feminist Thought.* Boston, MA: Beacon Press.

Tack, M.W., and C.L. Patitu. 1992. *Faculty Satisfaction: Women and Minorities in Peril.* ASHE-ERIC Higher Education Report No. 4. Washington, DC: The George Washington University, School of Education and Human Development.

Williams, C.L. 1992. "The Glass Escalator: Hidden Advantages for Men in the 'Female' Professions." *Social Problems* 39: 253-267.

Wyche, K.F., and S.B. Graves. 1992. "Minority Women in Academia: Access and Barriers to Professional Participation." *Psychology of Women Quarterly* 16: 429-437.

Yoder, J.D. 1991. "Rethinking Tokenism: Looking Beyond Numbers." *Gender & Society* 5: 178-192.

Yoder, J.D., and P. Aniakudo. 1997. "'Outsider Within' the Firehouse: Subordination and Difference in the Social Interactions of African American Women Firefighters." *Gender & Society* 11: 324-341.

# WOMEN PHYSICISTS:
## WHERE ARE THEY?

Rosalie G. Genovese and Sylvia F. Fava

## ABSTRACT

This chapter begins with background information on women in physics, reviewing their numbers, marital and family status, and employment experiences over time. Although strides have been made and their numbers have increased, women face considerable problems establishing academic careers, owing to both global factors that limit opportunities for all physicists and factors specific to their gender. The prevalence of physicist couples adds dual-career strains to a difficult employment environment for many women. Suggestions for making the chilly academic climate more hospitable to women are reviewed. However, the authors question whether young women are counseled adequately about the future awaiting them in academic physics. They also suggest that the employment focus be shifted to settings offering more options and advancement potential, especially career paths in industry and entrepreneurship. The experiences of some successful women in these settings are offered as examples.

Current Research on Occupations and Professions, Volume 10, pages 259-282.
Copyright © 1998 by JAI Press Inc.
All rights of reproduction in any form reserved.
ISBN: 0-7623-0034-5

# INTRODUCTION

This paper deals with people in their occupational settings. For women, that occupational setting always includes home, family, and children in a way that does not apply for men. The integration of these spheres is especially complex for women in nontraditional fields like the physical sciences.

Many of the first successful women scientists solved the work-family problem by remaining single. They adjusted their lives to the system. For example, Rita Levi-Montalcini, who won the Nobel Prize in medicine with Stanley Cohen for discovering the Nerve Growth Factor, chose to forgo marriage and children to devote herself to her work (see Rossiter 1982). In 1920, the first time women were included in the statistics on physicists, none were married.

In recent decades, the lives of women in physics have reflected the general trend of combining careers and marriage. They often marry other scientists. Fay Ajzenberg-Selove (1994), one of the first women to obtain a Ph.D. in nuclear physics, married the physicist Walter Selove. In her autobiography, she describes the discrimination and other barriers she faced in a male-dominated field. These were not the only major obstacles that she overcame, since she survived life in France during World War II before escaping to the United States.

In the mid-1970s, a little more than 62% of both women and men receiving doctorates in physics were married (Kistiakowsky 1980, p. 35). Women who continued their careers often had only one child. They also tended to devise careers that meshed with their family situation, since they continued to assume major responsibility for children and household (Kistiakowsky 1980, p. 39).

These changes in marital status are reflected in approaches to the work of women scientists. The focus of several recent articles and books is on the careers of scientific women within the framework of their family and other relationships (see Abir-Am and Outram 1987). More than two dozen scientists are profiled in *Creative Couples in the Sciences* (Pycior, Slack, and Abir-Am 1996). Married women scientists in the late nineteenth and early twentieth century tended to have unpaid jobs (if they were lucky), to receive little recognition, and to face assumptions that their male partners did the "real" work.

To obtain data on how women scientists were faring in the mid-1980s, we surveyed members of the Committee on the Status of Women in Physics (CSWP) of the American Institute of Physics. Two thousand questionnaires were sent to members in the United States, with the cooperation and assistance of the Committee (Fava and Deierlein 1988, 1989). About 44% responded to questions about their education, careers, and marital status.

Some of these physicists reported situations similar to those of their earlier counterparts: unpaid positions or being paid sporadically when extra funds were available. They reported facing still-prevalent departmental attitudes that they did not *need* a paid position because their spouse had a job. An unspoken view is that hiring a married woman takes a job away from a man with a family to support.

Respondents also found university search committees often unwilling to hire young married women because they believed that the candidates would have children and then either leave or stay, but not concentrate completely on their work. Even if interviewers did not directly ask about a candidate's marital status or family plans because of affirmative action requirements, they assumed that a woman scientist of childbearing age planned a family in the future.

Research confirms that scientific women continue to have primary responsibility for managing the household and child care (Rayman and Burbage 1989; Sonnert 1995a, 1995b). Couples who have egalitarian arrangements about jobs and child care are in the minority.

As for children, pioneering women may have had some advantages over their modern counterparts, since household and child care help often was more available and affordable then. Some scientific women had relatives who cared for their children (see Pycior et al. 1996). They also sometimes could do research at home, perhaps even in a home laboratory. That option is unlikely today, when physicists or astronomers need such highly sophisticated, expensive technology as accelerators and telescopes. Scientists have to vie with each other for time on these scarce machines and they cannot schedule experiments to fit their household schedules. Moreover, even if they could keep current in their field, independent researchers rarely get grants or funding; awards usually go to institutional employers.

## Rationale for Focusing on Women Physicists

There are several reasons for our focus on women in one field of science. In many studies of women scientists, the social and physical sciences are grouped together, even though the fields and status of women in them are very different. Women represent the majority of doctorate holders in some social sciences. They also are more prominent in the biological sciences than in physics or engineering. Women represented fewer than 11% of the physics doctorates awarded in 1990 (Dresselhaus 1994, p. 4).

Physics always presented strong barriers to women. Many remain, although in more subtle form today. A widespread and still common belief has been that women cannot do physics. As Kistiakowsky indicated in a 1980 article, it was considered not a "suitable" field for women, who were viewed as "unnecessary, injurious, and out of place." When she proposed starting a committee on the status of women in physics, a male physicist retorted that it was not needed since there were only two women in physics and they were not unhappy (Kistiakowsky 1992, p. 6). (She was not considered to be one of the two!) In the October 1992 *CSWP Gazette*, physicists Engle and Cladis wrote about their experiences in this male-dominated field. Engle provides vignettes and commentary on what it was like to be one of the few women at physics meetings and the predatory attitudes of male colleagues toward her.

The masculine model of science has been graphically described in many studies (see, e.g., Sonnert 1995a, 1995b). Trawek (1988) studied the culture of high-energy physics in the United States, Japan, Switzerland, and Germany. The traits considered to be essential for admission into the community—"aggressiveness," "haughty" self-confidence, and "competitiveness"—are defined as masculine in our society. She concluded that other behavior styles were likely to be associated with failure in physics (Trawek 1988, p. 87). Not surprisingly, many women are discouraged from entering this world because of these personality qualities and work styles associated with success. In a similar vein, McIlwee and Robinson (1992) explored the workplace culture in engineering, also masculine, and analyzed how women often experienced an incompetence in applied work settings that they had not felt in college.

Another reason for focusing on women physicists is that a high proportion is married to male physicists. This arrangement complicates a woman's career in science, as the discussion in the section on work-family integration makes clear.

## ASSESSING WOMEN'S FUTURE IN PHYSICS

Brush (1991, p. 404) calls physics "the coldest science" for women. The loss of women at various stages on the way to the doctorate has been termed a "leaky pipeline" and is more pronounced in physics than in other sciences. About 150 women receive Ph.D.s in physics annually. This figure represents 11% of the total, the lowest percent except for engineering (Dresselhaus 1994, p. 4).

Although recent studies show some progress, the number of women in top academic positions in research universities remains small. An American Physical Society member survey in 1990 found that women occupy only 3% of the academic ranks in physics departments, with a sharp drop as rank increases. Only 42% of women were tenured in 1990, compared to 70% of men (*CSWP Gazette*, January 1993). On the other hand, more women than men (33% versus 18%) reported that they were in tenure-track positions, reflecting their younger age. (However, these findings are based on a small number of responses.)

Another disturbing fact is that a critical mass of women in physics departments still does not exist in many universities. The Massachusetts Institute of Technology (MIT) has been one exception, a highly rated educational institution with a consistent history of tenured women professors in the physics department. Comparisons with other countries do not reflect favorably on the United States either. The United States has the lowest percentage of women in physics faculty ranks of any country (Dresselhaus 1994, p. 5).

Women's future prospects in physics are not promising because of both global factors affecting physicists as a whole and the "chilly climate" for women specifically (Dresselhaus 1994). Job opportunities are limited further for women in dual-career scientific marriages.

Both women and men are grappling with the dearth of jobs. In academia, jobs have dwindled, budgets are tight, mandatory retirement has ended, and significant cutbacks have occurred in research funding. In industry, the job market has been affected by decreased defense and other government spending. The cancellation of the super-collider project was a significant blow to scientists in university, industry and government settings.

## Dwindling Opportunities in Academia

Poor employment prospects in academia mean that the only job for many physicists after their Ph.D. is a postdoctoral fellowship. These jobs, commonly referred to as "postdocs," are temporary appointments for a specified period of time, with or without the possibility of renewal. For many men and women in the sciences, these are the only available academic jobs, resulting in two classes of employees in science departments. One group, tenured faculty, enjoy high salaries, job security, and the resources needed to establish reputations early in their careers. The other group consists of postdocs who are fully qualified but have little or no prospect of tenure. They are paid less and are supervised by tenured scientists. In fact, they may earn little more than half as much as their tenure-track colleagues do. Their chance to establish a track record of success in obtaining grants and supervising projects is hindered by their status.

Added competition for the limited number of jobs comes from foreign doctoral students. According to National Science Foundation (NSF) statistics, the increase in science and engineering doctoral degrees came more from non-U.S. citizens on permanent visas (63%) than from U.S. citizens (25%). Non-U.S. citizens on temporary visas accounted for the remaining 12% (Hill 1996). Moreover, immigrants represented 30% of U.S. residents with doctorates in physics and astronomy in 1993 (Regets 1995).

The job market in physics is worse than in other sciences. Employment data indicate that physics, along with chemistry, is depressed and that Ph.D.s earn less than their counterparts in computer science or electrical engineering. Fox and Stephan (1996) used data from a 1993-1994 national survey of doctoral students and from the National Research Council's (NRC) Survey of Doctoral Recipients to study students' preferences and job choices in chemistry, computer science, electrical engineering, microbiology, and physics.

When asked about their prospects, physics respondents were pessimistic. Not without cause, as it turns out. Professional societies in physics, as in chemistry and math, report that increased percentages of recent graduates are taking longer to find their first job. In 1995, more than half of the recent graduates (U.S. citizens and permanent residents) with definite plans had accepted post-doctoral positions (Gady, Babco, and Pearson 1996).

At the national level, policymakers are asking whether students are being encouraged to obtain Ph.D.s without being informed of the precarious future awaiting them. Many will not find adequate full-time jobs in their specialization. A 1995 study by Rand Corporation and Stanford University researchers estimated that 22% more doctorates are being turned out in science and engineering than there are jobs available (Lancaster 1997, p. B1).

Economic security after the post-doctoral stage may prove elusive:

> Put differently, the concern is that students are myopic and easily recruited into graduate school without having a good understanding of their future job prospects; and that since faculty have strong incentives to recruit students and post docs to work in their labs, they may advise them poorly (Fox and Stephan 1996, p. 11).

Alan Hale, astronomer and co-discoverer of the Hale-Bopp comet, put out an e-mail message to the effect that "career opportunities for scientists were so limited that he had been unable to find work to adequately support his family, and that he couldn't in good conscience encourage students to pursue science careers" (Lancaster 1997, p. B1). He is cobbling together a career of sorts by running a nonprofit foundation devoted to astronomy education. He also is lecturing, writing a book, and consulting, so far without pay.

The announcement that Manpower, the largest temporary agency in the country, has begun to recruit physicists with advanced degrees shows the precariousness of the job market (Zachary 1996, p. A2). Corporate clients, mostly in the computer and electronics industries, can hire temporary workers with advanced physics degrees. Many of the jobs are in physics-related fields—for example, developing computer chips or software programs. This "temping" of physicists came about because so many graduate students could not find permanent jobs and experienced physicists have lost jobs in the downsizing of both government and industrial labs. With resumes supplied by the American Institute of Physics, Manpower is trying to find applicants corporate jobs for periods ranging from six months to two years. This development may represent the equivalent in industry of the temporary post-doctoral appointment in academia—the physicist as nomad.

If graduate science enrollments slow in response to the unfavorable job market, even more tenured positions could be lost. Budget concerns already have led universities to cut full-time positions. Instead, adjunct faculty are hired on a temporary, term-by-term basis to meet staffing needs. The educational institution saves on benefits as well as salaries and avoids long-term personnel commitments. Competition for the few tenured positions becomes even fiercer in such an environment.

In addition to these global factors, women physicists also face gender-related hurdles in employment. According to the study by Fox and Stephan (1996), students in physics and microbiology preferred academic careers in research universities. However, prospects for all careers are "strikingly low in physics compared to other fields" (Fox and Stephan 1996, p. 5). Moreover, physics is notable for gender differences. Men prefer nonacademic careers where there are more jobs and higher salaries. Perhaps surprisingly, prospects for nonacademic careers actually are higher for women in physics. Moreover, it is "the one field in which women do as well—or better—than men" (Fox and Stephan 1996, p. 6).

However, women's prospects really are not that rosy. Physics, along with chemistry, has the greatest proportion of women either not in the labor force, unemployed, or working in unrelated fields. They also are more likely than men to work part-time. Additionally, although women scientists are almost as likely as men to obtain prestigious faculty positions at research institutions, the *exceptions* are in physics or electrical engineering (Fox and Stephan 1996, p. 10).

In 1985, almost three-quarters of the physicists in our CSWP study were employed full-time and another 12% were full-time students. Fifteen percent were employed part-time, retired, or homemakers. Unemployment was common for many women Ph.D.s. in our survey, with one-third reporting at least one period of unemployment. Differences existed between married and non-married women. Ninety-one percent of those reporting unemployment were married. Major reasons for unemployment among married Ph.D.s related to child bearing and child care. For the unmarried, reasons related mainly to employment conditions, often lack of funding or job opportunities.

Therefore, after accomplishing the difficult goal of a doctorate, many women Ph.D.s seem destined for a succession of temporary post-doctoral positions, making it harder for them to obtain grants and direct research, critical factors for further advancement in science. The drive to eliminate affirmative action policies may make it even more difficult for women to be hired for tenure-track positions.

Potential scientists may react to this poor job situation in different ways. Men typically tend to show more career "savvy," reacting quickly to a changing job market by moving out of specializations with fading prospects in favor of growing fields. Fox and Stephan (1996) suggest that men may substitute working in industry at high salaries for more desirable but unavailable academic jobs. Women seem to prefer jobs in teaching universities, perhaps because that option seems open to them. The flexibility of teaching schedules also may play a role. Since men are less interested in teaching, Fox and Stephan suggest that women may be getting employment "left-overs."

Research shows that women make inroads into a field after men leave, but that prestige and pay levels then decrease (Fox and Stephan 1997; for a general analysis of this phenomenon, see Reskin and Roos 1990). On the other hand, when men move into a field previously dominated by women, pay and prestige tend to increase and women's representation drops. Again, there is the hint that women get the left-over jobs.

## Work-Family Issues

Women physicists' careers also are affected by their family roles. In addition to the work-family strains experienced by working women in general, the demands of physics careers exert even more pressure. Physics can be described as "a greedy occupation," swallowing up other aspects of people's lives (Coser and Coser 1974). The family also can be described as greedy, since the amount of time that could be devoted to the care and nurture of members is almost limitless. The lives and careers of women physicists are affected further by their tendency to marry physicists or other scientists.

Sonnert (1995b) views marriage and parenthood as a set of problems and opportunities for women scientists. They have to try to reconcile three (ticking) clocks: their own biological and career clocks and their spouses' career clock. Some women give priority to one clock and then another; others try to coordinate the two; still others choose one clock over the other. Choices have to be made and the outcomes may be unclear. At present, women are given one of two messages—either that they should not plan a career in science because they may need time for childrearing—or that they should not have children if they want to become scientists (Brush 1991, p. 412).

Work-family issues remain a major source of stress for women in fields like physics, characterized by long working hours and extreme pressure to publish. The rewards are worth it for the extremely dedicated women with outstanding abilities who are committed to a demanding career. There always will be exceptional women who choose a career in science, fully aware of the sacrifices they may make. The careers of such prominent twentieth century women physicists as Nobelist Maria Goeppert Mayer illustrate the odds against which such eminent scientists succeeded (see Jones 1990). As we noted at the beginning of this chapter, pioneering women in the sciences often chose to forgo marriage and family for single-minded pursuit of their careers. But some women today question whether they want to give up time with their families and other interests for the single-minded pursuit of a physics career.

Therefore, the consequences of policies to encourage women's commitment to science careers may be employment uncertainty, economic insecurity, and strains on family life for many. The rewards and pleasures of science that lead many women and men to be dedicated to this work are well-documented, but the economic and social costs may be too high for some.

## Dual-Career Considerations

Women physicists are a special case of dual-career couple, because such a high proportion is married to physicists. In our CSWP study, about half of the doctoral respondents were married to physicists. A physicist couple has to job hunt in a extremely tight employment market. (For a general discussion of issues facing dual-career couples, see Fava and Genovese 1983.) Finding two jobs in the same field is extremely difficult; finding one is hard enough. Even when the dual-career partners are in different academic fields, finding two positions in the same geographic area can be challenging. Most departments do not want to hire spouses, partly because of the political issues that arise with regard to tenure and other sensitive decisions.

The careers of astrophysicist Neil Gehrels and his spouse Ellen Williams, a physics professor, illustrate this "two-body problem." They have twice compromised on job opportunities so that they could live in the same city (Gibbons 1992a, p. 1380). Travel plans and other schedules are complicated and take considerable planning because of their two preschool children. The impression left by the special *Science* issue on women in science is that few couples have made the postdoctoral situation work "with both members staying in science" (Weaver 1993, p. 3).

Geographic mobility often becomes a way of life for scientist couples, especially if both are in physics and one or both have untenured positions. Marriage often restricts opportunities for women in both graduate training and employment. Some universities try to find a job for the trailing spouse; others do not.

Moving from place to place takes its toll on the whole family. No one really talks about the impact of frequent moves, precarious jobs, and economic insecurity on children who are repeatedly uprooted from familiar neighborhoods and schools. Aging parents also may be affected by their adult children's frequent moves.

Women respondents in our CSWP study also reported involuntary unemployment following moves to advance a spouse's career. Periods of unemployment, underemployment, or temporary jobs sometimes lasted for years. Mobility or immobility affects women scientists in other countries as well. The tendency for scientists in the same field to marry each other is widespread.

Two Australian physicists, Joan Freeman and Rachel Makinson, each married a physicist and fashioned careers largely determined by their husbands' career decisions. Each received her degree in 1939. Rachel Makinson moved to Australia to marry an Australian but did not expect to remain there. It took her 20 years to become resigned to staying in Australia. On the other hand, Joan Freeman was born in Australia but did her graduate work in nuclear physics at the Cavendish Laboratory in Cambridge, England. She would have liked to return to Australia, but found much more opportunity in England. Even when more work possibilities opened up in her homeland, she knew that her husband would not agree to move.

Makinson's job opportunities were limited because unwritten anti-nepotism rules kept her from becoming a permanent employee in the Physics Department of the University of Sydney. When she left for another job, she was prevented from becoming a permanent staff member because she was married. Instead, she worked for 20 years as a temporary employee who needed special Ministerial permission annually for reappointment (Allen 1990, pp. 75-85).

A woman's career still tends to take second place, often because her husband's career is at a more advanced stage and is given precedence for economic reasons. Some couples develop an egalitarian pattern and alternate the priority given to each partner's career. Although that arrangement sounds great in the abstract, the approach is not always easy to put into practice. Opportunities may not arise in the desired sequence. It also may be difficult and costly for the more established partner to leave a tenured position.

Family responsibilities also affected women's careers. Whether married or single, respondents expressed concerns about combining professional advancement and a family. Some described how childrearing had affected their careers: "I took time off from work for child care. Also, I had five years of under-employment— half time or part time—because there were no other opportunities where my husband's job was. I waited 7 years after receiving my Ph.D. for a 'proper' job—as a tenured assistant professor" (Fava and Deierlein 1989, p. 7).

As we noted in the Introduction, women today often postpone a family or decide not to have children at all to give priority to their careers. Their decision may be influenced by the still-pervasive belief that women cannot have it all—children and a career. As one respondent wrote, "I have been warned not to get pregnant. My boss has said it in a menacing tone [of voice] several times. Family is not an acceptable distraction for a woman scientist" (Fava and Deierlein 1989, p. 13). Respondents also faced institutionalized penalties if pregnancy or a long illness interrupted their career.

Respondents often expressed concern about whether they were doing a good enough job in the work place and at home when they tried to combine the two. (Fava and Deierlein 1989, p. 12). Child care remained a major concern. As one woman in our survey noted, "I do not think it is fair that highly educated and highly motivated women should be forced to neglect their own children just because this is the only option left (if they are) to keep their professional training alive."

Dual-career couples also face issues at work. There is considerable overlap when they work together or in the same organizational setting. Their common experiences in one sphere carry over to the other. Since the partners also interact with many of the same people at work and perhaps socially as well, they are thought of as a couple.

Therefore, such marriages have an impact on how women are perceived as scientists. It often is harder for the woman to establish her independent professional identity, especially when her husband is older and in a more senior position. She has to worry about how her work (research, papers, grants, productivity in general) is assessed, both when she collaborates with her spouse and when she works independently. Today, the male scientist is frequently the senior author, just as in the past,

The danger is that women get tagged as the "helper" (Hughes 1973) or part of the two-person, one-career syndrome (Papanek 1974) We all know, however, that women often do work for which men get credit, whether they collaborate with spouses or are research assistants on projects directed by men.

However, couples emphasize advantages as well as problems in dual-career science marriages. The pluses include understanding, encouragement, and someone with whom to share their work and concerns. Women scientists often say that only someone in the same or a related field could understand the schedules, pressures, and excitement of their work. A spouse who works in the same field may serve as a sounding board (Sonnert 1995b, p. 159).

Additionally, an older and more established spouse may serve as his wife's mentor, providing her with information and encouragement and introducing her into his networks of scientific colleagues. That arrangement can also have costs, as the woman may find it difficult to establish her own reputation.

Women continue to run up against policies prohibiting them from being hired in the same department or university as their spouse. If only one partner can be hired, it is more likely to be the male. Some women in our CSWP study expressed frustration about their stalled careers. One physicist was a full-time adjunct professor who put in 50 hours a week, commuting 40 miles each way for an unpaid position. She commented: "I love physics. I hate what has happened to me." Another woman had a post-doctorate appointment without pay. She said that it was her only choice if she wanted to live with her husband. The university paid her occasionally, when extra funds were available or when she taught a course (Fava and Deierlein 1989, p. 15). Such an arrangement makes a scientist feel like a charity case or an afterthought, not a researcher whose work is respected.

Some respondents in our CSWP study left physics for such reasons as lack of jobs or chances for advancement, lack of acceptance or harassment, or the geographic mobility problems of dual-career couples. They faced the task of building a career in a new, perhaps unrelated, field. On an institutional level, their withdrawal wastes human capital and the resources of the universities that trained them.

## Solutions to Work-Family Strains and Conflicts

At present, most solutions to combining work and family are individual ones, requiring women to modify their schedules or priorities to fit into a work organization's structure. Rayman and Burbage (1989) found that some women made a tradeoff to have more time for their children, choosing to work part-time with limited job satisfaction. Others shared household responsibilities with their husband.

At a recent conference honoring nine women scientists, speakers offered advice on combining career and family. One honoree, physicist and MIT Institute Professor Mildred Dresselhaus, advised women interested in a career in science, technology, or engineering to "choose your spouse or partner carefully." She also recommended that "if you want a family, be willing to invest in a good babysitter" (Solomon et al. 1997, p. 23). Shirley Malcom, Head of the Directorate for Education and Human Resources Programs of the American Association for the Advancement of Science, also advised a woman to choose the right partner and to "balance your life by making choices" (Solomon et al. 1997, p. 42).

The gist of recommendations concerning demands and time pressures on women in early career stages seemed to be that they cannot have it all. Dresselhaus advised women to postpone community activities until their careers were established. Another honoree, Florence Hazeltine, Director of the Center for Population Research at the National Institutes of Health, made a similar recommendation: "When you're young, stay focused on your career. When you are established professionally, then it is time to give something back to the professional community, perhaps 'pay back' through mentoring" (Solomon 1997, p. 23).

Some women suggest "serial obsession" as a way to deal with career and biological clocks, devoting time to children when they are young and then returning full-time to the lab when they are older. Such a routine was adopted by many successful women scientists who now are trying to convince their students that it is possible to have a career and a family using this approach (Angier 1995).

But other women disagree with expecting the *individual* to change to meet organizational requirements. Instead, they think that the work structure and environment should change to be more accepting of and friendly to women (and families). Employer policies that recognize the family responsibilities of both women and men can help those juggling multiple responsibilities (see Fehrs and Czujko 1992, pp. 33-40). Recommendations center on anti-discrimination, promotion, and tenure policies. For example, career paths can be structured to take into account childbearing and child-rearing responsibilities.

The tenure system receives much criticism for its rigid timetable. Brush (1991) recommends that universities stretch out the tenure time, provide paid family care leave for a defined period (one term or two quarters), institute an agreed-upon time of unpaid family leave for up to two years, provide subsidized day care either on campus or nearby, and offer temporary non-tenure-track positions for spouses when outside positions are not found. Such policies would benefit men as well, allowing the university to compete for the most qualified young scholars. Another solution is to eliminate the restriction against a department hiring its own graduates, especially important when the region has only one or two research universities. (Etzkowitz et al. 1994).

Benefits such as family leave help employees meet their family needs, whether these involve young children, ill family members, or elderly relatives. The issue of elder care almost never gets raised in discussions of work-family issues facing women scientists, but it needs to addressed since more women and men at midlife have aging parents who need care (see Genovese 1997). Such responsibilities may come just as women are reaching the peak of their careers.

## Should We Ask Different Questions about Science Careers?

The issues raised in the foregoing discussion leads us to raise two important questions about the future of women in physics.

1.  Perhaps we should ask not why so few women go into physics, but rather *why so many do*, given the employment picture.

2.  If women continue to choose to go into physics and other sciences, what kind of productive careers can they develop?

First, we should ask whether women who choose physics will be satisfied with their career choice, before encouraging more women to get advanced degrees. It may be unfair to recruit young women for traditional careers in scientific fields like physics when prospects are so poor. They may become nomads and second-class citizens.

Research shows that young women in high school tend to lose interest or get lower grades in math and science at adolescence. They may turn away from the sciences because they do not want to be considered "weird" (see Hanson 1996). But perhaps some are making a rational decision. They may reject science because they know how hard such a career will be, especially when combined with raising a family. Perhaps they see the future and do not like it. High school and college students who are encouraged to aim for a doctorate in science should be helped by advisors to assess the benefits and costs of such choices compared to other careers. Part of this process is a consideration of economic and work-family issues.

Much attention is focused on trying to increase doctoral degrees in science awarded to women annually. However, these efforts may not result in their subsequent career satisfaction. Fox (1993, p. 17) studied the experiences of women in several fields of science and found that the rewards they expected after reaching their educational goals did not materialize. She concludes that it is time to go beyond a preoccupation with the numbers of women in science: "*It makes little sense to increase the pools of trained women in science if that training does not translate into rank and reward*" (italics in the original).

Therefore, the admirable goal of gender equality should not keep advisors from emphasizing the investment of time, energy and money, as well as commitment, required for a scientific career. A student also should be informed about the dearth of jobs in many areas of science. Otherwise, after joining a select group of women and men in the sciences, she may find herself starting over in a different field. This career crisis may occur during the post-doctoral period, at the tenure stage, or even later. Weaver (1993, p. 3) notes that a special issue on women in *Science* magazine (1992) did not confront the problems faced by the scientist who has been laid off: "Trying to find a job long after graduate school when one is in her forties or beyond is especially dismaying and depressing."

Brush (1991, p. 416) suggests that perhaps women who leave the pipeline "are behaving more intelligently than those who want to recruit them but refuse to provide adequate incentives, such as reasonable working conditions and promotion opportunities."

The "pipeline" metaphor, so often used in describing science education, points to a problem in recruitment and career development. It suggests a factory-management attitude that views people as raw material to be made into products, without considering their views or well-being. Other professions, newly open to them, may be more attractive. Brush (1991, p. 416) suggests that if "those leaders who are trying to *push* women into science and engineering, would devote the same energy to creating conditions that would *pull* them into technical careers, everyone's interest would be better served" (italics in the original).

Thus, a second major question is whether women physicists can develop satisfying careers, despite the shrinking opportunities in academia and government. Their challenge is to find settings in which they can use their scientific knowledge while gaining more control over their lives, more recognition, and perhaps a more secure and comfortable life. For many scientists, this career route means developing their own niches in nontraditional settings. The next two sections explore some possibilities and briefly describe a few successful career paths.

## CAREER PATHS OUTSIDE ACADEMIA

If settings outside the university are to attract women, long-standing negative attitudes toward applied careers and jobs in industry will have to change. Some changes already have occurred, as evidenced by more favorable articles in scientific publications and positive attitudes toward nonacademic careers expressed by professional associations. Nonacademic scientists are becoming more visible in professional associations and meetings. Their careers and success are noted and applauded in scientific magazines.

In part, the change reflects the self-interest of the associations and policymakers in the sciences. Industry is the sector where the jobs are and emphasis on applied research is growing in society. If academic careers are the only ones valued, then the science pipeline will leak even more female, and male, physicists at every stage. Yet, as of 1990, only 3.9% of employed women doctoral physicists worked in industry or were self-employed (Dresselhaus 1993, p. 6).

Women often are employed in what are considered less desirable academic jobs—for example, teaching at a smaller college or working in nontenured positions. Yet, leaving academic teaching or research for a nonacademic research or nonresearch position has been considered a sign of "failure, disappointment, compromise, abandonment of the desire to be a research scientist, throwing away many years of training and hard work" (Sherman-Gold 1992, p. 12).

When men make a similar career move, they tend to see and interpret it as "a natural career evolution, an upward move in a successful career path," whereas women consider it an admission that they were not successful in their previous work (Sherman-Gold 1992, p. 12). These negative attitudes clearly will have to change, since they keep women from exploring job possibilities or tailoring innovative careers to match their interests. One way to raise the prestige of nonacademic careers is to stop labeling them "alternative" careers, implying that these are second choices to be considered only when more desirable academic careers do not materialize.

## Careers in Industry

Some women have begun to recognize clear advantages to careers in industry, compared to academia (see, e.g., Lindbeck 1997; Wallander 1993). A respondent in the CSWP survey indicated how she had weighed an academic versus an industry career and concluded:

> The moving and uprooting that this [an academic career] would entail seemed prohibitive for a two-career family with a small child who would have to adjust to several new places and daycare situations. . . . The stability of having a permanent job in industry rather than temporary positions was very appealing, not to mention of course the huge difference in salary. The point I'm trying to make is that many women may avoid academic careers because they can't see dragging their families around with them for a series of temporary positions.

The number of women scientists in industry is still small (Davis 1996). Some take jobs in industry if their scientist spouse is employed by the major university in the area. However, a dual-career couple faces another kind of constraint in the industrial sector. When one partner is employed by a company, a competitor may refuse to hire the other. This "competitor exclusion rule" is the corporate equivalent of the anti-nepotism rule in academia (Rayman and Brown 1996, p. 5).

Nevertheless, women with scientific backgrounds may find greater opportunities in industry than in academia, partly because affirmative action legislation has made more of a difference for hiring in industry. Affirmative action requirements are written into and enforced in government contracts. Additionally, some companies have made concerted efforts to recruit and retain women scientists, offering defined career paths, mentoring, and favorable work-family policies. In these settings, they tend to find a critical mass of women colleagues and avenues for networking within the corporation.

Despite the lingering stigma attached to applied research, a career in industry has advantages. Some have been noted above. In addition, the hours may not be so long that a scientist has practically no life outside the lab—an important consideration for women in the child-bearing and child-rearing years.

Women with or without doctorates also can choose to pursue managerial careers in industrial companies or work for consulting firms. Admittedly, physicists who are dedicated to research are less receptive to a management career track. To go into management, women scientists need more than technical expertise and problem-solving ability. They also should be "inventive, innovative and entrepreneurial" (Hamm 1994, p. 12; also see National Research Council 1994).

Women sometimes limit their career options in industry by narrowing their job search to the applied physics research area. Other opportunities exist. For example, high-tech firms may recruit women with physics backgrounds because their technical skills are an asset in this rapidly changing area.

The experience of women scientists in industry has been mixed. Some companies with male-oriented cultures have presented women scientists with a hostile environment (Culotta 1993b; Friedmann 1993). In many settings, women scientists meet the same obstacles to promotion as they do in academic situations.

Retention has been a problem. The numbers of women leaving jobs in industry has been higher than those of men. But jobs are plentiful and career advancement potential exists. Some women scientists may try several settings before settling on a situation that suits them. Job changing has become so acceptable today that it should not hamper an applicant's prospects.

However, although women and men may start out with comparable salaries, men outpace them in 10 or 15 years, largely because more men go into management (Brush 1991, p. 412). The number of women in upper management levels is still small, although some progress has been made.

Women's science groups are working to develop policies to improve the climate in industry. A recent National Research Council (NRC) 1994 report, *Women Scientists and Engineers Employed in Industry: Why So Few?* made a range of recommendations, starting with recruitment efforts to identify qualified women. Suggested on-the-job policies and practices include: career development, flexible work schedules, and other programs to meet work-family needs. Equitable job assignments and salaries of course are essential. Women can help improve their situation by using such strategies as mentoring and women's networks.

These approaches are much more effective when supported and encouraged by top management. And more women in executive positions and corporate boards would foster this goal. Many corporations have only one token woman director or none at all.

Some companies have designed programs to keep and promote women. Corning developed a successful approach to retaining women scientists after many recruits left soon after being hired. Xerox has made a commitment to hiring women in such fields as physics, computer science, and electrical engineering. Esther Conwell, a physicist and Fellow of the Corporation at Xerox, became the first woman to receive the Edison Medal from the Institute of Electrical and Electronic Engineers for her work on semiconductors. Other companies with programs to recruit and retain women—including Bell Labs (now split between AT&T and Lucent), ALCOA, Aerospace, along with Xerox—are profiled in *Women Scientists and Engineers Employed in Industry: Why So Few?* (NRC 1994).

Shirley Jackson, a physicist and MIT's first black female Ph.D., has had a varied career in industry and academia and now has a top job in government. She worked for Bell Labs as a condensed matter theorist and then moved to Rutgers so that she could work with students. Now she is in a high government position as head of the Nuclear Regulatory Commission. She has testified to the "double bind" for black women in science and technology. During her education, she experienced isolation and began working to combine self-help with helping others, now her trademark (Gibbons 1993, p. 393).

However, opportunities for government employment have decreased in recent years. The end of the Cold War and a smaller government role in fostering research and development in defense has resulted in fewer jobs.

## Entrepreneurship and Other Creative Responses

Another career route for women scientists is to become entrepreneurs. Some young women are starting their own companies in technical fields (see, e.g., Lawman 1997; Ouellette 1997; Dogar 1997). If women can overcome the barriers to financing a new venture, the rewards of "being one's own boss" and starting a company can be great. Advantages include independence, the potential to develop personal abilities and watch their company grow, and financial independence (see Culotta 1993, p. 406). Flexibility in scheduling is also a plus, although hours are long, especially in the start-up stage.

The computer industry is one area in which women are making strides. For example, Kim Pollese was on *Time* magazine's (1997, p. 42) list of 100 influential people as the most influential Web entrepreneur of 1997. She left Sun Microsystems, where she had been instrumental in developing Java, a programming language, to start her own Silicon Valley company, Marimba. Her academic background is in biophysics.

Women with backgrounds in biotechnology and computer fields seem to be especially oriented to starting companies. Lawman's 1997 article, "Birthing a Biotech Company," compared the process of starting her company to the stages of giving birth. Cynthia Robbins-Roth founded a leading biotechnology newsletter and with a partner also started a consulting firm that offers advice to biotechnology companies. Her experience explains why going into industry or starting a company is attractive. She had been a postdoctoral fellow in biochemistry and immunology but left academia because "there were a lot of women scientists in academia. When we couldn't get promoted, we said, 'Screw you, guys,' and went into biotech" (Brownlee 1997, p. 62). After academia, Robbins-Roth went to work for Genentech and then started a company.

Although women are moving into the biotech industry, progress still needs to be made at the executive and board levels; their representation in these corporate settings continues to be small. One study showed that only 107 out of about 580 executive positions in the industry were filled by women (Brownlee 1997, p. 62).

A September 1997 *Working Woman* special technology report includes a list of the 10 most important women in technology. Their backgrounds include degrees in business or finance, doctorates in math or cellular biology, and advanced degrees in engineering or computer science.

Other examples of women entrepreneurs include Jill Shapiro, vice-president of CH2M HILL, who was one of nine women honored at a 1997 AWIS conference. An environmental planner with a Ph.D. in physical and inorganic chemistry (1960s), her career has included positions at the National Institutes of Health,

Institut Pasteur in Paris, and National Bureau of Standards. As a post-doctoral scientist at the University of Washington in the early 1970s, she was one of a group whose members lost their positions as a result of industry cutbacks in the region. She started an environmental consulting firm, later sold it to her staff, and started another firm before going to CH2M HILL.

Her advice to women who want to be scientists or engineers: Don't worry about your title if you are working in industry. Be honest about what you don't know. (She found women less willing to indicate their lack of knowledge.) Finally, she recommended that they network laterally and vertically at work. Take risks and do new things (Solomon et al. 1997, pp. 23-24).

Another entrepreneurial example is Marianne Hamm, a physicist who started a company with her spouse and two colleagues. It qualifies as a woman-owned small business, employs 40 employees, and was twice included in *Inc.* magazine's "Inc. 500 List." She and her husband, also a physicist, had already experienced numerous geographic moves and job searches after they earned doctorates; hers is in heavy ion physics, his is in accelerator physics.

Her advice to physics students who want to become entrepreneurs is: "broaden your background and experience," be flexible and alert to nontraditional opportunities; don't focus only on physics to the exclusion of other activities on campus, especially those that develop interpersonal and management skills; and "don't be afraid to take a calculated risk" (Hamm 1994, pp. 13-14)

Carol Latham took a different approach to starting a company. She became an entrepreneur by finding a niche that no one else had considered (Petzinger 1994, p. B1). Her work history is an interesting one. After receiving an undergraduate degree in chemistry, she went to Standard Oil of Ohio (Sohio) to work on developing oil-based products. Then she stayed home for the next 20 years as a housewife and mother. But after a divorce in 1981, she was back at Sohio surrounded by Ph.D.s The organization was "99.99999% male" and she felt "patronized and barely tolerated." When she began working in ceramics, she thought about combining them with plastics. Her coworkers did not wanted to cross boundaries or move away from their areas of specialization.

However, she saw an opportunity to fill the void and started a new company on a shoestring in an apartment. Her business cards identified her as "technical director," not president. Orders came from some big high technology companies. In 1996, the firm had 18 employees and sales of two million dollars, with no debt.

For some women, becoming self-employed may be a temporary solution to gain needed flexibility during early child rearing. For example, a survey of women chemists (Roscher 1990, p. 128) found that slightly fewer than 2% of women chemists are self-employed, compared to slightly more than 2% of the men. The availability of child care played an important role for some women. The self-employed president of a small company reported that her flexible schedule gave her more time with her children. Her employer, a large oil company, offered only childcare referrals and her proposal for an on-site day care center was rejected.

She was happier with her current work, even though it was unrelated to traditional chemistry. Another chemist had a home-based writing and editing business whose clients were primarily in the pharmaceutical industry. She had left a similar, better-paying position with a large company because of inadequate childcare (Roscher 1990, p. 129).

We might ask how women's approach to science research affects their success as entrepreneurs. Sonnert (1995b, p. 152) suggests that the styles and definitions of what constitutes "good" science typically may differ among women and men. He found that women stressed comprehensiveness and integrity and emphasized thoroughness and perfectionism. They focused on conformity to the traditional rules of science. They rejected "quick and dirty" work and were less willing to cut corners to succeed. Sonnert thinks this gives women less of a "careerist" approach to science. Whereas men stressed creativity and a good presentation, women tended to be more careful and prone to double checking everything they did. Perhaps they are reacting to the closer scrutiny and criticism their work receives, compared to their male counterparts' research.

Starting a new venture differs in many ways from working on a research project in a laboratory, despite similarities like creativity. Skills needed by entrepreneurs also include risk-taking, the ability to convince others to support your ideas, flexibility, and management ability. Women who decide to become entrepreneurs may already possess these qualities, or will develop them early in their venture. The 1994 NRC report on women in industry summarizes additional skills needed for management or entrepreneurial careers (for a general review of the literature on women entrepreneurs, see Starr and Yudkin 1996).

Women (and men) are taking creative approaches to career planning so that they can use their training, sometimes in untraditional ways. For example, Sandra Panem is a scientist who went from academia to a career with a venture capital firm that specializes in biotechnology. She attributes her failure to get tenure at the University of Chicago in the early 1980s to more than discrimination, although she believes that subtle discrimination is found at all major research institutions. Her research became an issue because her investigation of whether viruses cause tumors was controversial at the time. (Perhaps this experience helps to explain why Sonnert found that women scientists often seemed to "play it safe" in their research.) She also was not well-served by her advisor, who did not tell her about the opposition to granting her tenure. Now she uses her scientific background to review cutting- edge research as a venture capitalist (Zuckerman, Cole, and Bruer 1991, pp. 127-154).

Scientists may develop careers in other nontraditional settings. For example, scientists can work in many aspects of patent examination and law. A recent issue of the *AWIS Magazine* titled "Patents Protecting Intellectual Property" shows the importance of this field. Some scientists have established careers as examiners in the US Patent Office (Brown 1997, pp. 12-14). Law firms specializing in patent applications also hire scientists as experts or partners. Lily Rin-Laures (1997, pp.

5-7) is a patent attorney with a science background and medical degree as well. Moreover, because many patent lawsuits are so complex, some judges assigned to these cases have scientific backgrounds. An advanced science degree may become a requirement for deciding such patent cases in the future. These work settings can offer women a chance to use their scientific expertise in a different way from making discoveries in the laboratory.

When young women decide to pursue a doctorate in physics or other sciences, they should be aware that a traditional career in a university physics department or laboratory may be unlikely. Only a few outstanding women (and men) will succeed in a shrinking and extremely competitive field. Therefore, they would be well-advised to plan for more than one type of career after a doctorate. Women in the sciences who stop short of the doctorate also should be knowledgeable about the range of career paths and settings open to them.

## CONCLUSIONS

In this chapter, we have assessed the climate for women's careers in physics. The policy to encourage women to enter this traditionally male-dominated field has come up against the reality of a very difficult employment market. Tenure-track positions are decreasing in universities and women still face barriers to being hired. Dual-career scientific couples face even greater difficulties in establishing careers.

Young women should be made aware of the uncertainty and economic insecurity that may await them after earning a doctorate in physics. Some still will opt for this career path. However, other settings and types of work offer opportunities and potential for advancement. They might consider the advantages of careers in industry or as entrepreneurs. Before significant numbers of women are likely to move into these settings, such careers will have to be viewed as a good first choice, not something to fall back on if academic jobs are not available.

## REFERENCES

Abir-Am, P.G., and D. Outram, eds. 1987. *Uneasy Careers and Intimate Lives: Women in Science, 1789-1979*. New Brunswick, NJ: Rutgers University Press.
Ajzenberg-Selove, F. 1994. *A Matter of Choices: Memoirs of a Female Physicist*. New Brunswick, NJ: Rutgers University Press.
Allen, N. 1990. "Australian Women in Science: A Comparative Study of Two Physicists." *Metascience* 8: 75-85.
Alper, J. 1993. "The Pipeline is Leaking Women All Along the Way." *Science* 260: 409-411.
Amato, I. 1992. "Profile of a Field: Chemistry. Women Have Extra Hoops to Jump Through." *Science* (March 13): 1372-1373.
Angier, N. 1995. "Why Science Loses Women in the Ranks." *New York Times* (May 4): E5.
Astin, H.S., and J.F. Milem. 1996. "The Status of Academic Couples in U.S. Institutions." *AWIS Magazine* 25(4, Summer): 12-14.

Brown, K. 1997. "From Bench Science to Patent Examining." *AWIS Magazine* 26(5, September/October): 12-14.

Brownlee, S. 1997. "Same Song, Second Verse." *Working Woman* (September): 62.

Brush, S. 1991. "Women in Science and Engineering." *American Scientist* (September-October): 404-419.

Byrne, E.M. 1993. *Women and Science: The Snark Syndrome.* Washington, DC: The Falmer Press.

Civian, J., with P. Rayman, B. Brett, and L.M. Baldwin. 1997. *Pathways for Women in the Sciences. The Wellesley Report,* Part II. Wellesley, MA: Wellesley College.

Cladis, P. 1992. "Women in Physics: Where are We Now? Where Do We Go from Here?" *CSWP Gazette* 12(2, October): 5, 8-9.

Coser, L., and R. Coser. 1974. *Greedy Institutions: Patterns of Undivided Commitment.* New York: The Free Press.

Culotta, E. 1993a. "Women Struggle to Crack the Code of Corporate Culture." *Science* 260: 398-404.

Culotta, E. 1993b. "Work and Family: Still a Two-Way Stretch." *Science* 260(April): 401.

Culotta, E. 1993c. "Entrepreneurs Say: 'It's Better to Be the Boss.'" *Science* 260(April): 406.

Davidson, J. 1997. "The Tenure Trap." *Working Woman* (June): 36-41, 68.

Davis, A.C. 1996. "Women and Underrepresented Minority Scientists and Engineers Have Lower Levels of Employment in Business and Industry." Data Brief, Science Resources Studies Division, National Science Foundation, December 31.

Dogar, R. 1997. "Movers and Shakers and Soft-Ware Makers." *Working Woman* (September): 52-53, 59, 82-83.

Dresselhaus, M.S. 1994. ""Update on the Chilly Climate for Women in Physics." *CSWP Gazette* 14(1, Spring): 4-9, 24.

Dresselhaus, M.S., J.R. Franz, and B. Clark. 1994. "Interventions to Increase the Participation of Women in Physics." *Science* 263: 1392-1393.

"Dual Career Couples: Balancing Work and Family." 1996. *AWIS Magazine* 25(4, Special Issue).

Engle, I. 1992. "Contemporary Vignettes: Women Physicists. Where are We? What is Our Collective Goal? What is Our Direction? And How Fast are We Moving?" *CSWP Gazette* 12(2, October): 3, 7-8.

Etzkowitz, H., C. Kemelgor, M. Neuschatz, and B. Uzzi. 1994. "Barriers to Women's Participation in Academic Science and Engineering." Pp. 43-67 in *Who Will Do Science?* edited by W. Pearson and A. Fechter. Baltimore, MD: The Johns Hopkins Press.

Etzkowitz, H., C. Kemelgor, M. Neuschatz, B. Uzzi, and J. Alonzo. 1994. "The Paradox of Critical Mass for Women in Science." *Science* 266(October 7): 51-54.

Etzkowitz, H., C. Kemelgor, and J. Alonzo. 1995. *The Rites and Wrongs of Passage: Critical Transitions for Female Ph.D. Students in the Sciences.* Arlington, VA: Report to the NSF.

Evans, M. 1996. "High-Tech's Newest Resource." *The Financial Post* (July 27): 10.

Fava, S.F., and K. Deierlein. 1988. "Women Physicists in the U.S.: The Career Influence of Marital Status." *CSWP Gazette* 8(2, August): 1-3.

Fava, S.F., and K. Deierlein. 1989. "Women Physicists: Non-Traditional Occupations and Traditional Family Roles." Pp. 7-16 in *Contemporary Readings in Sociology,* edited by J. DeSena. Dubuque, IA: Kendall-Hunt Publishing.

Fava, S.F., and R.G. Genovese. 1983. "Family, Work and Individual Development in Dual Career Marriages: Issues for Research." Pp. 163-185 in *Research in the Interweave of Social Roles*, Vol. 3: *Jobs and Families,* edited by H.Z. Lopata and J. Pleck. Greenwich, CT: JAI Press.

Fehrs, M., and R. Czujko. 1992. "Women in Physics: Reversing the Exclusion." *Physics Today* (August): 33-40.

Fox, M.F. 1993. "Women, Men, and the Social Organization of Science." *AWIS Magazine* 22(1): 17.

Fox, M.F., and P.E. Stephan. 1996. "Careers in Science: Preferences, Prospects, Realities." Paper presented at the "Science Careers, Gender Equity and the Changing Economy" Conference, co-sponsored by the Commission on Professionals in Science and Technology and the Radcliffe Public Policy Institute, October.

Friedmann, L. 1993. "More Women in Industry is Good for Business." *AWIS Magazine* 22(5, September/October): 10-11.

Gady, C.D., E.L. Babco, and W. Pearson, Jr. 1996. "Employment Trends for Doctorates in Science and Engineering." Paper presented at the "Science Careers, Gender Equity and the Changing Economy" Conference, co-sponsored by the Commission on Professionals in Science and Technology and the Radcliffe Public Policy Institute, October.

"Gender Survey of the APS Membership—1990." 1993. *CSWP Gazette* 12(3, (January): 5-10.

Genovese, R.G. 1997. *Americans at Midlife: Caught Between Generations.* Westport, CT: Bergin and Garvey.

Gibbons, A. 1992a. "Key Issue: Two-Career Science Marriage." *Science* 255(March 13): 1380-1381.

Gibbons, A. 1992b. "Key Issue: Tenure." *Science* 255(March 13): 1386.

Gibbons, A. 1993. "Gaining Standing—By Standing Out." *Science* 260(April 16): 393

Hamm, M.E. 1994. "Life Beyond Research." *CSWP Gazette* 14(1, Spring): 12-14.

Hanson, S.L. 1996. *Lost Talent: Women in the Sciences.* Philadelphia, PA: Temple University Press.

Hill, S.T. 1996. "Science and Engineering Doctorate Awards are at an All-Time High." Data Brief No. 4, Science Resources Studies Division, National Science Foundation, June 19.

Hughes, H.M. 1973. "Maid of All Work or Departmental Sister-in-Law? The Faculty Wife Employed on Campus." Pp. 5-10 in *Changing Women in a Changing Society,* edited by J. Huber. Chicago, IL: University of Chicago Press.

Jones, L.M. 1990. "Intellectual Contributions of Women to Physics." Pp. 188-215 in *Women of Science: Righting the Record,* edited by G. Kass-Simon and P. Farnes. Indianapolis, IN: Indiana University Press.

Kass-Simon, G., and P. Farnes, eds. 1990. *Women of Science: Righting the Record.* Indianapolis, IN: Indiana University Press.

Kilborn, P.T. 1993. "The Ph.D.'s Are Here, but the Lab Isn't Hiring." *New York Times* (July 18): E3.

"Kim Polese: Web Entrepreneur." 1997. *Time* (April 21): 42.

Kistiakowsky, V. 1980. "Women in Physics: Unnecessary, Injurious, and Out of Place?" *Physics Today* 33(February): 32-40.

Kistiakowsky, V. 1992. "The Origins of the Committee on Women in Physics: How Much Has Changed and How Little." *CSWP Gazette* 12(2, October): 1, 6-7.

Lancaster, H. 1997. "Landing a Job in the Sciences Takes Some Creativity." *Wall Street Journal* (April 29): B1.

Lawman, P. 1997. "Birthing a Biotech Company: A Labor-Intensive Process." *AWIS Magazine* 26(2): 8-10.

Lindbeck, A.C. 1997. "From Academia to Industry: Portrait of a Process Chemist." *AWIS Magazine* 26(2): 6-7, 11.

Long, J.S. 1993. "Sex Differences in Rank Advancement in Academic Careers." *American Sociological Review* 58: 703-722.

Mann, C.C. 1991. "Wanted: Wayward Particles." *Discover* (December): 50-55.

Matyas, M.L., L. Baker, and R. Goodell. 1989. *Marriage, Family, and Scientific Careers: Institutional Policy Versus Research Findings.* Proceedings of a Symposium, American Association for the Advancement of Science Annual Meeting, January 16, San Francisco, CA.

Matyas, M.L., and L.S. Dix, eds. 1992. *Science and Engineering Programs: On Target for Women?* Washington, DC: National Research Council.

McIlwee, J.S., and J.G. Robinson. 1992. *Women in Engineering: Gender, Power, and Workplace Culture.* Albany, NY: State University of New York Press.

National Research Council. 1994. *Women Scientists and Engineers Employed in Industry: Why So Few?* Washington, DC: National Academy Press.

National Science Foundation. 1994. *Nonacademic Scientists and Engineers: Trends from the 1980 and 1990 Censuses.* Arlington, VA: SRS, Directorate for Social, Behavioral and Economic Sciences.

National Science Foundation. 1995. *Science and Engineering Degrees 1966-1993: Detailed Statistical Tables.* Arlington, VA: SRS, Directorate for Social, Behavioral and Economic Sciences.

National Science Foundation. 1996 (September). *Women, Minorities, and Persons with Disabilities in Science and Engineering: 1996.* Arlington, VA: National Science Foundation.

Ouellette, L.A. 1997. "The Journey Toward Independence and Business Ownership." *AWIS Magazine* 26(2): 20-21.

Papanek, H. 1974. "Men, Women and Work: Reflections on the Two-Person Career." *American Journal of Sociology* 78(4): 852-872.

Petzinger, T., Jr. 1994. "Carol Latham Knows the Spoils Go to Those Who Cross Boundaries." *Wall Street Journal* (May 16): B1.

Planning Group to Assess Possible OSEP Initiatives for Increasing the Participation of Women in Scientific and Engineering Careers, Office of Scientific and Engineering Personnel. 1989. *Responding to the Changing Demography: Women in Science and Engineering.* Washington, DC: National Science Foundation, January.

Pycior, H.M., N.G. Slack, and P.G. Abir-Am, eds. 1996. *Creative Couples in the Sciences.* New Brunswick, NJ: Rutgers University Press.

Rayman, P., and H. Burbage. 1989. "Professional Families: Falling Behind While Getting Ahead." In *Marriage, Family and Scientific Careers: Institutional Policy Versus Research Findings.* San Francisco: American Association for the Advancement of Science.

Rayman, P., and B. Brett. 1993. *Pathways for Women in the Sciences. The Wellesley Report*, Part I. Wellesley, MA: Wellesley College.

Rayman, P.M., and S.V. Brown. 1996. "Women Scientists: Can They Work and Have a Life?" Paper presented at the "Science Careers, Gender Equity and the Changing Economy" Conference, co-sponsored by the Commission on Professionals in Science and Technology and the Radcliffe Public Policy Institute, October.

Regets, M.C. 1997. "Immigrants Are 23 Percent of U.S. Residents With S and E Doctorates." Issue Brief NSF 97-321, Science Resources Studies Division, National Science Foundation, September 23.

Rin-Laures, L. 1997. "Science and the Law: Patent Law as a Career." *AWIS Magazine* 26(5, September/October): 5-7.

Roscher, N.M. 1990. *Women Chemists: 1990.* A Supplementary Report to the American Chemical Society's 1990 Survey of Members' Salary and Employment. Washington, DC: American Chemical Society. December.

Rossiter, M.W. 1982. *Women Scientists in America: Struggles and Strategies to 1940.* Baltimore, MD: Johns Hopkins University Press.

Schwartz, A.E. 1996. "Work Vs. Family—Too Many Secrets." *AWIS Magazine* 25(4, Summer): 16.

Sherman-Gold, R. 1992. "Alternative Careers for Women Scientists: Career Choice or Refuge?" *AWIS Magazine* 21(5, September/October): 12-13.

Solomon, F., K. Stremler, K. Atkinson, and A. Robinson. 1997. "Shaping Our Future: A Leadership Conference for Women in Science, Technology, and Engineering." *AWIS Magazine* 36(3, May/June): 22-25, 42.

Sonnert, G., with G. Holton. 1995a. *Who Succeeds in Science: The Gender Dimension.* New Brunswick, NJ: Rutgers University Press.

Sonnert, G., with G. Holton. 1995b. *Gender Differences in Science Careers: The Project Access Study.* New Brunswick, NJ: Rutgers University Press.

Starr, J., and M. Yudkin. 1996. *Women Entrepreneurs: A Review of Current Research.* Center for Research on Women Special Report. Wellesley, MA: Center for Research on Women.

"The Ten Most Important Women in Tech." *Working Woman* (September): 54-59.

Wallander, D. 1993. "A Science Career: There Are Options . . ." *AWIS Magazine* 22(4): 8-9.
Weaver, E. 1993. "The Case for a Career in Science." *AWIS Magazine* 22(1, January/February): 2-3.
Zachary, G.P. 1996. "Looking for a Real Rocket Scientist? Manpower to Offer Physicists as Temps."
   *Wall Street Journal* (November 27): A7.
Zuckerman, H., J. Cole, and J.T. Bruer. 1991. *The Outer Circle: Women in the Scientific Community.*
   New York: W.W. Norton and Co.

# BIOGRAPHICAL SKETCHES OF CONTRIBUTORS

**Nandini Narain Assar** is a doctoral candidate in the Department of Sociology at Virginia Polytechnic Institute and State University. Her areas of interest include race/ethnicity, gender, immigration, and qualitative methods. Her current research examines the political economy of immigration policy, cultural traditions, and gender as they play out among Asian Indian immigrants in the United States, and the relationship between the researcher and participants.

**Toni M. Calasanti** is an associate professor of sociology at Virginia Polytechnic Institute and State University (Virginia Tech), where she is also affiliated with Women's Studies and the Center for Gerontology. Much of her research is concerned with the intersection of gender with work/retirement, and increasingly with race and class. Her examination of work is broad, including paid and unpaid labor in both formal and informal sectors. Sole and co-authored work has appeared in a variety of outlets, including *Social Problems, The Journal of Gerontology: Social Sciences, The Gerontologist*, and *The Sociological Quarterly.*

**Paula Dempsey** is a graduate student in the Department of Sociology and Anthropology, Loyola University Chicago. Research interests in addition to work and occupations include religious movements and the sociology of knowledge.

**Sylvia F. Fava** is professor emerita, Brooklyn College and the Graduate Center, City University of New York. Educated at Queens College—CUNY and Northwestern University, she has authored and co-authored numerous books and articles on the community, urban and suburban society, housing, world urbanism and, most recently, the interplay between gender and marital status in occupations such as sociology and physics. She has been vice-president of the Eastern Sociological

Society and elected chair of the Community and Urban Section of the American Sociological Association.

**Ann Barry Flood** is a professor and Director of Policy Studies for the Evaluative Clinical Sciences, Dartmouth Medical School, Hanover, NH. She received her doctorate in organizational sociology from Stanford University in 1977. She has also served on the staff of the U.S. Senate Finance Committee during 1989 as a Robert Wood Johnson Policy Fellow. Her area of theoretical and policy expertise involve understanding the professional and organization factors that influence physician's styles of practice, patients' outcomes, and the cost of health care.

**Allen M. Fremont**, M.D., Ph.D., is a resident in internal medicine at Beth Israel Deaconess Medical Center and a clinical fellow at Harvard Medical School. His research interests include nonclinical determinants of physician practice styles, social psychological factors in health and illness, and health outcomes research. He is currently co-editing the fifth edition of the *Handbook of Medical Sociology*.

**Rosalie Genovese** is an affiliated scholar with the University of Rochester's Susan B. Anthony University Center. She is the author of *Americans at Midlife: Caught Between Generations* (1997) and *Families and Change: Social Needs and Public Policies* (1984). Her research interests include dual-career couples, self-help networks, programs to improve the economic status of women, and community-based planning.

**Jona Goldschmidt** is an associate professor in Loyola University Chicago's Department of Criminal Justice. He was formally the assistant executive director for programs of the American Judicature Society. He has also taught at Arizona State University and Northern Arizona University. A member of the Illinois and California bars, he is a graduate of the University of Illinois-Urbana (B.S. 1968), DePaul University (J.D. 1972), and Arizona State University (Ph.D. 1990). His areas of research and publication include pro se litigation, unauthorized practice of law, alternative dispute resolution, sociology of professions, judicial selection, and judicial ethics.

**Elizabeth Goodrick** is an assistant professor in the Department of Organizations and Human Resources in the School of Management, State University of New York at Buffalo. Her research interests focus on the development and impact of socially created meaning systems, particularly within the health care sector. Recent publications include "Organizational Discretion in Responding to Institutional Practices: Hospitals and Cesarian Births" with Gerald R. Salancik (*Administrative Science Quarterly*, 41: 1-28) and "Business as Usual: The Adoption of Managerial Ideology by U.S. Hospitals" with James R. Meindl and Ann Barry Flood (*Research in the Sociology of Health Care*, Vol. 14, JAI Press). She received her Ph.D. in organizational theory from the University of Illinois at Urbana-Champaign.

**Dai Kejing** is a professor at the Institute of Sociology, the Chinese Academy of Social Sciences, Beijing. Her publications focus on rural labor resources and agriculture, marriage and the family, the life course, and the history of Chinese sociology. She was active in the 1995 World Women's Conference in Beijing.

**Helena Z. Lopata** became Professor Emerita of Sociology at Loyola University Chicago in 1997. She obtained her Ph.D. from the University of Chicago in 1954. Her areas of interest are occupations and professions, social development and social roles, and immigration and ethnicity, all from a symbolic interactionist perspective. Her continued research focus is on the Cosmopolitan Community of Scholars.

**Kandace Pearson Schrimsher** is currently teaching sociology at Diablo Valley College, Pleasant Hill, CA. She obtained her Ph.D. in sociology from Loyola University Chicago in 1996. She has also taught at Oakland University and John Carroll University and has conducted research for the Public Affairs Research Lab at Oakland University and Case Western Reserve University School of Law. Her areas of interest include occupations and professions, work and family issues, and gender. Her current research focuses on corporate relocations, trailing spouses, and corporate wives.

**Kathleen F. Slevin** is an associate professor and Chair of the Department of Sociology at the College of William and Mary in Virginia. A feminist sociologist, she writes and teaches about gender, race, class, and age. She has recently co-authored a book on retired professional African-American women which is scheduled for publication in 1998 by New York University Press.

**Janice Witt Smith** is an assistant professor of management in the School of Business and Economics at North Carolina Agricultural and Technical State University in Greensboro, North Carolina. She is actively engaged in research on examining the experiences of institutional and social isolation of women and minorities in the workplace, as well as affective organizational commitment, intent to turnover, and mentoring. She has presented her research at a number of different conferences, including the National Academy of Management, Southern Management Association, Southeastern Institute for Operations Research and the Decisions Sciences (SE InfORMS), Entrepreneurial Development Program, and Commonwealth Professional Business Women's Organization. She serves as a reviewer for the Women in Management Division of the Academy of Management; is a member of the editorial review board of *Strategic and Organizational Leadership Journal*; is track chair for the Management and Strategy Track of SE InfORMS; serves as a textbook reviewer for Prentice-Hall; was a Fellow, ABD African-American Faculty Fellow Mentoring Program, Virginia Tech; and participated in the Doctoral Student Consortia, Academy of Management, 1993-1994.

**Richard Travisano** teaches and writes sociology at the University of Rhode Island. His present research interests are lobstermen and jazz musicians. He is also writing poetry and a memoir of his Italian-American childhood. His two favorite recent publications are "Meaning without Symbols: Toward Revising Symbolic Interaction" and "Lobstering Out of Narrow River: Minding One's Grasp." (These appear in Norman Denzin's *Studies in Symbolic Interaction*, volumes 14 and 16, respectively.

**Carol S. Wharton** is an associate professor of sociology at the University of Richmond, Virginia. Her current research interests concern the interrelations among gender, work, and family, and how the arrangements women make to accommodate each sphere change over the life course.

**C. Ray Wingrove** is a professor of sociology and occupies the Irvin May Chair in Human Relations at the University of Richmond. He has conducted a number of studies involving older women and has paid particular attention to the techniques effective in interviewing the elderly. He recently co-authored a book on retired professional African-American women which is scheduled for publication in 1998 by New York University Press.

**Jing Zhang** received her Ph.D. in sociology from Loyola University Chicago in 1997. She has conducted research for the School of Social Service Administration at the University of Chicago and has been teaching at DePaul University and Truman College.Her research interests are community development, cultural diversity, and economics-related issues. She is also interested in research methodology.

# Current Research on Occupations and Professions

Edited by **Helena Z. Lopata,**
*Department of Sociology and Anthropology,*
*Loyola University, Chicago*

**Volume 9,** 1996, 296 pp.                    $78.50/£49.95
ISBN 1-55938-877-3

Edited by **Helena Z. Lopata** and **Anne E. Figert,**
*Department of Sociology and Anthropology,*
*Loyola University, Chicago*

**CONTENTS:** Introduction, *Helena Z. Lopata and Anne E. Figert.* Professionals, Consumers, and the European Medicines Agency: Policy Making in the European Union, *Louis H. Orzack.* Professional Work in Public and Private Settings: The Use and Evaluation of the DSM in Psychiatric Units, *Bernice A. Pescosolido, Anne E. Figert, and Keri M. Lubell.* Car Saleswomen: Expanding the Scope of Salesmanship, *Helen M. Lawson.* Professionalism Among Indian Executives, *S.L. Hiremath, Raghavendra Gudagunti and Jayashree Kulkarni.* "Pitching" Images of the Community to the Generalized Other: Promotional Strategies of Economic Development Officers, *Robert Prus and Augie Fleras.* To Be or Not To Be Your Own Boss? A Comparison of White, Black and Asian Scientists and Engineers, *Joyce Tang.* Work in the Family Business, *John Ward and Drew S Mendoza.* "Hard Cases": Prosecutorial Accounts for Filing Unconvictable Sexual Assault Complaints, *Lisa Frohmann.* Interior Architectural Design: Conventions and Innovations, *Susie Fehrenbacher-Zeiser.* The Changing Terrain of Sex Stratification in the Legal Profession, *Jo Dixon and Jordana Pestrong.* Competition and Cooperation Among Pharmacists and Physicians in Ontario, 1920-1940, *Linda Muzzin and Roy Hornosty.* Illegal Home Care Workers: Polish Immigrants Caring for American Elderly, *Mary Erdmans.* Biographical Sketches of Contributors.

Also Available:
**Volumes 1-8** (1980-1993)                    $78.50/£49.95 each

**JAI PRESS INC.**
55 Old Post Road No. 2 - P.O. Box 1678
Greenwich, Connecticut 06836-1678
Tel: (203) 661- 7602    Fax: (203) 661-0792

# Research in the Sociology of Organizations

Edited by **Samuel B. Bacharach,** *New York State School of Industrial and Labor Relations, Cornell University*

**Volume 14,** 1996, 360 pp.  $78.50/£49.95
ISBN 0-7623-0019-1

Edited by **Samuel B. Bacharach,** *New York State School of Industrial and Labor Relations, Cornell University* **Peter A. Bamberger,** and **Miriam Erez,** *Faculty of Industrial Engineering and Management, Techion-Israel Institute of Technology*

**CONTENTS:** The Cross-Cultural Analysis of Organizations: Bridging the Micro-Macro and Ideographic-Nomethetic Gaps, *Samuel B. Bacharach, Peter A. Bamberger, and Miriam Erez.* Effect of Cultural Diversity in Work Groups, *David C. Thomas, Elizabeth C. Ravlin, and Alan W. Wallace.* Nested Cultures and Identities: A Comparative Study of Nation and Profession/ Occupation Status Effects on Resource Allocation Decisions, *Janet M. Dukerich, Brain R. Golden, and Carol K. Jacobson.* Profile Analysis of the Sources of Meaning Reported by U.S. and Japanese Local Government Managers, *Mark F. Peterson, James R. Elliott, Paul D. Bliese, and Mark H.B. Radford.* Underlying Assumptions of Agency Theory and Implications for Non-U.S. Settings: The Case of Japan, *Allan Bird and Margarethe F. Wiersema.* Organizational Action and Institutional Reforms in China's Economic Transition: A Comparison of Two Industries, *Douglas Guthrie.* Rediscovering the Individual in the Formation of International Joint Ventures, *Paul Olk and P. Christopher Earley.* Cultural Contingencies and Leadership in Developing Countries, *Rabindra N. Kanungo and Manuel Mendonca.* Planned Change in Organizations: The Influence of National Culture, *Anne-Wil Harzing and Geert Hofstede.*

Also Available:
**Volumes 1-13** (1982-1995)  $78.50/£49.95 each

## JAI PRESS INC.

55 Old Post Road No. 2 - P.O. Box 1678
Greenwich, Connecticut 06836-1678
Tel: (203) 661- 7602   Fax: (203) 661-0792

# Research in Social Stratification and Mobility

Edited by **Kevin Leicht,** *Department of Sociology, University of Iowa*

**Volume 15,** 1997, 274 pp.                    $78.50/£49.95
ISBN 0-7623-0048-5

**CONTENTS:** Introduction, *Michael Wallace.* PART I. STUDIES IN SOCIAL MOBILITY. Class Mobility in Israeli Society: A Comparative Perspective, *John H. Goldthorpe, Meir Yaish, Vered Kraus.* The Mobility Patterns of Part-Time Workers, *Jerry A. Jacobs and Zhenchao Qian.* The Impact of Mothers' Occupations on Children's Occupational Destinations, *Aziza Khazzoom.* How Reliable are Studies of Social Mobility: An Investigation into the Consequences of Errors in Measuring Social Class, *Richard Breen and Jan O. Jonsson.* PART II. THE ISRAELI LABOR MARKET. On the Cost of Being an Immigrant in Israel: The Effects of Tenure, Origin, and Gender, *Moshe Semyonov.* Job Search, Gender, and the Quality of Employment in Israel, *Sigal Alon and Haya Stier.* PART III. JAPAN, THE UNITED STATES, AND JAPANESE-AMERICANS. Effects of Ascribed and Achieved Characteristics on Social Values in Japan and the United States, *Catherine B. Silver and Charlotte Muller.* Wages Among White and Japanese-American Male Workers, *Arthur Sakamoto and Satomi Furuichi.* PART IV. OTHER TOPICS. Trends in Male and Female Self-Emplouyment: Growth in a New Middle Class or Increasing Marginalization of the Labor Force?, *Richard Arum.* Revisiting Broom and Cushing's "Modest Test of an Immodest Theory", *Micahel Wallace.* Bibliographic Sketches of Contributors to Volume 15. Index.

Also Available:
**Volumes 1-14** (1981-1995)                    $78.50/£49.95 each

---

**FACULTY/PROFESSIONAL** discounts are available in the U.S. and Canada at a rate of 40% off the list price when prepaid by personal check or credit card and ordered directly from the publisher.

---

## JAI PRESS INC.
55 Old Post Road No. 2  - P.O. Box 1678
Greenwich, Connecticut 06836-1678
Tel: (203) 661- 7602    Fax: (203) 661-0792

**JAI PRESS**

# Research in the Sociology of Work

Edited by **Randy Hodson,** *Department of Sociology, The Ohio State University*

**Volume 6, The Globalization of Work**
1997, 372 pp.                                    $78.50/£49.95
ISBN 0-7623-0020-5

*The Globalization of Work* will focus on various facets of work in the global economy including the impact of global competition on workers in core economies, comparative studies of work in developed and developing nations, international production systems, and the work experiences of international migrants. Each contributor will propose solutions to the current challenges and problems associated with the globalization of work. This volume will be an invaluable resource to students and researchers across the social sciences and management sciences.

Also Available:
**Volumes 1-5** (1981-1994)                      $78.50/£49.95 each